GÖDEL '96

Lecture Notes in Logic
A Publication of
The Association for Symbolic Logic

LECTURE NOTES IN LOGIC 6

GÖDEL '96

Logical Foundations of Mathematics, Computer Science, and Physics — Kurt Gödel's Legacy

Edited by

Petr Hájek
Institute for Computer Science
Academy of Sciences of the Czech Republic
Prague, Czech Republic

CRC Press
Taylor & Francis Group
Boca Raton London New York

CRC Press is an imprint of the
Taylor & Francis Group, an **informa** business
AN A K PETERS BOOK

CRC Press
Taylor & Francis Group
6000 Broken Sound Parkway NW, Suite 300
Boca Raton, FL 33487-2742

First issued in hardback 2017

ISBN-13: 978-1-5688-1153-6 (pbk)
ISBN-13: 978-1-1384-6686-9 (hbk)

Visit the Taylor & Francis Web site at
http://www.taylorandfrancis.com

and the CRC Press Web site
http://www.crcpress.com

Library of Congress Cataloging-in-Publication Data

Gödel '96 (1996 : Brno, Czech Republic)
Gödel '96 : logical foundations of mathematics, computer science, and physics -- Kurt Gödel's Legacy / edited by Petr Hájek.
p. cm. -- (Lecture notes in logic ; 6)
Originally published : Berlin : Springer-Verlag, 1996.
Includes bibliographical references and index.
ISBN 1-56881-153-5 (pbk. : alk. paper)
1. Logic, Symbolic and mathematical--Congresses. 2. Mathematics--Philosophy--Congresses. 3. Computer science--Congresses. 4. Mathematical physics--Congresses. I. Hájek, Petr. II. Title. III. Series.

QA9.A1 G6 1996a
511.3--dc21 2001016534

Preface

The aim of the conference *Logical Foundations of Mathematics, Computer Science and Phycics - Kurt Gödel's Legacy (GÖDEL'96)*, organized to the occasion of the ninetieth anniversary of the birth of Kurt Gödel, is to pay tribute to Kurt Gödel by arranging a scientific event presenting a forum for papers relevant to foundational aspects of Logic in Mathematics, Computer Science, Philosophy and Physics – areas influenced by Kurt Gödel's work.

The conference has been organized in Brno, the birthplace of Gödel, by Masaryk University Brno in co-operation with the Institute of Computer Science of the Academy of Sciences of the Czech Republic, Prague, and with the international Kurt Gödel Society based in Vienna (Organizing Committee chaired by J. Zlatuška). The Association for Symbolic Logic has recognized the conference as an ASL sponsored meeting.

The Program Committee has consisted of Z. Adamowicz, Warsaw; J. Bičák, Prague; L. Bukovský, Košice; D. de Jongh, Amsterdam; J. Grygar, Prague; E. Köhler, Vienna; J. Krajíček, Prague; P. Hájek (chair), Prague; A. Leitsch, Vienna; D. Mundici, Milano; G. Müller, Heidelberg; J. Paris, Manchester; C. Parsons, Harvard.

The Program Committee has selected the invited speakers and undertook the refereeing process. From 52 submitted papers, 14 have been selected for both presentation at the conference and publication in the Proceedings, 8 other papers have been selected only for presentation at the conference (mainly due to preliminary character of the paper). A few authors, whose papers have been accepted, decided not to publish them. Two invited authors (D. Isaacson, M. Magidor) were unable to finish their papers in time, but have remained invited speakers. It is regretted that two invited speakers (G. Kreisel and A. MacIntyre) had to decide not to attend the meeting. Nevertheless, I feel that the papers in this volume cover the domains touched by Kurt Gödel's work reasonably well and that they contribute to our understanding of the present state of knowledge in these domains.

I would like to thank to the editors of the series Lecture Notes in Logic for accepting this volume of Proceedings into the series in spite of the fact that it also contains papers from physics and philosophy of mathematics. The presence of these papers is clearly given by the interests of Kurt Gödel and without them the picture would be incomplete.

A financial support is very important for any conference such as this one and financial sponsorship by the following institutions is highly appreciated:

- EACSL – European Association for Computer Science Logic,
- IUHPS/DLMPS – International Union of History and Philosophy of Science / Division Logic, Methodology and Philosophy of Science,
- UNU-IIST – The United Nations University / International Institute for Software Technology,
- Hewlett-Packard s.r.o.,

- ZPA-CZ Trutnov, Czech Republic.

The hard work of M. Daniel, D. Harmancová and A. Štědrý from the Institute of Computer Science, Academy of Sciences of the Czech Republic, in converting all the papers into a unified LaTeX shape is highly recognized.

Before closing this preface, I would like to express my anticipation that the conference will be a successful celebration of the ninetieth anniversary of the birth of Kurt Gödel, a man whose work influenced the XX century science and many of us so deeply.

Prague, May 8, 1996 Petr Hájek

Table of Contents

Part II. Contributed Papers

Part I

Invited Papers

Gödel's program for new axioms: Why, where, how and what?

Solomon Feferman*

Department of Mathematics
Stanford University
Stanford, CA 94305 USA
email: sf@csli.stanford.edu

Summary. From 1931 until late in his life (at least 1970) Gödel called for the pursuit of new axioms for mathematics to settle both undecided number-theoretical propositions (of the form obtained in his incompleteness results) and undecided set-theoretical propositions (in particular CH). As to the nature of these, Gödel made a variety of suggestions, but most frequently he emphasized the route of introducing ever higher axioms of infinity. In particular, he speculated (in his 1946 Princeton remarks) that there might be a uniform (though non-decidable) rationale for the choice of the latter. Despite the intense exploration of the "higher infinite" in the last 30-odd years, no single rationale of that character has emerged. Moreover, CH still remains undecided by such axioms, though they have been demonstrated to have many other interesting set-theoretical consequences.

In this paper, I present a new very general notion of the "unfolding" closure of schematically axiomatized formal systems S which provides a uniform systematic means of expanding in an essential way both the language and axioms (and hence theorems) of such systems S. Reporting joint work with T. Strahm, a characterization is given in more familiar terms in the case that S is a basic system of non-finitist arithmetic. When reflective closure is applied to suitable systems of set theory, one is able to derive large cardinal axioms as theorems. It is an open question how these may be characterized in terms of current notions in that subject.

1. Why new axioms?

Gödel's published statements over the years (from 1931 to 1972) pointing to the need for new axioms to settle both undecided number-theoretic and set-theoretic propositions are rather well known. They are most easily cited by reference to the first two volumes of the edition of his *Collected Works.*[1] A number of less familiar statements of a similar character from his unpublished essays and lectures are now available in the third volume of that edition.[2]

* Invited opening lecture, Gödel '96 conference, Brno, 25-29 August 1996. This paper was prepared while the author was a fellow at the Center for Advanced Study in the Behavioral Sciences, Stanford, CA, whose facilities and support are greatly appreciated.

[1] Cf. in Gödel [1986] the items dated: *1931*(p.181, ftn.48a), *1934*(p.367), *1936*(p.397), and in Gödel [1990] those dated: *1940*(p.97, ftn.20[added 1965]), *1946*(p.151), *1947*(pp.181-183), *1964*(pp.260-261 and 268-270), and *1972a*, Note 2 (pp.305-306).

[2] Cf. in Gödel [1995] the items dated: **1931?*(p.35), **1933o* (p.48), **1951*(pp.306-307), **1961/?*(p.385) and **1970a,b,c*(pp.420-425).

Given the ready accessibility of these sources, there is n o need for extensive quotation, though several representative passages are singled out below for special attention.

With one possible exception (to be noted in the next section), the single constant that recurs throughout these statements is that the new axioms to be considered are in all cases of a set-theoretic nature. More specifically, to begin with, axioms of higher types, extended into the transfinite, are said to be needed even to settle undecided arithmetical propositions.[3] The first and most succinct statement of this is to be found in the singular footnote 48a of the 1931 incompleteness paper, in which Gödel states that "...the true reason for the incompleteness inherent in all formal systems of mathematics is that the formation of ever higher types can be continued into the transfinite...[since] the undecidable propositions constructed here become decidable whenever appropriate higher types are added". In an unpublished lecture from that same period Gödel says that analysis is higher in this sense than number theory and set theory is higher than analysis: "...there are number-theoretic problems that cannot be solved with number-theoretic, but only with analytic or, respectively, set-theoretic methods" (Gödel [1995], p.35). A couple of years later, in his (unpublished) 1933 lecture at a meeting of the Mathematical Association of America in Cambridge, Massachusetts, Gödel said that for the systems S to which his incompleteness theorems apply "...exactly the next higher type not contained in S is necessary to prove this arithmetical proposition...[and moreover] there are arithmetic propositions which cannot be proved even by analysis but only by methods involving extremely large infinite cardinals and similar things" (Gödel [1995], p.48). This assertion of the necessity of axioms of higher type — a.k.a. axioms of infinity in higher set theory — to settle undecided arithmetic (Π^0_1) propositions, is repeated all the way to the final of the references cited here in footnotes 1 and 2 (namely to 1972).

It is only with his famous 1947 article on Cantor's continuum problem that Gödel also pointed to the need for new set-theoretic axioms to settle specifically *set-theoretic* problems, in particular that of the Continuum Hypothesis CH. Of course at that time one only knew through his own work the (relative) consistency of AC and CH with ZF, though Gödel conjectured the falsity of CH and hence its independence from ZFC. Moreover, it was the question of determining the truth value of CH that was to preoccupy him almost exclusively among all set-theoretic problems — except for those which might be ancillary to its solution — for the rest of his life. And rightly so: the continuum problem — to locate $2^{\aleph_0}$ in the scale of the alephs whose existence is forced on us by the well-ordering theorem — is the very first chal-

[3] The kind of proposition in question is sometimes referred to by Gödel as being of "Goldbach type" i.e. in Π^0_1 form, and sometimes as one concerning solutions of Diophantine equations, of the form $(P)D = 0$, where P is a quantifier expression with variables ranging over the natural numbers; cf. more specifically, the lecture notes **193?* in Gödel [1995].

lenging problem of Cantorian set theory, and settling it might be considered to bolster its conceptual coherence. In his 1947 paper, for the decision of CH by new axioms, Gödel mentioned first of all, axioms of infinity:

> The simplest of these ... assert the existence of inaccessible numbers (and of numbers inaccessible in the stronger sense) $> \aleph_0$. The latter axiom, roughly speaking, means nothing else but that the totality of sets obtainable by exclusive use of the processes of formation of sets expressed in the other axioms forms again a set (and, therefore, a new basis for a further application of these processes). Other axioms of infinity have been formulated by P. Mahlo. [Very little is known about this section of set theory; but at any rate][4] these axioms show clearly, not only that the axiomatic system of set theory as known today is incomplete, but also that it can be supplemented without arbitrariness by new axioms which are only the natural continuation of those set up so far. (Gödel [1990], p.182)

However, Gödel goes on to say, quite presciently, that "[a]s for the continuum problem, there is little hope of solving it by means of those axioms of infinity which can be set up on the basis of principles known today...", because his proof of the consistency of CH via the constructible sets model goes through without change when such statements are adjoined as new axioms (indeed there is no hope in this direction if one expects to prove CH false):

> But probably [in the face of this] there exist other [axioms] based on hitherto unknown principles ... which a more profound understanding of the concepts underlying logic and mathematics would enable us to recognize as implied by these concepts. (*ibid.*)

Possible candidates for these were forthcoming through the work of Scott [1961] in which it was shown that the existence of measurable cardinals (MC) implies the negation of the axiom of constructibility, and the later work of Hanf [1964] and of Keisler and Tarski [1964] which showed that measurable cardinals and even weakly compact cardinals must be very much larger than anything obtained by closure conditions on cardinals of the sort leading to hierarchies of inaccessibles. But as we now know through the extensive subsequent work on large cardinals as well as other strong set-theoretic principles such as forms of determinacy, none of those considered at all plausible to date settles CH one way or the other (cf. Martin [1976], Kanamori [1994]). Gödel himself offered only one candidate besides these, in his unpublished 1970 notes containing his "square axioms" concerning so-called scales of functions on the $\aleph_n$'s. The first of these notes (***1970a** in Gödel [1995]) purports to prove that the cardinality of the continuum is $\aleph_2$ while the second (***1970b**, op.cit.) purports to prove that it is $\aleph_1$. However, there are essential gaps in

[4] The section enclosed in brackets was deleted from the 1964 reprinting of the 1947 article (cf. Gödel [1990], p. 260).

both proofs and in any case the axioms considered are far from evident (cf. the introductory note by R.M. Solovay to **1970a,b,c* in Gödel [1995], pp. 405-420).

Gödel's final fall-back position in his 1947 article is to look for axioms which are "so abundant in their verifiable consequences...that quite irrespective of their intrinsic necessity they would have to be assumed in the same sense as any well-established physical theory" (Gödel [1990], p.183). It would take us too far afield to look into the question whether there are any plausible candidates for these. Moreover, there is no space here to consider the arguments given by others in pursuit of the program for new axioms; especially worthy of attention are Maddy [1988, 1988a], Kanamori [1994] and Jensen [1995] among others.

My concern in the rest of this paper is to concentrate on the consideration of axioms which are supposed to be "exactly as evident" as those already accepted. On the face of it this excludes, among others, axioms for "very large" cardinals (compact, measurable, etc.), axioms of determinacy, axioms of randomness, and axioms whose only grounds for accepting them lies in their "fruitfulness" or in their simply having properties analogous to those of $\aleph_0$. Even with this restriction, as we shall see, there is much room for reconsideration of Gödel's program.

2. Where should one look for new axioms?

While the passage to higher types in successive stages, in one form or another, is sufficient to overcome incompleteness with respect to number-theoretic propositions because of the increase in consistency strength at each such stage, it by no means follows that this is the *only* way of adding new axioms in a principled way for that purpose. Indeed, here a quotation from Gödel's remarks in 1946 before the Princeton Bicentennial Conference is very apropos:

> Let us consider, e.g., the concept of demonstrability. It is well known that, in whichever way you make it precise by means of a formalism, the contemplation of this very formalism gives rise to new axioms which are exactly as evident and justified as those with which you started, and this process of extension can be iterated into the transfinite. So there cannot exist any formalism which would embrace all these steps; but this does not exclude that all these steps (or at least all of them which give something new for the domain of propositions in which you are interested) could be described and collected together in some non-constructive way. (Gödel [1990], p.151)

It is this passage that I had in mind above as the one possible exception to Gödel's reiterated call for new set-theoretic axioms to settle undecided number-theoretic propositions. It is true that he goes on immediately to say that "[i]n set theory, e.g., the successive extensions can most conveniently be represented by stronger and stronger axioms of infinity". But note that here he is referring to set theory as an *example* of a formalism to which the general idea of expansion by "new axioms exactly as evident and justified as those with which you started" may be applied as a special case. That idea, in the case of formal systems S in the language of arithmetic comes down instead to one form or another of (proof-theoretic) reflection principle, that is a formal scheme to the effect that whatever is provable in S is correct. In its weakest form (assuming the syntax of S effectively and explicitly given), this is the collection of statements

$$(\mathrm{Rfn}_S) \qquad Prov_S(\#(A)) \rightarrow A$$

for A a closed formula in the language of S, called the *local reflection principle.*[5] This is readily generalized to arbitrary formulas A uniformly in the free variables of A as parameters, in which case it is called the *uniform reflection principle* RFN_S. The axioms Rfn_S, and more generally, RFN_S may indeed be considered "exactly as evident and justified" as those with which one started. Moreover, as shown by Turing [1939], extension by such axioms may be effectively iterated into the transfinite, in the sense that one can associate with each constructive ordinal notation $a \in O$ a formal system S_a such that the step from any one such system to its successor is described by adjunction of the reflection principle in question, and where all previous adjunctions are simply accumulated at limit s by the formation of their union. These kinds of systematic extensions of a given formal system were called *ordinal logics* by Turing; when I took them up later in 1962, I rechristened them *(transfinite) recursive progressions of axiomatic theories* (cf. Feferman [1962, 1988]). While Turing obtained a completeness result for Π_1^0 statements via the transfinite iteration in this sense of the local reflection principle, and I obtained one for all true arithmetic statements via the iteration of the uniform reflection principle, both completeness results were problematic because they depended crucially on the judicious choice of notations in O, the selection of which was no more "evident and justified" in advance than the statements to be proved.

What was missing in this first attempt to spell out the general idea expressed by Gödel in the above quotation was an explanation of which ordinals — in the constructive sense — ought to be accepted in the iteration procedure. The first modification made to that end (Kreisel [1958], Feferman [1964]) was to restrict to *autonomous* progressions of theories, where one advances to a notation $a \in O$ only if it has been proved in a system S_b, for some b which precedes a, that the ordering specifying a is indeed a well-ordering. It was with this kind of procedure in mind that Kreisel called in his paper [1970] for the study of *all principles of proof and ordinals which are implicit in given concepts.* However, one may question whether it is appropriate at

[5] Note that the consistency statement for S is an immediate consequence of the local reflection principle for S.

all to speak of the concept of ordinal, in whatever way restricted, as being implicit in the concepts of, say, arithmetic. I thus began to pursue a modification of that program in Feferman [1979], where I pr oposed a characterization of that part of mathematical thought which is implicit in our conception of the natural numbers, without any prima-facie use of the notions of ordinal or well-ordering. This turned out to yield a system proof-theoretically equivalent to that proposed as a characterization of *predicativity* in Feferman [1964] and Schütte [1965]. Then in my paper [1991], I proposed more generally, a notion of *reflective closure* of arbitrary schematically axiomatized theories, which gave the same result (proof-theoretically) as the preceding when applied to Peano Arithmetic as initial system. That made use of a partial self-applicative notion of truth, treated axiomatically. The purpose of the present article is to report a new general notion of reflective closure of a quite different form, which I believe is more convincing as an explanation of *everything that one ought to accept if one has accepted given concepts and principles*. In order not to confuse it with the earlier proposal, I shall call this notion that of the *unfolding* of any given schematically formalized system. This will be illustrated here in the case of non-finitist arithmetic as well as the case of set theory. Exact characterizations in more familiar terms have been obtained for the case of non-finitist arithmetic in collaboration with Thomas Strahm; these will be described in Section 4 below. However, there is no space here to give any proofs.

3. How is the unfolding of a system defined?

As we shall see, it is of the essence of the notion of unfolding that we are dealing with schematically presented formal systems. In the usual conception, *formal schemata* for axioms and rules of inference employ *free predicate variables* $P, Q, \ldots$ of various numbers of arguments $n \geq 0$. An appropriate substitution for $P(x_1, \ldots, x_n)$ in such a scheme is a formula $A(x_1, \ldots x_n \ldots)$ which may have additional free variables. (Thus if P is 0-ary, any formula may be substituted for it.) Familiar examples of *axiom schemata* in the propositional and predicate calculi are

$$\neg P \to (P \to Q) \qquad \text{and} \qquad (\forall x)P(x) \to P(t) .$$

Further, in non-finitist arithmetic, we have the *Induction Axiom Scheme*

(IA) $$P(0) \wedge (\forall x)[P(x) \to P(x')] \to (\forall x)P(x) ,$$

while in set theory we have the *Separation* and *Replacement Schemes*

(Sep) $$(\exists b)(\forall x)[x \in b \leftrightarrow x \in a \wedge P(x)], \text{ and}$$

(Repl) $$(\forall x \in a)(\exists! y)P(x, y) \to (\exists b)(\forall y)[y \in b \leftrightarrow (\exists x \in a)P(x, y)] .$$

Familiar examples of *schematic rules of inference* are, first of all, in the propositional and predicate calculi,

$$P, P \rightarrow Q \Rightarrow Q \quad \text{and} \quad [P \rightarrow Q(x)] \Rightarrow [P \rightarrow (\forall x)Q(x)] \text{ (for } x \text{ not free in } P\text{)},$$

while the scheme for the *Induction Rule* in finitist arithmetic is given by

$$\text{(IR)} \qquad P(0), P(x) \rightarrow P(x') \Rightarrow P(x)\ .$$

It is less usual to think of schemata for axioms and rules given by *free function variables* $f, g, \ldots$ But actually, it is more natural to formulate the Replacement Axiom Scheme in functional form as follows:

$$\text{(Repl)}' \quad (\forall x \in a)(\exists y)[f(x) = y] \rightarrow (\exists b)(\forall y)[y \in b \leftrightarrow (\exists x \in a) f(x) = y]\ .$$

Note that here, and for added compelling reasons below, our function variables are treated as ranging over *partial functions.*

The informal philosophy behind the use of schemata here is their *open-endedness.* That is, they are not conceived of as applying to a specific language whose stock of basic symbols is fixed in advance, but rather as applicable to *any* language which one comes to recognize as embodying meaningful basic notions. Put in other terms, *implicit in the acceptance of given schemata is the acceptance of any meaningful substitution instances.* But *which* these instances are need not be determined in advance. Thus, for example, if one accepts the axioms and rules of inference of the classical propositional calculus given in schematic form, one will accept all substitution instances of these schemata in any language which one comes to employ. The same holds for the schemata of the sort given above for arithmetic and set theory. In this spirit, we do not conceive of the function, resp. predicate variables as having a fixed intended range and it is for this reason that th ey are treated as *free* variables. Of course, if one takes it to be meaningful to talk about the totality of partial functions, resp. predicates, of a given domain of objects, then it would be reasonable to bind them too by quantification. In the examples of unfolding given here, it is only in set theory that the issue of whether and to what extent to allow quantification over function variables is unsettled.

Now our question is this: *given a schematic system* S, *which operations and predicates — and which principles concerning them — ought to be accepted if one has accepted* S? The answer for operations is straightforward: *any operation from and to individuals is accepted in the unfolding of* S *which is determined (in successive steps) explicitly or implicitly from the basic operations of* S. Moreover, the *principles* which are added concerning these operations are just those which are derived from the way they are introduced. Ordinarily, we would confine ourselves to the *total operations* obtained in this way, i.e. those which have been proved to be defined for all values of their arguments, but it should not be excluded that their introduction might depend in an essential way on prior *partial operations*, e.g. those introduced by recursive definitions of a general form.

We reformulate the question concerning predicates in operational terms as well, i.e.: *which operations on and to predicates — and which principles concerning them — ought to be accepted if one has accepted* S? For this, it is necessary to tell at the outset *which logical operations on predicates are taken for granted in* S. For example, in the case of non-finitist classical arithmetic these would be (say) the operations $\neg, \wedge$ and $\forall$, while in the case of finitist arithmetic, we would use just $\neg$ and $\wedge$. It proves simplest to treat predicates as propositional functions; thus $\neg$ and $\wedge$ are operations on propositions, while $\forall$ is an operation on functions from individuals to propositions. Now we can add to the operations from individuals to individuals in the unfolding of S also *all those operations from individuals and/or propositions to propositions which are determined explicitly or implicitly (in successive steps) fro m the basic logical operations of* S. Once more, the principles concerning these operations which are included in the expansive closure of S are just those which are derived from the way they are introduced. Finally, *we include in the expansive closure of* S *all the predicates which are generated from the basic predicates of* S *by these operations*; the principles which are taken concerning them are just those that fall out from the principles for the operations just indicated.

This notion of unfolding of a system is spelled out in completely precise terms in the next section for the case of non-finitist arithmetic. But the following two points ought to be noted concerning the general conception described here. First of all, one should not think of the unfolding of a system S as delimiting the range of applicability of the schemata embodied in S. For example, the principle of induction is applicable in every context in which the basic structure of the natural numbers is recognized to be present, even if that context involves concepts and principles not implicit in our basic system for that structure. In particular, it is applicable to impredicative reasoning with sets, even though (as will be shown in the next section) the unfolding closure of arithmetic is limited to predicative reasoning. Secondly, we may expect the language and theorems of the unfolding of (an effectively given system) S to be effectively enumerable, but we should not expect to be a ble to decide which operations introduced by implicit (e.g. recursive fixed-point) definitions are well defined for all arguments, even though it may be just those with which we wish to be concerned in the end. This echoes Gödel's picture of the process of obtaining new axioms which are "just as evident and justified" as those with which we started (quoted in Section 2 above), for which we cannot say in advance exactly what those will be, though we can describe fully the means by which they are to be obtained.

4. The expansive closure of non-finitist arithmetic: what's obtained

Here the starting schematic system NFA (Non-Finitist Arithmetic) has language given by the constant 0, individual variables $x, y, z, \ldots$, the operations

Sc and Pd for successor and predecessor, a free unary predicate variable P and the logical operations $\neg, \wedge$ and $\forall$.

Assuming classical logic, $\wedge, \rightarrow$ and $\exists$ are defined as usual.[6] We write t' for $Sc(t)$ in the following. The axioms of NFA are:

Ax 1. $x' \neq 0$
Ax 2. $Pd(x') = x$
Ax 3. $P(0) \wedge (\forall x)[P(x) \rightarrow P(x')] \rightarrow (\forall x)P(x)$.

Ax 3 is of course our scheme (IA) of induction. Before defining the full unfolding $\mathcal{U}$(NFA) of this system, it is helpful to explain a subsystem $\mathcal{U}_0$(NFA) which might be called the *operational unfolding* of NFA, i.e. where we do not consider which predicates are to be obtained. Basically, the idea is to introduce new operations via a form of generalized recursion theory (g.r.t.) considered axiomatically. The specific g.r.t. referred to is that developed in Moschovakis [1989] and in a different-appearing but equivalent form in Feferman [1991a] and [1996]; both feature *explicit definition* (ED) and *least fixed point recursion* (LFP) and are applicable to arbitrary structures with given functions or functionals of type level ≤ 2 over a given basic domain (or domains). The basic structure to consider in the case of arithmetic is $\langle \mathrm{N}, Sc, Pd, 0\rangle$, where N is the set of natural numbers. To treat this axiomatically, we simply have to enlarge our language to include the terms for the (in general) partial functions and functionals generated by closure under the schemata for this g.r.t., and add their defining equations as axioms. So we have terms of three types to consider: *individual terms, partial function terms* and *partial functional terms*. The types of these are described as follows, where, to allow for later extension to the case of $\mathcal{U}$(NFA), we posit a set Typ_0 of types of level 0; here we will only need it to contain the type ι of individuals, but below it will be expanded to include the type ι of propositions:

Typ 1. $\iota \in Typ_0$, where ι is the type of individuals. In the following κ, ν range over Typ_0 and $\bar{\iota}$, resp. $\bar{\kappa}$ range over types of finite sequences of individuals, resp. of objects of Typ_0.
Typ 2. τ, σ range over the types of partial functions of the form $\bar{\iota} \stackrel{\sim}{\rightarrow} \nu$, and $\bar{\tau}$ ranges over the types of finite sequences of such.
Typ 3. $(\bar{\tau}, \bar{\kappa} \stackrel{\sim}{\rightarrow} \nu)$ is used as types of partial functionals.

Note that objects of partial function type take only individuals as arguments; this is to insure that propositional functions, to be considered below, are just such functions. On the other hand, we may have partial functionals of type described under Typ 3 in which the sequence $\bar{\tau}$ is empty, and these reduce to partial functions of any objects of basic type in Typ_0.

[6] All our notions and results carry over directly to NFA treated in intuitionistic logic; the only difference in that case is that we take the full list of logical operations, $\neg, \wedge, \vee, \rightarrow, \forall$, and $\exists$ as basic.

The terms $r, s, t, u, \ldots$ of the various types under Typ 1 – Typ 3 are generated as follows, where we use $r : \rho$ to indicate that the term r is of type ρ.

Tm 1. For each $\kappa \in Typ_0$, we have infinitely many variables $x, y, z, \ldots$ of type κ.
Tm 2. $0 : \iota$.
Tm 3. $Sc(t) : \iota$ and $Pd(t) : \iota$ for $t : \iota$.
Tm 4. For each τ we have infinitely many partial function variables $f, g, h, \ldots$ of type τ.
Tm 5. $Cond(s, t, u, v) : (\bar{\tau}, \bar{\kappa}, \iota, \iota \overset{\sim}{\to} \nu)$ for $s, t : (\bar{\tau}, \bar{\kappa} \overset{\sim}{\to} \nu)$ and $u, v : \iota$.
Tm 6. $s(\bar{t}, \bar{u}) : \nu$ for $s : (\bar{\tau}, \bar{\kappa} \overset{\sim}{\to} \nu)$, $\bar{t} : \bar{\tau}$, $\bar{u} : \bar{\kappa}$.
Tm 7. $\lambda \bar{f}, \bar{x}.t : (\bar{\tau}, \bar{\kappa} \overset{\sim}{\to} \nu)$ for $\bar{f} : \bar{\tau}, \bar{x} : \bar{\kappa}, t : \nu$.
Tm 8. LFP $(\lambda f, \bar{x}.t) : (\bar{\iota} \overset{\sim}{\to} \nu)$ for $f : \bar{\iota} \overset{\sim}{\to} \nu, \bar{x} : \bar{\iota}$, $t : \nu$.

We now specialize this system of types and terms to just what is needed for $\mathcal{U}_0$(NFA), by taking $Typ_0 = \{\iota\}$. The formulas $A, B, C, \ldots$ of $\mathcal{U}_0$(NFA) are then generated as follows:

Fm 1. The atomic formulas are $s = t, s \downarrow$, and $P(s)$ for $s, t : \iota$.
Fm 2. If A, B are formulas then so also are $\neg A, A \wedge B$, and $\forall x A$.

As indicated above, formulas $A \vee B, A \to B$, and $\exists x A$ are defined as usual in classical logic. We write $s \simeq t$ for $[s \downarrow \vee t \downarrow \to s = t]$. Below we write $t\ [\bar{f}, \bar{x}]$, resp. $A\ [\bar{f}, \bar{x}]$ for a term, resp. formula, with designated sequences of free variables $\bar{f}, \bar{x}$; it is not excluded that t, resp. A may contain other free variables when using this notation. Since we are dealing with possibly undefined (individual) terms t, the underlying system of logic to be used is the *logic of partial terms* (LPT) introduced by Beeson [1985], pp. 97-99, where $t \downarrow$ is read as: t is defined. Briefly, the changes to be made from usual predicate logic are, first, that the axiom for $\forall$-instantiation is modified to

$$\forall x A(x) \wedge t \downarrow \to A(t) .$$

In addition, it is assumed that $\forall x(x \downarrow)$, i.e. only compound terms may fail to be defined (or put otherwise, non-existent individuals are not countenanced in LPT). It is further assumed that if a compound term is defined then all its subterms are defined ("strictness" axioms). Finally, one assumes that if $s = t$ holds then both s, t are defined and if $P(s)$ holds then s is defined. Note that $(s \downarrow) \leftrightarrow \exists x(s = x)$, so definedness need not be taken as a basic symbol.

The axioms of $\mathcal{U}_0$(NFA) follow the obvious intended meaning of the new compound terms introduced by the clauses Tm 5-8:

Ax 4. $(Cond(s, t, u, u))(\bar{f}, \bar{x}) \simeq s(\bar{f}, \bar{x}) \wedge [u \neq v \to (Cond(s, t, u, v))(\bar{f}, \bar{x}) \simeq t(\bar{f}, \bar{x})]$.
Ax 5. $(\lambda \bar{f}, \bar{x}.s[\bar{f}, \bar{x}])(\bar{t}, \bar{u}) \simeq s[\bar{t}, \bar{u}]$.
Ax 6. For $\varphi = \mathrm{LFP}(\lambda f, \bar{x}.t[f, \bar{x}])$, we have:
(i) $\varphi(\bar{x}) \simeq t[\varphi, \bar{x}]$

(ii) $\forall \bar{x}\{f(\bar{x}) \simeq t[f,\bar{x}]\} \rightarrow \forall \bar{x}\{\varphi(\bar{x})\downarrow \rightarrow \varphi(\bar{x}) = f(\bar{x})\}$.

Finally, the predicate substitution rule for $\mathcal{U}_0$(NFA) is:

(Subst) $$A[P] \Rightarrow A[B/P]$$

where in the conclusion of this rule, B is any formula with a designated free variable x, $B[x]$, and we substitute $B[t]$ for each occurrence of $P(t)$ in A. This completes the description of $\mathcal{U}_0$(NFA).

In the following we shall write

$$\{\text{if } y = 0 \text{ then } s[\bar{f},\bar{x}] \text{ else } t[\bar{f},\bar{x}]\} \text{ for } (Cond(\lambda\bar{f},\bar{x}.s, \lambda\bar{f},\bar{x}.t, y, 0))(\bar{f},\bar{x}),$$

in order to meet the strictness axioms of LPT; this piece of notation has the property that the compound term is defined when $y = 0$ if s is defined, even if t is not defined, while it is defined when $y \neq 0$ and t is defined if s is not defined.

We shall use capital letters F for closed terms of function type such that NFA proves $\forall \bar{x}(F(\bar{x})\downarrow)$, i.e. for which F is proved to be total. Suppose given such terms G, H of arguments $(\bar{x})$ and $(\bar{x}, y, z)$, resp. Then we can obtain an F with

$$\begin{aligned} F(\bar{x},0) &= G(\bar{x}) \\ F(\bar{x},y') &= H(\bar{x},y,F(\bar{x},y)) \end{aligned}$$

provable in $\mathcal{U}_0$(NFA). This is done by taking

$$\varphi = \text{LFP}[\lambda f,\bar{x},y.\{\text{if } y = 0 \text{ then } G(\bar{x}) \text{ else } H(\bar{x},Pd(y),f(\bar{x},Pd(y)))\}] .$$

It is then proved by induction on y that $\varphi(y)\downarrow$; this is by an application of the substitution rule to the schematic induction axiom IA (Ax 3) together with part (i) of the LFP axiom (Ax 6). Then we can take F to be the term φ. It follows that $\mathcal{U}_0$(NFA) serves to define all primitive recursive functions, and so by IA and the substitution rule, we see that $\mathcal{U}_0$(NFA) contains the system of Peano Arithmetic PA in its usual first order (non-schematic) form. I believe this argument formalizes the informal argument (usually not even consciously expressed) which leads us to accept PA starting with the bare-bones system NFA.

Conversely, $\mathcal{U}_0$(NFA) is interpretable in PA, by interpreting the function variables as ranging over (indices of) partial recursive functions, and then the function(al) terms are interpreted as (indices of) partial recursive function(al)s. It follows that we have closure under the LFP scheme. Finally, one shows that if $A[P]$ is provable in $\mathcal{U}_0$(NFA) and B is any formula, and if A^*, B^* are their respective translations, then $A^*[B^*/P]$ is provable in PA. Thus we conclude:

Theorem 1. *$\mathcal{U}_0$(NFA) is proof theoretically equivalent to PA and conservatively extends PA.*

Now to explain the full expansive closure of NFA we treat (as already mentioned) *predicates as propositional functions*, more or less following Aczel with his notion of Frege structures (Aczel [1980]). For this purpose we add a new basic type π, the type of *propositions*, and explain propositional functions as *total* functions f of type $\bar{\iota} \xrightarrow{\sim} \pi$. To fill out the language and axioms of $\mathcal{U}_0(\text{NFA})$ we thus begin by taking $Typ_0 = \{\iota, \pi\}$. As before, κ, ν range over Typ_0, τ, σ over types of the form $\bar{\iota} \xrightarrow{\sim} \nu$ (and thus are either types of partial functions from individuals to individuals or partial functions from individuals to propositions), and $(\bar{\tau}, \bar{\kappa} \xrightarrow{\sim} \nu)$ ranges over the types of partial functionals (of partial function, individual and propositional arguments, to individuals or propositions). Now the closure conditions on terms are expanded t o include the logical operations on and to propositions. These are given by the additional symbols $Eq, Pr, Neg, Conj$ and Un with the following clauses:

Tm 9. $Eq(s,t) : \pi$ for $s,t : \iota$.
Tm 10. $Pr(s) : \pi$ for $s : \iota$.
Tm 11. $Neg(s) : \pi$ for $s : \pi$.
Tm 12. $Conj(s,t) : \pi$ for $s,t : \pi$.
Tm 13. $Un(s) : \pi$ for $s : \iota \xrightarrow{\sim} \pi$.

The intended meaning of these symbols is elucidated by Ax 7-11 below.

The formulas $A, B, C, \ldots$ of $\mathcal{U}(\text{NFA})$ are generated as follows, where $T(x)$ is an additional predicate which expresses that x is a true proposition:
Fm 1.
(a) $s = t, s\downarrow$, and $P(s)$ are atomic for $s,t : \iota$.
(b) $s = t, s\downarrow$, and $T(s)$ are atomic for $s,t : \pi$.
Fm 2. If A, B are formulas, so also are $\neg A, A \wedge B, \forall x A$.

The axioms of $\mathcal{U}(\text{NFA})$ are now as follows (in addition to Ax 1-6 above), where we reserve $x, y, \ldots$ as variables of type ι and $a, b, \ldots$ as variables of type π:

Ax 7. $Eq(x,y)\downarrow \wedge [T(Eq(x,y)) \leftrightarrow x = y]$.
Ax 8. $Pr(x)\downarrow \wedge [T(Pr(x)) \leftrightarrow P(x)]$.
Ax 9. $Neg(a)\downarrow \wedge [T(Neg(a)) \leftrightarrow \neg T(a)]$.
Ax 10. $Conj(a,b)\downarrow \wedge [T(Conj(a,b)) \leftrightarrow T(a) \wedge T(b)]$.
Ax 11. $(\forall x)(fx\downarrow) \rightarrow Un(f)\downarrow \wedge [T(Un(f)) \leftrightarrow (\forall x) T(f(x))]$, for $f : \iota \xrightarrow{\sim} \pi$.

Because propositional terms in general implicitly depend on the predicate parameter P, we must restrict the rule $A[P] \Rightarrow A[B/P]$ to formulas A which do not contain any such terms. We write $Pred_n(t)$ for $(\forall \bar{x})(t(\bar{x})\downarrow)$ when $t : \bar{\iota} \xrightarrow{\sim} \pi$ and $\bar{\iota}$ is of length n. Now the usual way of thinking of a *sequence* of n-ary predicates is as a function f of type $\iota \xrightarrow{\sim} (\bar{\iota} \xrightarrow{\sim} \pi)$ such that for each x, $f(x)\downarrow$ and $Pred_n(f(x))$. However, we do not have these types in our set-up (although that is easily modified to include them). Instead, a sequence of n-ary predicates is treated as being represented by a g of type $\iota, \bar{\iota} \xrightarrow{\sim} \pi$ such that for each $x, \bar{y}$ we have $g(x, \bar{y})\downarrow$, in other words so that for each x, $Pred_n(\lambda \bar{y} \cdot g(x, \bar{y}))$. Such g can, at the same time, be considered as an

$(n+1)$-ary predicate and in that guise g is simply the *join* of the sequence it represents: $J(g) = g$.

Now the main result about proof-theoretic strength of $\mathcal{U}$(NFA) is the following theorem, obtained in collaboration with Thomas Strahm.

Theorem 2. *$\mathcal{U}$(NFA) is proof-theoretically equivalent to the system of ramified analysis up to but not including Γ_0, and conservatively extends that system.*

The system of ramified analysis up to and including level β is denoted RA_β, and the union of these for $\beta < \alpha$ is denoted $\mathrm{RA}_{<\alpha}$. For $\alpha = \omega \cdot \alpha$ this is proof-theoretically equivalent to the iteration of $(\Pi^0_1 - \mathrm{CA})$ through all levels $\beta < \alpha$. Using Kreisel's proposed characterization of predicative analysis in terms of the autonomous progression of ramified systems, the least impredicative ordinal was determined to be Γ_0 in Feferman [1964] and, independently, Schütte [1965]. Theorem 2 thus re-characterizes *predicativity* as *what ought to be accepted concerning operations and predicates if one has accepted the basic notions and principles of* NFA, *including the logical operations* $\neg, \wedge$ *and* $\forall$ *applied to variables for the natural numbers.* The proof of this theorem is rather involved and full details will be presented elsewhere; the following merely gives an indication of ho w to embed $\mathrm{RA}_{<\Gamma_0}$ in $\mathcal{U}$(NFA), by means of the methods of Feferman [1979], sec.3.3. Basically, one shows for each initial segment $\prec_\alpha$ of the standard primitive recursive well-ordering of order type Γ_0 how to establish in $\mathcal{U}$(NFA) the principle of transfinite induction up to α applied to arbitrary formulas A, in symbols, $\mathrm{TI}(\prec_\alpha, A)$. For this it suffices to prove $\mathrm{TI}(\prec_\alpha, P)$ and then apply the substitution rule. Now with the full scheme at hand, one can define the jump (Π^0_1) hierarchy relative to P along $\prec_\alpha$ by LFP recursion and prove that it defines a predicate by induction on this ordering. Note that the definition of this hierarchy makes use of arithmetical steps at successor stages, guaranteed by the axioms Ax 7-11, and of join at limit stages, guaranteed by the use of the J operator as explained above. As is shown in the reference loc.cit., by use of this hierarchy relative to P up to α , one can prove $\mathrm{TI}(\prec_\gamma, P)$ for $\gamma = \kappa^{(\alpha)}(0)$ in the Veblen hierarchy of critical functions. Define $\gamma_0 = 0, \gamma_{n+1} = \kappa^{(\gamma_n)}(0)$; then $\Gamma_0 = \lim_n \gamma_n$, so by this means we can embed RA_α in $\mathcal{U}$(NFA) for each $\alpha < \Gamma_0$. The proof that $\mathcal{U}$(NFA) is no stronger than $\mathrm{RA}_{<\Gamma_0}$ requires some interesting new arguments from infinitary proof theory. However, it is worth noting that in this proof, partial functions of type $\bar{\iota} \stackrel{\sim}{\rightarrow} \iota$ are still interpreted as partial recursive functions. Indeed the same holds for functions of type $\bar{\iota} \stackrel{\sim}{\rightarrow} \pi$ when propositions are treated intensionally.

Remarks

1. **Implicit definability of functions.** Another way of introducing partial functions given by implicit defining conditions is if we associate with each partial $f : \bar{\iota}, \iota \stackrel{\sim}{\rightarrow} \iota$ a $g : \bar{\iota} \stackrel{\sim}{\rightarrow} \iota$ with

(ID) $\forall\bar{x},y,z[f(\bar{x},y)\simeq 0\wedge f(\bar{x},z)=0\rightarrow y=z]$
$\rightarrow\forall\bar{x}[(\exists y)f(\bar{x},y)=0\rightarrow f(\bar{x},g(x))=0]$.

Adding (ID) as an axiom to $\mathcal{U}_0$(NFA) and $\mathcal{U}$(NFA) does not affect Theorems 1 and 2. It is plausible to include (ID) in the unfolding process applied to any system with a distinguished constant 0.

2. **Predicate types in place of the type of propositions.** We can treat predicates directly, instead of in terms of propositional functions, by introducing a basic type of n-ary predicates π_n for each $n\geq 1$. Then the atomic formulas to be used in the $\mathcal{U}$ process for this symbolism are of the form $s=t$ for $s,t:\pi_n, s\downarrow$ for $s:\pi_n$ and $(t_1,\ldots,t_n)\in s$ for $s:\pi_n$ and $t_j:\iota$ $(j=1,\ldots,n)$. The axioms provide for suitable operations corresponding to atomic predicates and for the effect of Neg and $Conj$ on each π_n and Un on π_{n+1} to π_n for each n. In addition, we include the *Join operator* J for each n, which when applied to a sequence of n-ary predicates, i.e. a total $f:\iota\rightarrow\pi_n$, produces the join predicate $J(f):\pi_{n+1}$. In the language, so modified, the rule of substitution $A[P]\Rightarrow A[B/P]$ is restricted to A which do not contain terms of predicate type. Then Theorem 2 h olds as before. An advantage of the predicate type over the propositional type approach is that we can separate out the role of the Join operator from that of the logical operations while, as we saw, J is forced on us in the propositional type approach. Strahm has shown that if J is omitted, then the resulting system $\mathcal{U}^-$(NFA) is proof-theoretically equivalent to $RA_{<\omega}$.

3. **Quantifying function variables.** It was argued in Section 3 that for the general notion of unfolding, (partial) function variables in their schematic role ought not to be quantified. However, when we come to set theory and examine informal arguments that lead us to accept its basic principles and their immediate extensions, it is plausible to allow some degree or other of function quantification. Proof-theoretical strength there is sensitive to the decision as to whether to allow such quantification, and, if so, to what extent, as will be seen in the next section. Interestingly, it happens that in the case of NFA, even if we allow full function quantification in the language of $\mathcal{U}_0$(NFA), resp. $\mathcal{U}$(NFA), with suitable restrictions on the hypothesis $A[P]$ of the substitution rule as above, we do not alter proof-theoretic strength, i.e. Theorems 1 and 2 continue to hold as stated.

4. **The unfolding of finitist arithmetic.** Clearly the starting point for the study of this notion would be a *quantifier-free* system FA based on Axs 1 and 2 and, in place of Ax 3, the *induction rule*

$$P(0), P(x)\rightarrow P(x')\Rightarrow P(x).$$

Beyond this, there are various notions of unfolding to be considered, related to various informal and formal explanations of finitism in the literature, due especially to Hilbert, Kreisel and Tait. Research on these notions is in progress.

5. The unfolding of set theory

This section is largely programmatic and, given the limitations of space, necessarily sketchy. On the face of it, set theory offers a prime candidate for the study of what is implicit in given notions and principles by means of the unfolding procedure, both for ZF as a schematic theory and for Gödel's program for new axioms. We begin with the former.

In the spirit of the functional formulation of the $\mathcal{U}_0$ and $\mathcal{U}$ procedures, we take the basic language of set theory to have individual variables a, b, c, x, y, z, $\ldots$, variables $f, g, h, \ldots$ for partial functions, the constants 0 and ω, the operation symbols $\{,\}, \bigcup, \wp$, and E (the characteristic function of the $\in$ relation) and the relation symbols $=$ and $\in$. In addition we have functionals **S**, **R** and **A** whose meaning will be explained in a moment. The axioms of the system ST are, besides Extensionality, the expected ones for $0, \omega, \{,\}, \bigcup, \wp$, and E, and the following four function and predicate schemata:

$$(\mathbf{S}) \quad \forall x \in a[\, f(x)\downarrow] \rightarrow \mathbf{S}(f,a)\downarrow \wedge \forall x[x \in \mathbf{S}(f,a) \leftrightarrow x \in a \wedge f(x) = 0]$$

$$(\mathbf{R}) \quad \forall x \in a[\, f(x)\downarrow] \rightarrow \mathbf{R}(f,a)\downarrow \wedge \forall y[\, y \in \mathbf{R}(f,a) \leftrightarrow \exists x \in a(\, f(x) = y)\,]$$

$$I(\in) \qquad \forall x[\, (\forall y \in x)P(y) \rightarrow P(x)\,] \rightarrow \forall x(P(x))$$

$$(\mathbf{A}) \qquad \forall x[f(x)\downarrow] \rightarrow \mathbf{A}(f)\downarrow \wedge [\,\mathbf{A}(f) = 0 \leftrightarrow \forall x(f(x) = 0)\,].$$

Thus **S** gives Separation, **R** gives Replacement, $I(\in)$ is the positive (inductive) schematic form of the Axiom of Foundation, and **A** serves to represent every definable class by means of a characteristic function. This last allows (**S**) and (**R**) to take the place of the expected schemata:

$$(\text{Sep}) \qquad \exists b \forall x[\, x \in b \leftrightarrow x \in a \wedge P(x)\,], \text{ and}$$

$$(\text{Repl}) \qquad (\forall x \in a)\exists! y P(x,y) \rightarrow \exists b \forall y[\, y \in b \leftrightarrow (\exists x \in a)P(x,y)\,].$$

The point of doing it by the above function schemata instead is that we can treat a wide variety of set theories uniformly, with the only changes being the deletion or addition (with appropriate axioms) of various individual, function and functional constants. For example, if we omit $\omega, \wp$, and **A**, we obtain a functional schematic form AST of *Admissible Set Theory*. To be more precise KP (taken with Δ_0-Replacement instead of Δ_0-Collection) is contained in $\mathcal{U}_0(\text{AST})$, and the latter is interpretable in the constructible sets of the former by taking the function variables to range over the $\Sigma_1^{(L)}$ partial functions. It would be of interest to determine the strength of $\mathcal{U}(\text{AST})$.

Quite a few useful general principles and functional constructions can be derived in $\mathcal{U}_0(\mathrm{AST})$ and $\mathcal{U}(\mathrm{AST})$, which then carry over to (the respective unfolding) of any set theory S extending AST. In particular, we can derive principles of induction for various classes C with an ordering $<_C$ in the form:

$$I(<_C) \qquad \forall x \in C[\forall y(y <_C x \to P(y)) \to P(x)] \to \forall x \in C[P(x)].$$

Here $<_C$ might be much "longer" than the ordinals, for which we have $I(<)$ by the axiom $I(\in)$. Taking Ω as a symbol for the class of ordinals, we can define, for example, the lexicographic ordering $<_{\Omega^2}$ on pairs of ordinals by $\langle \xi, \eta\rangle <_{\Omega^2} \langle \alpha, \beta\rangle \leftrightarrow \xi < \alpha \vee \xi = \alpha \wedge \eta < \beta$, and prove $I(<_{\Omega^2})$ in $\mathcal{U}_0(\mathrm{AST})$. From this and the LFP construction we can derive a principle of recursion for hierarchies of functions $\lambda\alpha, \beta.\, f_\alpha(\beta)$ by means of any given functional G which determines each f_α in terms of $\langle f_\xi\rangle_{\xi<\alpha}$. More generally, I expect that we can establish $I(<_\rho)$ in $\mathcal{U}_0(\mathrm{AST})$ for each $\rho < \varepsilon_{\Omega+1}$ and similarly for each $\rho < \Gamma_{\Omega+1}$ in $\mathcal{U}(\mathrm{AST})$, where the ordering up to $\Gamma_{\Omega+1}$ is defined in AST on a suitable class of "notations" as in Feferman [1968]. We would then obtain related principles of recursion and construction of hierarchies as for $<_{\Omega^2}$ above. Note that the form of this ordering is independent of which set theory S we are in, but the interpretation in a standard model of S depends on what ordinal Ω turns out to be. What stronger S serve to do is supply a greater variety of functionals G for generating hierarchies associated with $<_\rho$ when $I(<_\rho)$ is provable.

Suppose S is an extension of our initial system ST to which we have added AC and the existence of arbitrarily large inaccessible cardinals. Then the preceding allows us to actually "name" specific large inaccessibles in the unfolding systems of S. In that sense, it already gives us some large cardinal axioms. But if we are to generate, e.g., hierarchies of Mahlo cardinals, we need to add to ST a new scheme which says in effect that *whatever holds in the universe of sets already holds in arbitrarily large transitive sets*, or what one would call a scheme of *Downwards Reflection*. This takes the following form:

$$\text{(D-Ref)} \qquad P \to \exists b[a \in b \wedge Trans(b) \wedge P^{(b)}].$$

If this scheme is denoted $A[P]$ and B is a statement which involves both quantified individual variables and (possibly) quantified function variables, when forming $B^{(b)}$ in $A[B/P]$ we relativize the former variables to b as usual, and the latter variables to partial functions from b to b. Write $Strans(b)$ for $\forall x \in b \forall y[y \subseteq x \to y \in b]$. We can infer

$$\text{(D-Ref)}' \qquad P \to \exists b[a \in b \wedge Strans(b) \wedge P^{(b)}]$$

by substituting $P \wedge \forall x \exists y[\wp(x) = y]$ for P in (D-Ref). Thus with the substitution rule $A[P] \Rightarrow A[B/P]$ taken to apply to any statement B in the unfolding language of ST in which function variables may be quantified unrestrictedly, we obtain a form of Bernays' downward second-order reflection principle (Bernays [1961], following on Levy [1960]). And as Bernays showed op.cit., the existence of hierarchies of Mahlo cardinals then follows from this principle. Briefly, one begins by substituting for P in (D-Ref) the statement that expresses that the universe is closed under power set and replacement, i.e.

$$\forall x \exists y[\wp(x) = y] \wedge \forall u \forall g[\forall x \in u\, \exists y(g(x) = y)]$$
$$\rightarrow \exists v[\mathbf{R}(g,u) = v \wedge \forall y(y \in v \leftrightarrow \exists x \in u(g(x) = y))].$$

Then the conclusion of this instance of D-Ref guarantees the existence of arbitrarily large inaccessible cardinals. It follows that any normal function on Ω has arbitrarily large inaccessible fixed-points. By substituting *that* statement for P in (D-Ref) we obtain the existence of arbitrarily large Mahlo cardinals — and so on.

Formulas involving (partial) function quantification are classified into the Π^1_n hierarchies as usual. The existence of Mahlo hierarchies follows from (D-Ref) by successively substituting suitable Π^1_1 statements for P. But if one is to obtain stronger large cardinal statements, e.g. the existence of weakly compact cardinals, it is necessary to make substitutions by more complicated formulas. For, as shown in the work of Hanf and Scott [1961], a cardinal κ is weakly compact iff it is Π^1_1 indescribable. The latter says that (D-Ref) holds in V_κ for all Π^1_1 statements, and saying *that* is Π^1_2. In general, we obtain the existence of arbitrarily large Π^1_n indescribables by suitably more complicated instances of (D-Ref). And that is *all* one can expect to follow from (D-Ref) in our languages using only function variables of type level 1 over the universe. And passing to higher types — however one were to argue for that — for substitution instances in (D-Ref), at most allows one to obtain the existence of Π^m_n indescribables for all m, n. But one certainly cannot obtain in this way the existence of measurable cardinals nor even some of its familiar consequences such as the existence of $0^\#$ (or even of some still weaker consequences from infinitary combinatorics, such as explained in Kanamori [1994] p.109).

However, as I see it, there is already a flat difference between the reasoning which leads us to the hierarchies of Mahlo cardinals, and that which leads, to begin with, to weakly compact cardinals. Here a quotation from Tarski is apropos:

> ... the belief in the existence of inaccessible cardinals $> \omega$ (and even of arbitrarily large cardinals of this kind) seems to be a natural consequence of basic intuitions underlying the "naïve" set theory and referring to what can be called "Cantor's absolute". On the contrary, we see at this moment no cogent intuitive reasons which could induce us to believe in the existence of cardinals $> \omega$ that are not strongly incompact, or which at least would make it very plausible that the hypothesis stating the existence of such cardinals is consistent with familiar axiom systems of set theory. As was pointed out at the end of Section 1, we do not know of any "constructively characterized"

> cardinal $> \omega$ of which we cannot prove that it is strongly incompact and for which therefore the problems discussed remain open. (Tarski [1962], p.134).

Gödel, commenting on this in a footnote (20) added in 1965 to his 1940 monograph (after referring to the work of Levy [1960] and Bernays [1961] leading to "all of Mahlo's axioms") said:

> Propositions which, if true, are extremely strong axioms of infinity of an entirely new kind have been formulated and investigated as to their consequences and mutual implications in Tarski [1962], Keisler and Tarski [1964] and the papers cited there. In contradistinction to Mahlo's axioms the truth (or consistency) of these axioms does not immediately follow from "the basic intuitions underlying abstract [sic] set theory" (Tarski [1962], p. 134), nor can it, as of now, be derived from them. However, the new axioms are supported by rather strong arguments from *analogy* ... (Gödel [1990] p. 97, italics mine).

What makes the separation of Mahlo from weakly compact cardinals reasonable is that when we substitute for P in (D-Ref) a Π^1_1 statement B, we may read B as asserting a *closure condition* in the ordinary sense on V under given function(al)s. But this reading is not plausibly extended to statements of higher function-quantifier complexity. From what Gödel says in the preceding quotation, it seems he would agree with this argument for demarcation.[7]

My personal attitude concerning the question of "actual" existence of various kinds of large cardinals, whether smaller or larger, is that *it is all pie in the sky*. This may make one wonder why I have even bothered with the present section. Well, the starting point was to see what one can say about which large cardinal statements are implicit in the basic notions and principles of set theory, *if* one accepts them, as Gödel and many other logicians certainly do, and to try to apply the unfolding procedure to begin to say something precise about that.[8] While that hypothetical acceptance does not apply to me, there are other potential values of great interest to me, which I hope will result from further pursuit of the present framework. The analogues to various large cardinal statements in admissible set theory are well-known.

[7] Tait [1990], p.76, ftn. 6, is puzzled by this view of Gödel's. But he says there that the existence of weakly compact cardinals follows from Π^1_1 reflection, which is mistaken, as we have seen.

[8] And if one is among the set theorists who believe there are reasons for accepting much larger cardinals than follow from $\mathcal{U}$(ST), it should be of interest to make explicit what are the basic notions and principles that lead one to such conclusions, rather than depend on arguments from analogy or fruitfulness. In this respect, a suggestion of Gödel in his 1946 Princeton remarks is most provocative: "It is cert ainly impossible to give a combinational and decidable characterization of what an axiom of infinity is; but there might exist, e.g., a characterization of the following sort: An axiom of infinity is a proposition which has a certain (decidable) formal structure and which in addition is true." (Gödel [1990], p. 151)

The work earlier in this section with AST suggests to me that there should be a way of stating these as part of a *common generalization* via the unfolding of S+(D-Ref) for S⊇AST, and not merely an *analogue*. Still further, there has been a surprising use of recursive ordinal notation systems employing "names" for very large cardinals in current proof-theoretic ordinal an alyses of formal systems (cf. e.g. Rathjen [1995]). What I would really hope comes out of this is a generalization which encompasses these as well, and helps explain how it is that they come to be employed at all for these purposes.

References

P. Aczel [1980], Frege structures and the notions of proposition, truth and set, in (J. Barwise, et al, eds.) *The Kleene Symposium*, North-Holland, Amsterdam. 31-59.

M. Beeson [1985], *Foundations of Constructive Mathematics*, Springer-Verlag, Berlin.

P. Bernays [1961], Zur Frage der Unendlichkeitsschemata in der axiomatischen Mengenlehre, in (Y. Bar-Hillel, et al, eds.) *Essays on the Foundations of Mathematics*, Magnes Press, Jerusalem, 3-49. (See also Bernays [1976].)

P. Bernays [1976], On the problem of schema of infinity in axiomatic set theory, in (G. Müller, ed.) *Sets and Classes*, North-Holland, Amsterdam, 121-172. (English translation of Bernays [1961].)

S. Feferman [1962], Transfinite recursive progressions of axiomatic theories, *J. Symbolic Logic* 27, 259-316.

S. Feferman [1964], Systems of predicative analysis, *J. Symbolic Logic* 29, 1-30.

S. Feferman [1968], Systems of predicative analysis, II: Representations of ordinals, *J. Symbolic Logic* 33, 193-220.

S. Feferman [1979], A more perspicuous formal system for predicativity, in (K. Lorenz, ed.) *Konstruktionen vs. Positionen*. Vol. I, Walter de Gruyter, Berlin, 68-93.

S. Feferman [1988], Turing in the land of $O(z)$, in (R. Herken, ed.) *The Universal Turing Machine. A Half-century Survey*, Oxford Univ. Press, Oxford, 113-147.

S. Feferman [1991], Reflecting on incompleteness, *J. Symbolic Logic* 56, 1-49.

S. Feferman [1991a], A new approach to abstract data types, II. Computation on ADTs as ordinary computation, in (E. Börger, et al, eds.) *Computer Science Logic*, Lecture Notes in Computer Science 626, 79-95.

S. Feferman [1996], Computation on abstract data types. The extensional approach, with an application to streams, to appear in *Annals of Pure and Applied Logic*.

K. Gödel [1986], *Collected Works, Vol. I. Publications 1929-1936*, Oxford Univ. Press, New York.

K. Gödel [1990], *Collected Works, Vol. II. Publications 1938-1974*, Oxford Univ. Press, New York.

K. Gödel [1995], *Collected Works, Vol. III. Unpublished Essays and Lectures*, Oxford Univ. Press, New York.

W. P. Hanf [1964], Incompactness in languages with infinitely long expressions, *Fundamenta Mathematicae* 53, 309-324.

W. P. Hanf and D. Scott [1961], Classifying inaccessible cardinals (abstract), *Notices A.M.S.* 8, 445.

R. Jensen [1995], Inner models and large cardinals, *Bull. Symbolic Logic* 1, 393-407.

A. Kanamori [1994], *The Higher Infinite*, Springer-Verlag, Berlin.

H. J. Keisler and A. Tarski [1964], From accessible to inaccessible cardinals, *Fundamenta Mathematicae* 53, 225-308. Corrections ibid. 57 (1965) 119.

G. Kreisel [1958], Ordinal logics and the characterization of informal concepts of proof, *Proc. International Congress of Mathematicians (Edinburgh 1958)*, Cambridge Univ. Press, New York, 289-299.

G. Kreisel [1970], Principles of proof and ordinals implicit in given concepts, in (J. Myhill, et al, eds.) *Intuitionism and Proof Theory*, North-Holland, Amsterdam, 489-516.

A. Levy [1960], Axiom schemata of strong infinity in axiomatic set theory, *Pacific Journal of Mathematics* 10, 223-238.

P. Maddy [1988], Believing the axioms, I. *J. Symbolic Logic* 53, 481-511.

P. Maddy [1988a], Believing the axioms, II. *J. Symbolic Logic* 53, 736-764.

D. A. Martin [1976], Hilbert's first problem: The continuum hypothesis, in (F. Browder, ed.) *Mathematical Developments Arising from Hilbert's Problems*, Proc. Symposia in Pure Math. 28, A.M.S., Providence, 81-92.

Y. Moschovakis [1989], The formal language of recursion, *J. Symbolic Logic* 54, 1216-1252.

M. Rathjen [1995], Recent advances in ordinal analysis: Π^1_2-CA and beyond, *Bull. Symbolic Logic* 1, 468-485.

K. Schütte [1965], Eine Grenze für die Beweisbarkeit der transfiniten Induktion in der verzweigten Typenlogik, *Archiv für Math. Logik und Grundlagenforschung* 7, 45-60.

D. Scott [1961], Measurable cardinals and constructible sets, *Bull. de l'Acad. Polonaise des Sciences* 9, 521-524.

W. Tait [1990], The iterative hierarchy of sets, *Iyyun, A Jerusalem Philosophical Quarterly* 39, 65-79.

A. Tarski [1962], Some problems and results relevant to the foundations of set theory, in (E. Nagel, et al, eds.) *Logic, Methodology and the Philosophy of Science* (Proc. of the 1960 International Congress, Stanford), Stanford Univ. Press, Stanford, 125-135.

A. Turing [1939], Systems of logic based on ordinals, *Proc. London Math. Soc.*, ser. 2, 45, 161-228.

Infinite-valued Gödel Logics with 0-1-Projections and Relativizations *

Matthias Baaz **

Institut für Algebra und Diskrete Mathematik E118.2,
Technische Universität Wien, A–1040 Vienna, Austria
Email address: baaz@logic.tuwien.ac.at

Summary. Infinite-valued Gödel logic, i.e., Dummett's **LC**, is extended by projection modalities and relativizations to truth value sets. An axiomatization for the corresponding propositional logic (sound and complete relative to any infinite set of truth values) is given. It is shown that certain simple infinite sets of truth values correspond to first-order Gödel logics which are not recursively axiomatizable.

1. Introduction

One of Gödel's main contributions to the study of nonstandard, in particular, many-valued and intuitionistic logics was his [4]. In that paper, he introduced a sequence of finite-valued propositional logics $\mathbf{G}_n$ intermediate in strength between classical and intuitionistic propositional logic. The definition of $\mathbf{G}_n$ is uniform, i.e., makes no explicit reference to the number of truth values. The only restrictions on the set of truth values V are that V is a (linearly ordered) subset of $[0, 1]$ and that $0, 1 \in V$. Dummett [3] subsequently showed that the infinite-valued Gödel logics are axiomatized by intuitionistic propositional calculus plus the axiom schema $(A \supset B) \vee (B \supset A)$. We extend this result to infinite-valued Gödel logics with the projection modalities on 0 and 1:

$$\nabla(A) = \begin{cases} 1 & \text{if } A \neq 0 \\ 0 & \text{if } A = 0 \end{cases} \qquad \Delta(A) = \begin{cases} 1 & \text{if } A = 1 \\ 0 & \text{if } A \neq 1 \end{cases}$$

Only the addition of Δ is of interest, as ∇ may be defined by $\nabla(A) \equiv \neg\neg A$.

The main result of the first part of this paper is the completeness theorem for all infinite sets of truth values V for the axiomatization consisting of the axiom schemas of intuitionistic propositional logic and of modal logic **S4** for Δ (including the necessitation rule $A/\Delta A$), plus the following schemas:

$$(A \supset B) \vee (B \supset A)$$
$$\Delta A \vee \neg\Delta A$$
$$\Delta(A \vee B) \supset \Delta A \vee \Delta B$$

* This paper is in its final form and no similar paper has been or is being submitted elsewhere.

** This work is supported by FWF grant P–10282 MAT.

The completeness result will be extended to relativizations to arbitrary truth value sets: A relativization to a subset $W \subseteq V$ is obtained by adding a new connective R_W with truth function

$$\mathsf{R}_W A = \begin{cases} 1 & \text{if } A \in W \\ 0 & \text{otherwise} \end{cases}$$

If V is closed under least upper and greatest lower bounds, we arrive naturally at first-order versions of the corresponding infinite-valued logics, taking as $\forall$ the inf and as $\exists$ the sup of the corresponding truth-value distributions. It is worth pointing out right away that, in contrast to the propositional case, which logic we get depends crucially on the order type of the set of truth values. So the Gödel logic based on $[0, 1]$ is axiomatizable while the one based on $\{1/k : k \in \omega \setminus \{0\}\} \cup \{0\}$ is not. The main result of the second part of the present paper is that *none* of the infinite-valued first-order Gödel logics with projection modalities (i.e., independent of the order type of the set of truth values) are recursively axiomatizable.

2. Propositional Gödel logics

We work in the language L_p of propositional logic containing a countably infinite set Var of propositional variables (X, Y, Z, ...), the constants $\top$ (true) and $\bot$ (false), as well as the connectives $\wedge$, $\vee$, $\supset$, and Δ. We introduce $\neg$ and ∇ as abbreviations: $\neg A \equiv A \supset \bot$ and $\nabla A \equiv \neg\neg A$. The set of formulas of L_p is denoted $\mathrm{Frm}(L_p)$.

Definition 2.1. *Let $V \subseteq [0, 1]$ be some set of truth values which contains* 0 *and* 1. *A* valuation $\mathfrak{V}$ based on V *is a function from* Var *to* V.. *The valuation for formulas is defined as follows:*

1. $A \equiv \top$: $\mathfrak{V}(A) = 1$.
2. $A \equiv \bot$: $\mathfrak{V}(A) = 0$.
3. $A \equiv B \wedge C$: $\mathfrak{V}(A) = \min(\mathfrak{V}(B), \mathfrak{V}(C))$.
4. $A \equiv B \vee C$: $\mathfrak{V}(A) = \max(\mathfrak{V}(B), \mathfrak{V}(C))$.
5. $A \equiv B \supset C$:
$$\mathfrak{V}(A) = \begin{cases} \mathfrak{V}(C) & \text{if } \mathfrak{V}(B) > \mathfrak{V}(C) \\ 1 & \text{if } \mathfrak{V}(B) \leq \mathfrak{V}(C). \end{cases}$$
6. $A \equiv \Delta B$:
$$\mathfrak{V}(A) = \begin{cases} 1 & \text{if } B = 1 \\ 0 & \text{if } B \neq 1 \end{cases}$$

$\mathfrak{V}$ satisfies *a formula* A, $\mathfrak{V} \models A$, *if* $\mathfrak{V}(A) = 1$. *The* propositional Gödel logic based on V, $\mathbf{GP}(V)$, *is the set of formulas A s.t. $\mathfrak{V}(A) = 1$ for every $\mathfrak{V}$ based on V. We write* $\mathbf{GP}(V) \models A$ *for* $A \in \mathbf{GP}(V)$.

It is easily verified that, for any $\mathfrak{V}$,

$$\mathfrak{V}(\neg B) = \begin{cases} 0 & \text{if } \mathfrak{V}(B) \neq 0 \\ 1 & \text{otherwise.} \end{cases} \qquad \mathfrak{V}(\triangledown B) = \begin{cases} 1 & \text{if } B \neq 0 \\ 0 & \text{if } B = 0 \end{cases}$$

Also, $\triangle$ cannot be defined by the other connectives. (To see this, suppose that $B(X)$ takes on only 0 or 1. Then, if $\mathfrak{V}(X) \neq 0 \neq \mathfrak{V}(Y)$, $\mathfrak{V}(B(X)) = \mathfrak{V}(B(Y))$ *independently* of whether $\mathfrak{V}(X) = 1$.)

Definition 2.2. *Let* **LGP** *be the calculus obtained by adding to the calculus for intuitionistic propositional logic [5] the following axioms*

$$\begin{array}{l} \top \\ (A \supset B) \vee (B \supset A) \\ \triangle A \vee \neg\triangle A \\ \triangle(A \vee B) \supset \triangle A \vee \triangle B \\ \triangle A \supset A \\ \triangle A \supset \triangle\triangle A \\ \triangle(A \supset B) \supset \triangle A \supset \triangle B \end{array}$$

as well as the rule

$$\frac{A}{\triangle A}$$

Remark 2.1. Maehara [6, Ch. 1] gave a sequent calculus for intuitionistic logic where the restriction to at most formula in the succedent applies not generally but only in the case of applications of $\supset$:right. If we use such a calculus, the axioms and rules involving $\triangle$ may be subsumed under the two rules:

$$\frac{A, \Gamma \to \Delta}{\triangle A, \Gamma \to \Delta} \qquad \frac{\triangle\Gamma \to \Delta}{\triangle\Gamma \to \triangle\Delta}$$

Proposition 2.1. **LGP** *is sound for* **GP**(V), *i.e., if* **LGP** $\vdash S$ *then* **GP**$(V) \models S$.

Proof. By induction on length of derivations.

Proposition 2.2. *The deduction theorem obtains in the form: If* $A_1, \ldots, A_n \vdash B$, *then* $\vdash \triangle A_1 \wedge \ldots \wedge \triangle A_n \supset B$.

Note that the usual deduction theorem, which is true in ordinary (infinite-valued) Gödel logic, is false here: In general, $A \supset \triangle A$ is not valid.

Proposition 2.3. **LGP** *proves the following formula:*

$$\begin{array}{l} (A \supset B) \wedge (B \supset A) \vee \\ (A \supset B) \wedge ((B \supset A) \supset A) \vee \\ (A \supset B) \wedge ((A \supset B) \supset B) \end{array}$$

Proof. Derive $(B \supset A) \vee ((B \supset A) \supset A)$ from $(B \supset (B \supset A)) \vee ((B \supset A) \supset B)$, $B \supset (B \supset A) \supset B \supset A$, $((B \supset A) \supset B), (B \supset A) \supset (B \supset A) \supset ((B \supset A) \supset A)$, and similarly $(A \supset B) \vee ((A \supset B) \supset B)$. The result follows from $(A \supset B) \vee (B \supset A)$.

3. Completeness of propositional Gödel logics with projections

To prove completeness of **LGP** we first show that an ordering of the variables w.r.t. $\supset$ induces an ordering on all formulas containing these variables only.

Lemma 3.1. *Let* $U = \{X_1, \ldots, X_n, \Delta X_1, \ldots, \Delta X_n, \top, \bot\}$, *and let* G *contain*

(a) $\{A \supset B, B \supset A\}$ *or* $\{A \supset B, (B \supset A) \supset A\}$ *or* $\{(A \supset B) \supset B, B \supset A\}$ *for all* $A, B \in U$.
(b) $\{\top \supset \Delta A\}$ *or* $\{\Delta A \supset \bot\}$ *for all* $\Delta A \in U$.

Then (a), (b) are derivable from **G** *for all formulas containing only* $X_1, \ldots,$ X_n *as propositional variables.*

Proof. By induction on the complexity of formulas:

1. $A \equiv C \vee D$: (a) By induction hypothesis, $\mathbf{G} \vdash C \supset D$ or $\mathbf{G} \vdash D \supset C$, i.e., $\mathbf{G} \vdash A \leftrightarrow D$ or $\mathbf{G} \vdash A \leftrightarrow C$. Apply induction hypothesis.
   (b) By (a) we have $\mathbf{G} \vdash \Delta A \leftrightarrow \Delta C$ or $\mathbf{G} \vdash \Delta A \leftrightarrow \Delta D$. Apply induction hypothesis.
2. $A \equiv C \wedge D$: Similarly.
3. $A \equiv C \supset D$: (a) By induction hypothesis, $\mathbf{G} \vdash C \supset D$ or $\mathbf{G} \vdash (C \supset D) \supset D$, so $\mathbf{G} \vdash A \leftrightarrow \top$ or $\mathbf{G} \vdash A \leftrightarrow D$. Apply induction hypothesis.
   (b) By (a), we have $\mathbf{G} \vdash \Delta A \leftrightarrow \top$ or $\mathbf{G} \vdash \Delta A \leftrightarrow \Delta D$.
4. $A \equiv \Delta C$: (a) By induction hypothesis (b) for C, $\mathbf{G} \vdash \Delta C \leftrightarrow \top$ or $\mathbf{G} \Delta C \leftrightarrow \bot$.
   (b) $\mathbf{G} \vdash \Delta\Delta C \leftrightarrow \Delta A$, so by (a), $\mathbf{G} \vdash \top \supset \Delta C$ or $\mathbf{G} \vdash \Delta C \supset \bot$, hence $\mathbf{G} \vdash \top \supset \Delta A$ or $\mathbf{G} \vdash \Delta A \supset \bot$. □

Definition 3.1. *Let* $U = \{X_1, \ldots, X_n, \Delta X_1, \ldots, \Delta X_n, \top, \bot\}$, *and let* G *be a set of formulas of the form* $A \supset B$ *or* $(B \supset A) \supset A$ *with* $A, B \in U$. G *is called a* complete $\supset$-order *if (a) and (b) of Lemma 3.1 are satisfied,* $\{A \supset \top, \bot \supset B, A \supset A\} \subseteq G$ *for all* $A, B \in U$, *and* $(\Delta X_i \supset X_i) \in G$ *for all* $i = 1, \ldots, n$.

Definition 3.2. *Let* G *be a complete* $\supset$*-order. The* stratification $[G^*, H^*]$ *of* G *is defined as follows: Let* G' *be the least set* $G' \supseteq G$ *s.t., if* $\{A \supset C, C \supset B\} \subseteq G'$, *then also* $(A \supset B) \in G'$ *(for all* A, B, $C \in U$*), and s.t., if* $(\top \supset X_i) \in G'$ *then also* $(\top \supset \Delta X_i) \in G'$ *(for all* $\Delta X_i \in U$*). Then*

$$\begin{aligned} G^* &= G' \setminus \{(A \supset B) \supset B : (A \supset B) \in G'\}, \text{ and} \\ H^* &= \{B : \{(A \supset B) \supset B, A \supset B\} \subseteq G'\} \end{aligned}$$

Proposition 3.1. *Let* G *be a complete* $\supset$*-order, and* $[G^*, H^*]$ *be its stratification. Then* $G \equiv \bigwedge G^* \wedge \bigwedge H^*$

Proof. Note that $\vdash (A \supset B) \supset (((A \supset B) \supset B) \leftrightarrow B)$.

Definition 3.3. *Let* $V = \{X_1, \ldots, X_n, \Delta X_1, \ldots, \Delta X_n, \top, \bot\}$, *let* G *be a complete* $\supset$*-order on* V, *and let* D, E, E' *and* F *be formulas in the variables* X_1, ..., X_n. *The* right reduction $R(E \supset D)$ *is defined by*

$$R(E \supset D) = \begin{cases} E \supset D_1 & \text{if } G \vdash C_1 \\ E \supset D_2 & \text{if } G \vdash C_2; \end{cases}$$

the left reduction $L(E \wedge D \wedge E' \supset F)$ *is defined by*

$$L(E \wedge D \wedge E' \supset F) = \begin{cases} E \wedge D_1 \wedge E' \supset F & \text{if } G \vdash C_1 \\ E \wedge D_2 \wedge E' \supset F & \text{if } G \vdash C_2, \end{cases}$$

where D_i, C_i *are as follows:*

D	D_1	C_1	D_2	C_2
$A \vee B$	B	$A \supset B$	A	$B \supset A$
$A \wedge B$	A	$A \supset B$	B	$B \supset A$
$A \supset B$	$\top$	$A \supset B$	B	$(A \supset B) \supset B$
ΔA	$\top$	$\top \supset \Delta A$	$\bot$	$\Delta A \supset \bot$.

Proposition 3.2. $G \vdash S \leftrightarrow R(S)$ *and* $G \vdash S \leftrightarrow L(S)$.

Proposition 3.3. *Let* G *be a complete and stratified* $\supset$*-order on* $U = \{X_1, \ldots, X_n, \Delta X_1, \ldots, \Delta X_n, \top, \bot\}$, *and let* $A \in U$, $H \subseteq U$. *Suppose that (a)* $(\top \supset \bot) \notin G$ *and (b)* $(B \supset A) \notin G$ *if* $B \in H$. *Then there is a valuation* $\mathfrak{V}$ *on* $V_{n+2} = \{1/k : 1 \leq k \leq n+1\} \cup \{0\}$ *s.t.* $\mathfrak{V}(G) = 1$ *and* $\mathfrak{V}(H) > \mathfrak{V}(A)$.

Proof. G determines an equivalence relation on U by: $A \in [B]$ iff $\{A \supset B, B \supset A\} \subseteq G$, and an order on the equivalence classes by: $[A] < [B]$ iff $(A \supset B) \in G$ but $(B \supset A) \notin G$. We have $[\top] \neq [\bot]$ and $[\bot] < [\top]$ by (a). ΔA is in $[\top]$ or $[\bot]$, and if $A \in [\top]$, then also $\Delta A \in [\top]$. Define $\mathfrak{V}$ according to the (at most $n+2$) equivalence classes on U, with $\mathfrak{V}([\top]) = 1$ and $\mathfrak{V}([\bot]) = 0$. Then $\mathfrak{V}(G) = 1$ and $\mathfrak{V}(H) > \mathfrak{V}(A)$ by (b).

Proposition 3.4. *Let* G, H, A *be as in the previous proposition. Then* $G \vdash H \supset A$ *if* $(\top \supset \bot) \in G$ *or* $(B \supset A) \in G$ *for some* $B \in H$.

Theorem 3.1. **LGP** *is complete for* **GP**(V) *for all infinite* D.

Proof. Suppose **LGP** $\nvdash H \supset A$. Let X_1, ..., X_n be the variables in H, A, and let $U = \{X_1, \ldots, X_n, \Delta X_1, \ldots, \Delta X_n, \top, \bot\}$. Let G_1, ..., G_r be all combinations of the sets $\{A \supset B, B \supset A\}$, $\{A \supset B, (B \supset A) \supset A\}$, $\{(A \supset B) \supset B, B \supset A\}$, and of $\{\top \supset \Delta X_i\}$, $\{\Delta X_i \supset \bot\}$, for all A, B, $\Delta X_i \in U$. Finally, let $G' = \{\Delta X_i \supset X_i : 1 \leq i \leq n\} \cup \{(A \supset \top), (\bot \supset A) : A \in U\}$. By Proposition 2.3, $G_i, G' \nvdash H \supset A$. Let $[G^*, H^*]$ be the stratification of $G_i \cup G'$. Then, by Propositions 2.2 and 3.1, $G^* \vdash \Delta H^* \wedge H \supset A$. The reduction rules yield H' and A' s.t. $H' \cup \{A'\} \subseteq U$ and $\mathfrak{V}(\Delta H^* \wedge H \supset A) = \mathfrak{V}(\Delta H^* \wedge H' \supset$

A') whenever $\mathfrak{V}(G^*) = 1$. By Proposition 3.3, there is such a valuation $\mathfrak{V}$ on V_{n+2} s.t. $\mathfrak{V}(G^*) = 1$ and $\mathfrak{V}(\Delta H^* \wedge H') > \mathfrak{V}(A')$. $\mathfrak{V}$ may naturally be taken to be an interpretation on any infinite set of truth values making $H \supset A$ not true.

Corollary 3.1. $\mathbf{GP}(V) = \bigcap_{n\in\omega} \mathbf{GP}(V_n)$ *for any infinite* V.

$\mathbf{GP}(V)$ for finite V is axiomatizable using suitable sequent calculi [1]. If $|V| = n + 2$ one may obtain an axiomatization also directly by adding the schema

$$(\bigvee_{1\leq i\leq n} Z \leftrightarrow X_i) \vee Z \leftrightarrow \top \vee Z \leftrightarrow \bot$$

to **LGP**. To see this, take a formula A valid in $\mathbf{GP}(V)$ and replace each variable occurring in it by one of $X_1, \ldots, X_n, \top, \bot$. All formulas thus obtained are valid in $\mathbf{GP}(V)$ and have $\leq n$ variables, hence, are provable in **LGP**. The schema then yields A.

We now proceed to extend the completeness result to relativizations to arbitrary subsets of the set of truth values.

Definition 3.4. *Let* $W \subseteq V$. *The language* L_p^W *is* L_p *plus a monadic operator* R_W. *The logic* $\mathbf{GP}(V, W)$ *is defined just like* $\mathbf{GP}(V)$ *with the truth function for* R_W *given by*

$$\mathfrak{V}(\mathsf{R}_W A) = \begin{cases} 1 & \text{if } \mathfrak{V}(A) \in W \\ 0 & \text{otherwise} \end{cases}$$

Corollary 3.2. $\mathbf{GP}(V, W)$ *is axiomatizable for arbitrary* W.

Proof. We distinguish cases according to whether W is s.t.

(a) there are $\{d_i : i \in \omega\} \subseteq W$, $\{e_i : i \in \omega\} \subseteq V \setminus W$ s.t. $d_i < e_i < d_{i+1}$ for all $i \in \omega$,

or there is a maximal k s.t. there are $\{d_1, \ldots, d_k\} \subseteq W$ and $\{e_1, \ldots, e_k\} \subseteq V \setminus W$ and

(b) $d_1 < e_2 < \ldots < e_{k-1} < d_k$,
(c) $d_1 < e_2 < \ldots < d_{k-1} < e_k$,
(d) $e_1 < d_2 < \ldots < e_{k-1} < d_k$,
(e) $e_1 < d_2 < \ldots < d_{k-1} < e_k$.

We extend **LGP** by

$$(A \leftrightarrow B) \supset (\mathsf{R}_W A \leftrightarrow \mathsf{R}_W B)$$
$$\mathsf{R}_W A \vee \neg \mathsf{R}_W A$$
$$\mathsf{R}_W U\ (\neg \mathsf{R}_W U) \text{ if } U \in W\ (U \notin W) \text{ for } U = \top, \bot.$$

plus the following formulas for cases (b)–(e):

(b) $\neg(\neg \mathsf{R}_W A_0 \wedge \mathsf{R}_W A_1 \wedge \neg \mathsf{R}_W A_2 \wedge \ldots \wedge \mathsf{R}_W A_k \wedge \bigwedge \neg\Delta(A_{i+1} \supset A_i))$ and $\neg(\mathsf{R}_W A_1 \wedge \neg \mathsf{R}_W A_2 \wedge \ldots \wedge \mathsf{R}_W A_k \wedge \neg \mathsf{R}_W A_{k+1} \wedge \bigwedge \neg\Delta(A_{i+1} \supset A_i))$

(c) $\neg(\neg R_W A_0 \wedge R_W A_1 \wedge \neg R_W A_2 \wedge \ldots \wedge \neg R_W A_k \wedge \bigwedge \neg\Delta(A_{i+1} \supset A_i))$ and $\neg(R_W A_1 \wedge \neg R_W A_2 \wedge \ldots \wedge \neg R_W A_k \wedge R_W A_{k+1} \wedge \bigwedge \neg\Delta(A_{i+1} \supset A_i))$

(d) $\neg(R_W A_0 \wedge \neg R_W A_1 \wedge R_W A_2 \wedge \ldots \wedge R_W A_k \wedge \bigwedge \neg\Delta(A_{i+1} \supset A_i))$ and $\neg(\neg R_W A_1 \wedge R_W A_2 \wedge \ldots \wedge R_W A_k \wedge \neg R_W A_{k+1} \wedge \bigwedge \neg\Delta(A_{i+1} \supset A_i))$

(e) $\neg(R_W A_0 \wedge \neg R_W A_1 \wedge R_W A_2 \wedge \ldots \wedge \neg R_W A_k \wedge \bigwedge \neg\Delta(A_{i+1} \supset A_i))$ and $\neg(\neg R_W A_1 \wedge R_W A_2 \wedge \ldots \wedge \neg R_W A_k \wedge R_W A_{k+1} \wedge \bigwedge \neg\Delta(A_{i+1} \supset A_i))$

Now suppose $\nvdash A$. Let $[G^*, H^*]$ be constructed as in the proof of Theorem 3.1 (i.e., disregarding R_W) and let $Y_1, \ldots, Y_\ell$ be representatives of the equivalence classes other than those of $\bot$ and $\top$. Let $\{R_W(Y_1)^{n_1}, \ldots, R_W(Y_\ell)^{n_\ell}\}$ (where $n_i \in \{0,1\}$ and $A^0 \equiv \neg A$, $A^1 \equiv A$) be a restriction on the $R_W(Y_i)$ consistent with the order of $Y_1, \ldots, Y_\ell$ and with W. Since $G^* \cup H^* \vdash B \leftrightarrow C$ for $C \in \{Y_1, \ldots, Y_\ell, \top, \bot\}$ for all formulas B containing only the original variables and $\vee, \wedge, \supset, \top, \bot$, R_W can be eliminated step-by-step using the additional axioms. The construction of a counterexample then works as before.

Conversely, if for every restriction, $\{R_W(Y_1)^{n_1}, \ldots, R_W(Y_\ell)^{n_\ell}\} \cup G^* \cup H^* \vdash A$ holds, we get $G^* \cup H^* \vdash A$ by Proposition 3.4 and $R_W Y_i \vee \neg R_W Y_i$.

4. First-order Gödel logics

In considering first-order infinite valued logics, care must be taken in choosing the set of truth values. In order to define the semantics of the quantifier we must restrict the set of truth values to those which are closed under infima and suprema. (Note that in propositional infinite valued logics this restriction is not required.) For instance, the *rational* interval $[0,1] \cap \mathbb{Q}$ will not give a satisfactory set of truth values. The following, however, do:

$$\begin{aligned} V_R &= [0,1] \\ V^0 &= \{1/k : k \in \omega \setminus \{0\}\} \cup \{0\} \\ V^1 &= \{1 - 1/k : k \in \omega \setminus \{0\}\} \cup \{1\} \end{aligned}$$

We work in a usual first-order language L extending L_p by individual variables x, y, z, …, predicate symbols P, Q, …, function symbols f, g, …, and the quantifiers $\forall$ and $\exists$.

Definition 4.1. *Let $V \subseteq [0,1]$ be some set of truth values which contains 0 and 1 and is closed under supremum and infimum. An* interpretation $\mathfrak{I} = \langle D, \mathsf{s}\rangle$ based on V *is given by the* domain D *and the* valuation function s *where s maps atomic formulas in* $\mathrm{Frm}(L^{\mathfrak{I}})$ *into V and n-ary function symbols to functions from D^n to D.*

s can be extended in the obvious way to a function on all terms. The valuation for formulas is defined as follows:

(1) $A \equiv \top$: $\mathfrak{I}(A) = 1$.

(2) $A \equiv \bot$*:* $\mathfrak{I}(A) = 0$.
(3) $A \equiv P(t_1, \ldots, t_n)$ *is atomic:* $\mathfrak{I}(A) = \mathbf{s}(P)(\mathbf{s}(t_1), \ldots, \mathbf{s}(t_n))$.
(4) $A \equiv B \wedge C$*:* $\mathfrak{I}(A) = \min(\mathfrak{I}(B), \mathfrak{I}(C))$.
(5) $A \equiv B \vee C$*:* $\mathfrak{I}(A) = \max(\mathfrak{I}(A), \mathfrak{I}(B))$.
(6) $A \equiv B \supset C$*:*

$$\mathfrak{I}(A) = \begin{cases} \mathfrak{I}(C) & \text{if } \mathfrak{I}(B) > \mathfrak{I}(C) \\ 1 & \text{if } \mathfrak{I}(B) \leq \mathfrak{I}(C). \end{cases}$$

(7) $A \equiv \Delta B$*:*

$$\mathfrak{I}(A) = \begin{cases} 1 & \text{if } \mathfrak{I}(B) = 1 \\ 0 & \text{if } \mathfrak{I}(B) \neq 1 \end{cases}$$

The set $\{\mathfrak{I}(A(d)) : d \in D\}$ *is called the* distribution *of* $A(x)$*, we denote it by* $\mathrm{Distr}_{\mathfrak{I}}(A(x))$*. The quantifiers are, as usual, defined by infimum and supremum of their distributions.*

(8) $A \equiv (\forall x)B(x)$*:* $\mathfrak{I}(A) = \inf \mathrm{Distr}_{\mathfrak{I}}(B(x))$.
(9) $A \equiv (\exists x)B(x)$*:* $\mathfrak{I}(A) = \sup \mathrm{Distr}_{\mathfrak{I}}(B(x))$.

$\mathfrak{I}$ satisfies *a formula* A, $\mathfrak{I} \models A$, *if* $\mathfrak{I}(A) = 1$.

The first-order Gödel logic based on V, $\mathbf{G}(V)$, *is the set of all formulas* $A(\bar{x})$ *s.t.* $\mathfrak{I} \models A(\bar{a})$ *for every interpretation based on* V *and every* $\bar{a} \in V^{<\omega}$.

While the set of tautologies of propositional infinite-valued Gödel logic is independent of the set of truth values, this is not the case in the first-order case. Here, the infinite-valued systems need not be equivalent.

Proposition 4.1. *Let*

$$\begin{aligned} C &= (\exists x)(A(x) \supset (\forall y)A(y)) \text{ and} \\ C' &= (\exists x)((\exists y)A(y) \supset A(x)) \end{aligned}$$

C' *is valid in both* $\mathbf{G}(V^0)$ *and* $\mathbf{G}(V^1)$*.* C *is valid in* $\mathbf{G}(V^1)$ *but not in* $\mathbf{G}(V^0)$*. Neither* C *nor* C' *are valid in* $\mathbf{G}(V_R)$*.*

Proof. See [2].

If $V \subseteq V'$, then $\mathbf{G}(V') \subseteq \mathbf{G}(V)$, i.e., $\mathbf{G}(V_R)$ is the logic with the fewest valid formulas. The next proposition shows that there are infinitely many infinite-valued first-order Gödel logics.

Proposition 4.2. *Let* $V_k = \{\frac{x}{k} + \frac{1}{2yk} : 0 \leq x < k, y \in \omega \setminus \{0\}\}$*. Then* $\mathbf{G}(V_k) \subset \mathbf{G}(V_\ell)$ *if* $k > \ell$*.*

Proof. Since V_ℓ can be embedded in V_k preserving the order structure if $k > \ell$, we have $\mathbf{G}(V_k) \subseteq \mathbf{G}(V_\ell)$.

Let

$$\begin{aligned} F_k^* \equiv & \bigwedge_{0\le i<j<k} (\forall x)[P_i(x) \ll (\forall x)P_j(x)] \wedge \\ \wedge & \bigwedge_{0\le i<k} (\exists x)[P_i(x) \supset (\forall y)P(y)] \supset (\forall y)P_i(y) \wedge \\ \wedge & \bigwedge_{0\le i<k} (\forall x)P_i(x) \ll R \wedge (\forall x)Q(x) \ll R \wedge (\forall x)P_i(x) \ll (\forall x)Q(x), \end{aligned}$$

where $A \ll B \equiv (A \supset B) \wedge ((B \supset A) \supset A)$, and let

$$F_k \equiv F_k^* \supset (\exists x)(Q(x) \supset (\forall y)Q(y)) \vee R$$

If one of the conditions in F_k^* is not satisfied, then the value of F_k equals 1, since then every conjunct in F_k^* gets a value $\le R$. If all conditions are satisfied, and the value of R is < 1, then F_k expresses: $(\forall x)Q(x)$ is an infimum which is different from k distinct infima none of which is a minimum. In $\mathbf{G}(V_k)$, it then must be a minimum, and so the value of $(\exists x)(Q(x) \supset (\forall y)Q(y)) \vee R$ equals 1. This need not be the case in general in $\mathbf{G}(V_\ell)$ with $\ell > k$.

In terms of complexity, there may be significant differences as well. It was shown, e.g., that $\mathbf{G}(V_\mathbb{R})$ is axiomatizable [7], but that $\mathbf{G}(V^0)$ is not [2].

As in the propositional case, we may extend $\mathbf{G}(V)$ to $\mathbf{G}(V, W)$ by adding an operator R_W. The definition of interpretation is extended by adding the clause

(10) $A \equiv \mathsf{R}_W B$:

$$\mathfrak{I}(A) = \begin{cases} 1 & \text{if } \mathfrak{I}(B) \in W \\ 0 & \text{otherwise} \end{cases}$$

and the other definitions amended accordingly.

5. Incompleteness of first-order Gödel logics with 0-1-projections and relativizations

In order to prove the main theorem of this section we need some tools from recursion theory.

Definition 5.1. *Let ψ be an effective recursive enumeration of the set PR_1^1 of all primitive recursive functions from ω to ω. We define a two place function φ (which enumerates a subclass of PR_1^1):*

$$\varphi_k(x) = \begin{cases} 0 & \text{if } x = 0 \\ 0 & \text{if } \psi_k(y) = 0 \text{ for } 1 \le y \le x \\ 1 & \text{otherwise} \end{cases}$$

The index set O_φ is defined as $\{k : (\forall y)\varphi_k(y) = 0\}$.

Proposition 5.1. *The index set O_φ is not recursively enumerable.*

Proof. By definition of φ, $\{k : (\forall y)\varphi_k(y) = 0\} = \{k : (\forall y)\psi_k(y) = 0\}$. But for every $g \in \mathrm{PR}^1_1$ the index set $\{k : (\forall y)\psi_k = g\}$ is Π_1-complete. Therefore O_φ is Π_1-complete and thus not recursively enumerable.

The essence of the incompleteness proof is represented by a sequence of formulas $(A_k)_{k\in\omega}$ constructed via φ s.t.

$$\mathbf{G}(V,W) \models A_k \iff k \in O_\varphi$$

i.e. O_φ is m-reducible to the validity problem of $\mathbf{G}(V,W)$.

We prove the incompleteness result separately for W which have a cumulation point which is approached from above and those with one approached from below. The idea is to write down axioms which express that the values of $P(s^n(0))$ form a decreasing sequence in W as long as $\varphi_k(n) = 0$, and that $P(s^n(0))$ gets the value of $P(0)$ if $\varphi_k(n) \neq 0$ Using Δ, we can force $\mathrm{Distr}(P(x))$ to have no minimum iff the decreasing sequence is infinite. Thus, the value of $(\exists x)\Delta(P(x) \supset (\forall y)P(y))$ will equal 1 if $\varphi_k \equiv 0$ and equal 0 otherwise. For W with a cumulation point approached from below the argument is similar with an increasing sequence.

Theorem 5.1. *Suppose V is a set of truth values and $W \subseteq V$ is infinite and there are only finitely many elements of W between any two elements of W. Then $\mathbf{G}(V,W)$ is incomplete.*

Proof. Obviously W cannot have both a minimum and a maximum. Suppose first that it has no minimum, i.e., contains an infinite descending sequence with no lower bound in W.

Let P be a one-place predicate symbol, s be the function symbol for the successor function and $\bar{0}$ be the constant symbol representing 0 (in particular, we choose a signature containing this symbol and all symbols from Robinson's arithmetic Q).

Let A_1 be a conjunction of axioms strong enough to represent every recursive function (e.g. the axioms of Q) and a defining axiom for the function φ and write Δ in front of every positively occurring atomic formula. This ensures that these formulas behave essentially as classical formulas. We define the formulas A_2^k, A_3^k, A_4^k for $k \in \omega$; for formulas representing the equality $\varphi_k(x) = 0$ we write $[\varphi_k(x) = 0]$ (these also contain Δ in front of atomic formulas).

$A_2^k \equiv (\forall x, y)(\neg[\varphi_k(x) = 0] \wedge \Delta(x \leq y) \supset \neg[\varphi_k(y) = 0])$

$A_3^k \equiv (\forall x)[\neg[\varphi_k(x)=0] \supset \Delta(P(\bar{0}) \supset P(s(x))) \wedge \Delta(P(s(x)) \supset P(0)) \wedge \mathsf{R}_W P(s(x))]$

$A_4^k \equiv (\forall x)[[\varphi_k(s(x)) = 0] \supset \neg\Delta(P(x) \supset P(s(x))) \wedge \mathsf{R}_W P(s(x))]$

$A_5 \equiv \mathsf{R}_W P(0)$

Finally we set

$$B_k \equiv A_1 \wedge A_2^k \wedge A_3^k \wedge A_4^k \wedge A_5$$

and

$$A_k \equiv B_k \supset (\neg(\exists x)\Delta[P(x) \supset (\forall y)P(y)].\square$$

Provided $\mathfrak{I}(B_k) = 1$, the sequence given by $\mathfrak{I}(P(s^n(0)))$ lies in W, hence cannot contain a cumulation point. Hence, the implication A_k is true iff this sequence is infinite, i.e., iff $\varphi_k(n) = 0$ for all $n \in \omega$.

Now suppose W contains an infinite increasing sequence without upper bound in W. We have to replace

$$A_4^k \quad \text{by} \quad (\forall x)[[\varphi_k(x) = 0] \supset \neg\Delta(P(s(x)) \supset P(x)) \wedge \mathsf{R}_W P(s(x))]$$

and set $A_k \equiv B_k \supset \neg(\exists x)\Delta((\exists y)P(y) \supset P(x))$.

6. Conclusion

The results of this paper establish that no extension of the infinite-valued first-order Gödel logic is recursively enumerable, if

1. the projection function Δ (or by a negation symmetric to $\neg$, i.e.,

$$\mathfrak{I}(\sim A) = \begin{cases} 0 & \text{if } \mathfrak{I}(A) = 1 \\ 1 & \text{if } \mathfrak{I}(A) \neq 1, \end{cases}$$

   which would make ΔA definable as $\sim\sim A$) and
2. a relativization operator based on a subset $W \subseteq V$ s.t. there are only finitely many elements of W between any two elements of W.

are present. This leaves open the question whether the result holds for all relativizations, and in particular, whether $\mathbf{G}([0,1],[0,1])$ (= $\mathbf{G}(V_R)$ extended by 0-1-projections) and $\mathbf{G}([0,1],[0,1]\cap\mathbb{Q})$ are r.e. Further investigations of 0-1-projections promise to shed light on these problems.

References

1. M. Baaz, C. G. Fermüller, and R. Zach. Elimination of cuts in first-order finite-valued logics. *J. Inform. Process. Cybernet.* **EIK**, **29**(6), 333–355, 1994.
2. M. Baaz, A. Leitsch, and R. Zach. Incompleteness of an infinite-valued first-order Gödel logic and of some temporal logics of programs. In *Computer Science Logic. Selected Papers from CSL'95*, 1996. to appear.
3. M. Dummett. A propositional calculus with denumerable matrix. *J. Symbolic Logic*, **24**, 97–106, 1959.
4. K. Gödel. Zum intuitionistischen Aussagenkalkül. *Anz. Akad. Wiss. Wien*, **69**, 65–66, 1932.
5. K. Gödel. Über eine bisher noch nicht benutzte Erweiterung des finiten Standpunktes. *Dialectica*, **12**, 280–287, 1958.
6. G. Takeuti. *Proof Theory*. Studies in Logic 81. (North-Holland, Amsterdam, 1987) 2nd ed.
7. G. Takeuti and T. Titani. Intuitionistic fuzzy logic and intuitionistic fuzzy set theory. *J. Symbolic Logic*, **49**, 851–866, 1984.

Contributions of K. Gödel to Relativity and Cosmology *

G. F. R. Ellis

University of Cape Town

Summary. K Gödel published two seminal papers on general relativity theory and its application to the study of cosmology. The first examined a non-expanding but rotating solution of the Einstein field equations, in which causality is violated; this lead to an in-depth examination of the concepts of causality and time in curved space-times. The second examined properties of a family of rotating and expanding spatially homogeneous solutions of the Einstein equations, which was a forerunner of many studies of such cosmologies. Together they stimulated examination of themes that were fundamental in the development of the Hawking-Penrose singularity theorems and in studies of cosmological dynamics. I review these two papers, and the developments that resulted from them.

1. Introduction

Gödel became interested in general relativity theory while he and Einstein were both on staff of the Institute for Advanced Studies in Princeton. Apparently they discussed the subject together often. His resultant two papers had a major impact:

> Curiously, the beginning of the modern studies of singularities in general relativity in many ways had its seeds in the presentation by Kurt Gödel (1949) of an exact solution of Einstein's equations for pressure-free matter, which could be thought of as a singularity-free, rotating but non-expanding cosmological model ... [this paper] was one of the papers presented in a special issue of *Reviews of Modern Physics* dedicated to Einstein on his 70th birthday. Gödel used this space-time as an example helping to clarify the nature of time in general relativity, for it is an exact solution of the Einstein equations in which there are closed timelike lines: an observer can travel into his own past, and (as an old man) stand alongside himself (as a young man). He shortly thereafter published a further paper (1952) discussing a family of exact solutions of Einstein's equations representing rotating and expanding spatially homogeneous universe models (and relying on the geometric results derived many decades earlier by Sophus Lie and Luigi Bianchi). As these permit non-zero redshifts, they could include realistic models of the observed universe.

* This paper is in its final form and no similar paper has been or is being submitted elsewhere

> These papers perhaps more than any other antecedents of later work particularly stimulated investigations leading to fruitful developments. (This may partly have been due to the enigmatic style in which they were written: literally for decades after, much effort was invested in giving proofs for some of the results stated without proof by Gödel).

(Tipler Clarke and Ellis 1980, pp. 111-112). This is what I will explore in the sequel. The discussion that follows cannot possibly consider all developments from these papers (according to the Science Citation Index, the first paper has been the subject of 220 citations between 1965 and 1993 and the second 47 citations in the same period); rather I will concentrate on main themes and arguments that have arisen.

2. Gödel's stationary universe

Gödel's paper of 1949 gave the first exact rotating fluid-filled cosmological solution of Einstein's gravitational field equations. It is uniquely characterized by its symmetry properties, for it is a highly symmetric space-time: it is the only perfect-fluid filled universe invariant under a G_5 of isometries multiply transitive on space-time, which is space-time homogeneous (there is a 4-dimensional subgroup of isometries simply transitive on space-time) and locally rotationally symmetric (there is a 1-dimensional isotropy group acting about each space-time point) (Ellis 1967). Thus every space-time point is equivalent to every other one, and the universe is axially symmetric about every event. However it is not spatially homogeneous, because there is no family of spatially homogeneous 3-surfaces in the space-time.

Because this universe is space-time homogeneous, the density μ and pressure p are the same everywhere, and hence (using the standard notation for kinematic variables, see Ehlers 1961, Ellis 1971) it does not expand ($\theta = 0$) and matter moves geodesically ($\dot{u}_a = 0$). It also has zero shear ($\sigma = 0$), so the matter velocity vector is a Killing vector field but is not hypersurface orthogonal:

$$u_{a;b} = u_{[a;b]} = \omega_{ab} \neq 0 \tag{2.1}$$

(i.e. it generates a timelike symmetry, making it stationary) and the only non-zero kinematic quantity is the vorticity ($\omega \neq 0$). The vorticity vector is covariantly constant:

$$\omega^a{}_{;c} = 0 \quad \Leftrightarrow \quad \omega_{ab;c} = 0\,. \tag{2.2}$$

The kinematic description (1),(2) uniquely characterizes these space-times (Ehlers 1961, Theorem 1.5.2 and 2.5.4). Thus the homogeneous substratum rotates uniformly relative to the local compass of inertia: ω^2 is constant everywhere.

From these properties it follows (Ellis 1971 Section 5.2) that the electric part of the Weyl tensor is non-zero and is given by

$$E_{ab} = -\omega_a\omega_b + h_{ab}\frac{1}{3}\omega^2\,, \; E_{ab;c} = 0 \tag{2.3}$$

but the magnetic part of the Weyl tensor is zero: $H_{ab} = 0$. Because of the rotational symmetry, the Weyl tensor is Petrov type D.

The matter source in the original solution is pressure-free matter, but there is a cosmological constant of negative sign (the opposite sign to that usually encountered). More generally one can regard the matter source as being a perfect fluid. The only non-trivial covariant field equation is the Raychaudhuri equation, which with restrictions (1), (2) becomes

$$\Lambda + 2\omega^2 = \frac{1}{2}\kappa(\mu + 3p). \tag{2.4}$$

The Bianchi identities give

$$E_s{}^{(m}\omega^{t)s} = 0\,, \tag{2.5}$$

$$-3E^t{}_s\omega^s = \kappa(\mu + p)\omega^t\,, \tag{2.6}$$

the first of which is identically satisfied and the second of which leads to the relation

$$2\omega^2 = \kappa(\mu + p)\,. \tag{2.7}$$

With the Raychaudhuri equation (4) this gives

$$\Lambda = \frac{1}{2}\kappa(-\mu + p). \tag{2.8}$$

Hence in the pressure-free case we get

$$\Lambda = -\frac{1}{2}\kappa\mu = -\omega^2 < 0. \tag{2.9}$$

One can alternatively represent it as a fluid or scalar field with

$$\Lambda = 0 \Rightarrow p = \mu = \omega^2/\kappa\,. \tag{2.10}$$

There are many ways to construct this solution, because of its high symmetry. Gödel himself apparently used a deformation of a metric (Klein's fundamental quadric) along a family of timelike lines at constant distance ('Clifford parallels') generating a space invariant under a 4-parameter simply transitive group of isometries.

Gödel used this exact solution of the Einstein equations to examine properties of time and causality in general relativity. Using axially symmetric comoving coordinates centered on a chosen world line, the metric tensor is

$$ds^2 = 2\omega^{-2}(-dt^2 + dz^2 + dr^2 - (\sinh^4 r - \sinh^2 r)d\phi^2 + 2\sqrt{2}\sinh^2 r\, d\phi\, dt)\,.$$

The light cones tip over more and more the further one moves out (Figure 31 in Hawking and Ellis 1973), so that for large enough r, the circles $\{r, t, x\,const\}$ are closed timelike lines. This demonstrates *causality violation*: an observer traveling on this path from some event P will end up, after some proper time has elapsed, at the same space-time event P; thus she can. as an old woman, stand next to herself as a young woman. Various paradoxes ensue (the old person can kill the young one at event P, for example, but then there will be no older person at that event who can kill the young one, for she will not have survived - in which case the young one survives after all until arriving at P, and is then able as an old person to kill the young one ...). Furthermore by traveling far enough away, any observer can reach an arbitrarily distant event in the past on her own world line, and so influence events in her own past history at an arbitrary early proper time in that history.

The essential point demonstrated is that the Einstein Field Equations, determining space-time curvature from the matter present, are compatible with such causal violation. Until this solution was discovered, it had been taken for granted this could not occur. Furthermore, because the universe is space-time homogeneous, there are closed timelike curves through every event (hence the causal violation is not localized to some small region). It must be emphasized that this breakdown of causality does not occur because of any multiple- connectivity of the space-time, such as happens for example in a 2-dimensional torus universe (it is easy to construct space-times with closed timelike lines if one allows 'cutting and pasting'). Rather the Gödel universe is simply connected (indeed it is homeomorphic to R^4).

A necessary condition that causal violation can occur is that there exist no cosmic time, that is, no time function which increases in the future direction along every (timelike) world line. Gödel demonstrated that no such time function exists in these models, indeed he showed there are no inextendible spacelike surfaces at all in this space-time (on attempting to extend them, they necessarily become null and then timelike). This is possible because of the cosmic rotation signaled by the non-zero vorticity (for if the vorticity were zero, there would be a potential function for the fluid flow vector field that would provide a cosmic time function). However not all rotating universes admit causal violation; it occurs here because of the uniform extent of the rotation (it does not die away at infinity).

Gödel did not describe the geodesic properties of this space-time, but may have investigated them (see pp.560-1 and footnote 11 in Gödel 1949a). Later investigations by Kundt (1956) and Chandrasekhar and Wright (1961) explicitly showed that there are no closed timelike geodesics in the Gödel universe. This is compatible with Gödel's results because the closed timelike lines he found are non-geodesic (some force would have to be exerted, for example by a rocket engine, for an observer to move on them and experience the violation of causality). The past null cone of each point on the coordinate axis, generated by the null geodesics through that point, diverges out from there to a maximum radius r_m where closed (non-geodesic) null lines occur and it experiences self-intersections, and then reconverges to the axis (Hawking and Ellis 1973). Thus the past light cone of each event is quite different than in flat space-time. No null geodesic every reach further from their starting point than r_m.

This study of geodesics also showed that these space-times are geodesically complete (and so singularity-free). This means that this universe is an example of an Anti-Mach metric. One of the still unsolved problems of gravitational theory was raised *inter alia* by Ernst Mach: what gives an explanation of the origin of inertia? and why is it that in the real universe, distant galaxies are apparently at rest in a local inertial rest-frame? This could simply be a coincidence, but cosmologists have sought for a causal explanation of this fact: hopefully in the form of a direct link of local inertial properties to the distribution of matter (Einstein 1949a).

The Gödel universe shows conclusively that this is not a necessary connection, for in that universe the metric and curvature are regular everywhere, and the space-time is complete (there is no boundary at finite distance from any space-time point), but the matter in the universe rotates relative to a local inertial rest- frame (because of the non-zero vorticity). Thus specification of matter by itself (the singularity-free condition is needed for a complete matter specification, as otherwise singularities can be regarded as limiting distributions of matter) does not guarantee the Machian property we observe in the real universe: some extra boundary conditions have to be imposed to guarantee this condition. The Gödel universe sparked considerable new discussion of this feature (e.g. Oszvath and Schücking 1962, Rindler 1977, Adler et al 1975).

At the end of his paper, Gödel related his solution to the rotation of galaxies, comparing observed rotation rates with the vorticity in his solution. He acknowledged that his solution was not a realistic universe model, in that it does not expand (and so could not explain the observed redshifts in the spectra of distant galaxies). Nevertheless it is interesting that he made some attempt to relate it to astrophysical observations of galactic rotation by Hubble, estimating ω from equation (9) and a value of 10^{-30} gm/cc for the density of matter, presumably obtained from Hubble's data (he specifically mentions Hubble's estimates of the rotation rates of galaxies). Gödel must have been led to these considerations and estimates by his conversations with Einstein. In any case this section clearly shows Gödel functioning in the mode of an applied mathematician (comparing observational data with model parameters to check the validity of a universe model).

Some while after the publication of this solution, Heckmann and Schücking showed there is an exact Newtonian analogue of the solutions (Heckmann and Schücking 1955), provided one adopts generalized boundary conditions (which are in fact needed for any Newtonian cosmology at all to be viable). In terms of suitably adapted coordinates the gravitational potential is $\Phi = \frac{1}{2}\omega^2(x_1^2 + x_2^2)$ which diverges at infinity, and generates a non-zero Newtonian tidal force field $E_{ab} = \Phi_{,ab} - \frac{1}{3}h_{ab}\Phi^a{}_{;a}$. In this case the only field equations to be satisfied is

$$2\omega^2 + \Lambda = \frac{1}{2}\rho, \tag{2.11}$$

correspondingt to (4). Thus the relation between these variables is less restricted than in EFE, when additionally (7) must be satisfied. Clearly there is in this case no implication of causal violation, but this now gives a Newtonian example of an anti-Mach metric where this effect results from the imposition of specific boundary conditions at infinity. The issue of boundary conditions for Newtonian cosmology is ongoing, and in a sense has still not been satisfactorily resolved; this solution provides a specific example that shows the significance of this issue.

In summary, Heckman and Schücking express the impact of the solution thus:

> From a theoretical point of view, Gödel's model is highly interesting in several respects. It shows that in an infinite space the matter can rotate absolutely. This is the first indication that Mach's ideas are not automatically contained in Einstein's theory of gravitation. On the other hand this model pointed out, as showed by Gödel, that there may arise considerable difficulties if one wants to introduce an absolute time coordinate into a model of the Universe. The existence of closed timelike lines in Gödel's model showed moreover that space-time structures in the large might be very complicated and that startling situations could arise, e.g. a person could travel into his own past.

(Heckmann and Schücking 1962)

3. Gödel's expanding universes

Gödel's stationary rotating universe is not a viable model of the real universe because in it the galaxies show no systematic redshifts (Gödel 1949). Apparently Gödel must now have put a great deal of effort into examining properties of more realistic universe models that both rotate and expand. The results were presented at an International Congress of Mathematics held at Cambridge (Massachusetts) from 30th August to 5th September 1950 (Gödel 1952). This represents the first explicit construction of spatially homogeneous expanding and rotating cosmological models. They are invariant under a non-abelian G_3 of isometries simply transitive on spacelike surfaces[1]. These are now called *Bianchi universes* (Heckmann and Schücking 1962, Ellis and MacCallum 1968, MacCallum 1980), because the classification of the 3-dimensional symmetry group transitive on the homogeneous 3-spaces is derived from that introduced much earlier by L Bianchi, which is based on an examination of the structure constants of the Lie algebra of the symmetry group.

The models examined by Gödel belong to the Bianchi IX family, invariant under the group $SO(3)$, and consequently with compact spacelike surfaces of homogeneity. Indeed this was his starting point. The matter content is taken to be pressure-free matter ('dust'). The space-times are rotating solutions ($\omega \neq 0$) with the usual space-time signature, satisfying the further conditions:

'I. The solution is to be homogeneous in space,

II. Space is to be finite,

III. The density is not constant'.

The last condition implies that the models are expanding. In order that vorticity be non-zero, the models are *tilted*, i.e. the matter flow lines are not orthogonal to the surfaces of homogeneity (King and Ellis 1973)[2]. The paper argues that these conditions allow only the Type IX group as the group of isometries, and introduces a decomposition of the metric tensor into projection tensors along and perpendicular to the fluid flow lines, that has become fundamental in later work, as well as the idea of an expansion quadric (what is now called the expansion tensor). Gödel stated, mainly without proof, a number of interesting properties of these space-times, which remain interesting cosmological models today.

On the one hand, he developed relations between vorticity and the local existence of time functions determining simultaneity for a family of observers:

> A necessary and sufficient condition for a spatially homogeneous universe to rotate is that the local simultaneity of the observers moving along with the matter be not integrable (i.e. do not define a simultaneity in the large).

Thus $\omega = 0$ implies the local existence of a time function defining simultaneity for all fundamental observers (Ehlers 1960, Ellis 1971), and so $\omega \neq 0$

[1] The much simpler Abelian case had been considered previously by Kasner (1921) and Lemaître (1933); but these cannot rotate, and their construction does not require explicit consideration of the group action and structure.

[2] Tilt is always a necessary condition for rotation; in Bianchi IX universes, it implies rotation.

implies tilt. This led him to an important observation: in such models, there is necessarily an anisotropy in source number counts: "for sufficiently great distances, there must be more galaxies in one half of the sky than in the other half". He estimated the size of this anisotropy (which is proportional to the vorticity), and which must occur in any tilted universe model whether rotating or not (King and Ellis 1973). He went on to develop vorticity conservation relations[3], and gave the condition for the vorticity vector to be parallel propagated along the matter flow lines (it must be an eigenvector of the shear tensor, cf. Ehlers 1960), relating this to the axes of rotation galaxies: "The fact that the direction of ω need not be displaced parallel to itself might be the reason for the irregular distribution of the directions of the axes of rotation of the galaxies (which at first sight seems to contradict an explanation of the rotation of galaxies from a rotation of the universe [together with conservation of angular momentum]"

Further, he linked these local studies to the global topology and the existence of closed timelike lines: "The precise necessary and sufficient condition for the non-existence of closed timelike lines (provided that the one-parameter manifold of the spaces $\rho = const$ is not closed) is that the metric in the spaces of constant density be spacelike". That is, provided the matter flowlines themselves do not close up, spatial homogeneity precludes closed timelike lines, but if the surfaces of homogeneity are timelike then closed timelike lines will occur (because these surfaces are compact). It follows that

> The non-existence of closed timelike lines is equivalent with the existence of a 'world-time', where by a world-time we mean an assignment of a real number t to every space-time point so that t always increases if one moves along a timelike line in its positive direction[4].

This is because if the surfaces of constant density are spacelike, a world-time can be defined by taking these 3-spaces as surfaces of constant time (and this is the only world time invariant under the group of transformations of the solution).

On the other hand, he gave some dynamical results that are deeper in that they involve a detailed study of the Einstein field equations (rather than just the kinematic identities that are the basis of the vorticity conservation results, see Ehlers 1960). First, he considered the locally rotationally symmetric ('LRS') cases, showing there exist no LRS cases satisfying the conditions above. Second, he stated that

> Under the additional assumption that the universe contains no closed timelike lines, the quadric of expansion, at no moment of time,

[3] partly implied in previous work by Synge (1937).

[4] Note that there is no requirement relating this function to measurements of simultaneity.

> can be rotationally symmetric around ω. In particular it can never be a sphere, i.e. the expansion is necessarily coupled with a deformation. This is true even for *all* solutions satisfying I-III, and gives another directly observable property of rotating universes of this type.

Gödel suggested that the result on rotational symmetry might be related to the spiral structure of galaxies. The somewhat convoluted last statement above means that there are no expanding and rotating spatially homogeneous type IX universes with vanishing shear. Third, he stated existence of stationary homogeneous rotating solutions with finite space, no closed timelike lines, and $\lambda > 0$, in particular such as differ arbitrarily little from Einstein's static universe; but that there exist no stationary homogeneous solutions with $\lambda = 0$. These results however are almost an afterthought; the reason is that such models are unrealistic, for they cannot expand on average.

Gödel gave only the briefest of hints as to how he proved the dynamic results. Because of the symmetry of these space-times, the Einstein Field equations reduce to a system of ordinary differential equations. He did not give those equations, but he gave a Lagrangean function from which they could be derived, and stated an existence theorem: "for any value of [the cosmological constant] λ (including 0), there exist ∞^8 rotating solutions satisfying all the conditions stated. The same is true if in addition it is required that a world-time should exist (or should not exist)". The latter is the requirement that initially the surfaces of homogeneity should be spacelike or timelike.

This paper by Gödel is enigmatic, because the proofs of some of the major results are only sketched in the briefest manner[5]; the material is presented in a somewhat random order; and it is sparse on references[6]. Nevertheless it was a profound contribution to theoretical cosmology.

4. Resulting studies of causality

Gödel's papers (1949,1952) lead to an in-depth reconsideration of the nature of time and causality in relativity theory. He had showed there were acausal simply connected exact solutions of EFE. One stream of development was looking at space-times that were not simply connected, for example Bass and Witten (1957) showed that a compact space-time was necessarily acausal; and various papers considered specific high-symmetry space-times where one

[5] E. Schücking asked Gödel how he had proved the statements made, and in essence the answer was by detailed calculation. Schücking suggests that Gödel did not give more details of the proofs because the method used was inelegent (private communication).

[6] Indeed the only reference is to his own paper, Gödel (1949).

could determine all possible connectivities. Perhaps most significant was the broad realization that one could not take either the topology or the causality of space-time for granted: one needed to consider multiply connected space-times and possibilities such as wormholes, for example, as well as the possibility of causal violations.

Resulting from this, a general analysis of the ideas of causality took place, developing also from two other directions. First, Zeeman's remarkable paper 'Causality implies the Lorentz group' (Zeeman 1964) showing that causal orderings induced by the metric of Minkowski space-time are preserved only by the Lorentz group and dilations. Second, the study of Causal domains and their boundaries by Penrose, arising out of work on the Cauchy development of initial data for space-time and the idea of global hyperbolicity (due to Leray and others) on the one hand, and studies of the conformal structure of space-time on the other.

A series of important ideas arose, developed particularly by Penrose, Carter, Geroch, and Hawking, that were crucial in the later studies of causality and singularities:

(1) the idea of causal domains: the domain of dependence of initial data, of Cauchy horizons bounding this domain of dependence, and of Cauchy surfaces (surfaces on which initial data determines the evolution of the entire space-time) in space-times where no such horizons exist. The latter case was shown to be equivalent to the condition of Global Hyperbolicity, and implied geodesic connectivity of the space-time (which is not true in general).

(2) a series of causality conditions of increasing strength (causality, future distinguishing, past distinguishing, strong causality) leading up to the strongest and physically most relevant, namely stable causality. The latter was shown to be true if and only if there is a cosmic time function, i.e. a function that increases along all timelike curves (which is not true in the Gödel stationary universe). This generalizes and completes Gödel's statement on the relation between time functions and causality.

(3) the broad idea of null boundaries of causal domains, and an understanding of their properties. These boundaries include Cauchy horizons, particle and event horizons, and causality horizons (where the nature of the causality conditions obeyed by space-time changes).

These ideas are discussed in broad outline in Tipler Clarke and Ellis (1980); they are presented in technical detail in Penrose (1972) and Hawking and Ellis (1973). I believe it is fair to say that Gödel's paper gave the impetus to a lot of this work by initiating a new round of questioning of the nature of time and causality in relativity theory.

5. Resulting studies of universe models

The papers also resulted in a series of studies that greatly expanded our understanding of the dynamics of universe models, extending and in many cases completing the work initiated by Gödel.

Firstly, they initiated systematic analysis of the family of Bianchi universe models. Taub (1951) gave an enlightening study of the equations and properties of empty Bianchi universes with arbitrary group type. This built on and extended the (largely unexplained) methods used by Gödel in his expanding universe study, and made the needed techniques accessible to workers in the field, based on local properties of symmetry groups and the classification of 3-d Lie algebras developed by Luigi Bianchi from Sophus Lie's work. Taub found some new 'anti-Mach' metrics, notably the remarkable space later known as Taub-NUT space (see Hawking and Ellis 1973 for a discussion).

Heckmann and Schücking (1962) extended the equations to a study of fluid-filled Bianchi models, initiating the systematic study of this class of models. This has become an important topic of study in terms of providing a parametrized set of alternative models to the standard Friedmann-Lemaître models of cosmology. The dynamical and observational properties of the Bianchi models have been extensively studied (see Ellis and MacCallum 1968, King and Ellis 1973, MacCallum 1980, 1993, Wainwright and Ellis 1996, and references therein). There is only space to mention here four aspects of this study. First, the 'mixmaster' universe studied by Misner (1968), which is in fact the same model studied by Gödel (1952), was shown by Misner and then by Lifshitz, Belinskii, and Khalatnikov to have complex oscillatory properties at early times, leading to chaotic-like behaviour. Whether or not the early epoch of this universe exhibits truly chaotic behaviour is still the object of investigation (Hobill et al, 1994). Second, the Hamiltonian methods introduced by Misner became the cornerstone of the Hamiltonian approach to cosmology (Ryan 1972), which in turn is the foundation of the study of quantum cosmology (Coleman et al, 1991). Third, a very interesting series of dynamical systems investigations of these universes has been undertaken, which is just coming to fruition and leading on to similar studies of inhomogeneous universe models (Wainwright and Ellis 1996). Fourth, these analyses were extended to the case of Newtonian cosmology by Heckmann and Schücking (1955,1956), see also Raychaudhuri (1957).

Secondly, the (observationally unrealistic) space-time homogeneous G_4 cases were studied and completely solved by Ozsvath, Schücking, Farnsworth, and Kerr, see the references in Ellis (1967).

Thirdly, the LRS cases were completely determined, see Ellis (1967) for the dust case, Stewart and Ellis (1968) for the fluid case, and van Elst and Ellis (1996) for a covariant approach to the fluid and dust cases. These include some of the simplest interesting anisotropic cosmological models with non-trivial properties (see e.g. Collins and Ellis 1979 for phase planes and the singularity structure of the tilted type V LRS models).

Fourthly, the local covariant analysis of dynamics of cosmological models developed from Gödel's second paper, utilising and extending his use of the projection tensors and his analyses of vorticity and the expansion tensor (Ehlers 1960, Ellis 1971). A proof of his theorem on shear-free motion was given for the general homogeneous case by Schücking (1957) and then extended to the general inhomogeneous dust case by Ellis (1967), who proved that in all cases, dust-filled solutions with vanishing shear must have either vanishing rotation or vanishing expansion. Extension of this result to various perfect fluid cases followed, see Collins (1986) for a summary.

Fifthly, and perhaps most significant of all, Gödel's paper seems to have been influential in the formulation of Raychaudhuri's fundamentally important equation, giving the rate of change of the volume expansion along fluid flow lines in terms of the fluid shear, rotation, and matter content (Raychaudhuri 1955, Ehlers 1961). This is the fundamental equation of gravitational attraction, playing a central role in the dynamics of all cosmological models. It underlies the instability of the Einstein static universe (Ellis 1971), and directly gives simple singularity theorems for both the dust case (Raychaudhuri 1955) and for perfect fluids (Ehlers 1960): neither anisotropy nor inhomogeneity can avoid a singularity in universe models where matter moves without rotation or acceleration. Together with its null analogue, obtained by Ehlers and Sachs[7], this equation is one of the pillars of the important Penrose-Hawking singularity theorems.

6. The singularity theorems

The point here is simple: Raychaudhuri's result shows that irrotational dust cannot avoid a singularity at the beginning of the universe. Can rotation or pressure avoid the singularity?

All efforts at a direct attack, based on the dynamical equations, failed. Many thought that it was only the symmetry of the FL models that led to the prediction of a start to the universe. A similar issue arose in the case of gravitational collapse. The resolution of this problem came in a brilliant paper by Roger Penrose (1965) who used a combination of arguments from the convergence properties implied by the null version of Raychaudhuri's equation and analysis of its implications for the boundaries of causal sets, to

[7] See Tipler Clarke and Ellis 1980 for a discussion, and Hawking and Ellis 1973 for a derivation.

show there must be a singularity (in the sense of existence of inextendible incomplete null geodesics) at the endpoint of realistic gravitational collapse. Stephen Hawking then extended this kind of argument to the cosmological case (the start of the universe), proving a series of theorems applicable in that context, and leading to the combined Hawking-Penrose theorem that applies in both cases (Hawking and Penrose 1970) and uses both the timelike and null versions of the Raychaudhuri equation.

The nature of these arguments, given in depth in Hawking and Ellis (1973), is summarized in Tipler Clarke and Ellis (1980), Section 3. The implication is that, classically considered, space-time has a beginning at the start of the universe. More realistically, a modern view would be that we cannot avoid a quantum gravity regime at the beginning of the universe (if gravity is indeed quantized), or in any case a quantum-field dominated era where energy violations take place. The issue of the nature of the beginning of the universe is still the subject of intense debate; the Hawking-Penrose theorems have set the parameters within which the discussion takes place. Those theorems owe much to Gödel's papers both in terms of the foundations they laid for analysis of causality in general relativity, and the initiation of dynamical analyses that clarified the role and nature of vorticity and led to the timelike and null versions of the Raychaudhuri equation.

7. Gödel's dialogue with Einstein

Because most of the interaction between Einstein and Gödel took place during their talks in the Institute, little is written down of that debate. However there is a brief public interchange between them resulting from Gödel's work. It is printed in the book edited by P A Schilpp (1949), produced for the occasionof Einstein's 70th birthday on 14th March 1949.

In his contribution to that book, Gödel (1949a) explains there are world models in which there exists no objective lapse of time. He then comments:

> It might be asked: Of what use is it if such conditions prevail in *possible* worlds? Does that mean anything for the question interesting us whether in *our* world there exists an objective lapse of time? I think it does. For (1) Our world, it is true, can hardly be represented by the particular solutions referred to above (because these solutions are static and therefore yield no redshift for distant objects); there exist however also *expanding* rotating solutions. In such a universe an absolute time might also fail to exist, and it is not impossible that our world is a universe of this kind. (2) The mere compatibility with the laws of nature of worlds in which there is no distinguished absolute time, and therefore no objective lapse of time can exist, throws some light on the meaning of time also in those worlds where

> an absolute time *can* be defined. For, if someone asserts that this absolute time is lapsing, he accepts as a consequence that, whether or not an objective lapse of time exists (i.e. whether or not time in the ordinary sense of the word exists), depends on the particular way in which matter and its motion are arranged in the world. This is not a straightforward contradiction; nevertheless, a philosophical view leading to such consequences can hardly be considered as satisfactory.

(Gödel 1949a, p.562). This article shows how Gödel was primarily concerned with the "non-objectivity of the present", and only secondarily with closed timelike lines. Einstein replies,

> Kurt Gödel's essay constitutes, in my opinion, an important contribution to the general theory of relativity, especially the analysis of the concept of time.... if [causal violations exist] the distinction 'earlier - later' is abandoned for world points which lie far apart in the cosmological sense, and those paradoxes, regarding the direction of the causal connections, arise, of which Mr Gödel has spoken. Such cosmological solutions of the gravitation equations (with not vanishing cosmological constant) have been found by Mr Gödel. It will be interesting to weigh whether these are not to be excluded on physical grounds.

(Einstein 1949). Later various causality assumptions were introduced to specifically exclude causal violations (see Hawking and Ellis 1973, Tipler Clarke and Ellis 1980), with a general assumption that this was necessary for physical resonableness of solutions; but that assumption has been challenged from time to time.

This debate has been renewed with vigour in the past couple of years, with the discovery of closed timelike lines associated with moving cosmic strings and 'wormholes', and the introduction by Hawking of the 'chronology protection conjecture'. An illuminating presentation of this new discussion may be found in Kip Thorne's splendid book *Black Holes and Time Warps* (Thorne 1994). The debate is not yet ended.

References

1. R. Adler, M. Bazin, M. Schiffer (1975). *Introduction to General Relativity.* (McGraw Hill Kogakusha, Tokyo), 437-448.
2. S. Chandrasekhar and J. P. Wright (1961). *Proc. Nat. Acad. Sci* **47**, 341.
3. S. Coleman, J. B. Hartle, T. Piran and S. Weinberg (Eds.) (1991). *Quantum Cosmology and Baby Universes.* (World Scientific, Singapore).
4. C. B. Collins (1986). Shearfree fluids in general relativity. *Can.J.Phys.* **64**, 191-199.

5. C. B. Collins and G. F. R. Ellis (1979). Singularities in Bianchi Cosmologies. *Physics Reports* **56**, 63-105.
6. J. Ehlers (1960): Beitrage zur mechanik kontinuerlichen medien. *Abh. Mainz. Akad. Wiss. u. Lit., Mat/Nat Kl.* Nr 11. English Translation: Contributions to the relativistic mechanics of continuous media. *Gen. Rel. Grav.* **25**, 1225-1266 (1993).
7. A. Einstein (1949). In Schilpp. (1949), 687-688.
8. A. Einstein (1949a). In Schilpp. (1949), 27-29, 65-67.
9. G. F. R. Ellis (1967). The dynamics of pressure-free matter in general relativity. *Journ. Math. Phys.* **8**, 1171-1194.
10. G. F. R. Ellis (1971). Relativistic Cosmology. In *General Relativity and Cosmology*, Proc. Int. School of Physics "Enrico Fermi", Course XLVII. Ed. R. K. Sachs (Academic Press, New York), 104-179.
11. G. F. R. Ellis and M. A. H. MacCallum (1969). A class of homogeneous cosmological models. *Comm. Math. Phys.* **12**, 108-141.
12. K. Gödel (1949). An example of a new type of cosmological solution of Einstein's field equations of gravitation. *Rev. Mod. Phys.* **21**, 447-450. [See Maths. Review **11**, 216].
13. K. Gödel (1949a). A Remark about the relationship between relativity theory and idealistic philosophy. In Schilpp. (1949), 557-462.
14. K. Gödel (1952). Rotating universes. In *Proc. Int. Cong. Math.* (Camb., Mass.). Ed. L. M. Graves et al. Volume 1, 175-181. [see Maths. Review **13**, 500].
15. S. W. Hawking and G. F. R. Ellis (1973). *The Large Scale Structure of Space-Time.* (Cambridge University Press, Cambridge).
16. S. W. Hawking and R. Penrose (1970). The Singularities of Gravitational Collapse and Cosmology. *Proc. Roy. Soc.* **A 314**, 529-548.
17. O. Heckmann and E. Schücking (1955). Remarks on Newtonian cosmology. I. *Zs.f.Ap.* **38**, 95-109.
18. O. Heckmann and E. Schücking (1955). Remarks on Newtonian cosmology. II. *Zs.f.Ap.* **40**, 75-92.
19. O. Heckmann and E. Schücking (1962). Relativistic cosmology. In *Gravitation*, Ed. L. Witten. (Wiley, New York), 438-469.
20. D. Hobill, A. Burd and A. Coley (Eds.) (1994). *Deterministic Chaos in General Relativity.* (Plenum Press, New York. Nato-ASI Series B: Physics).
21. E. Kasner (1921). Geometrical Theorems on Einstein's cosmological equations. *Am.J.Math.* **3**, 217.
22. A. R. King and G. F. R. Ellis (1973). Tilted homogeneous cosmologies. *Comm. Math. Phys.* **31**, 209-242.
23. W. Kundt (1956). Tragheitsbahnen in einem vom Gödel angegeben kosmologischen model. *Zs.f.Ap.* **145**, 611-620.
24. G. Lemaître (1933). L'Univers en expansion. *Ann. Soc. Sci Brux.* **53**, 51-85.
25. A. Lichnwerowicz (1955). *Theories relativistes de la Gravitation et d'lelectromagnetisme.* (Masson, Paris).
26. M. A. H. MacCallum (1980). In *General Relativity: An Einstein Centenary Survey.* Ed. S. W. Hawking and W. Israel (Cambridge University Press, Cambridge).
27. M. A. H. MacCallum (1993). Anisotropic and inhomogeneous cosmologies. In *The Renaissance of General Relativity and Cosmology.* Ed. G. F. R. Ellis, A. Lanza and J. Miller (Cambridge University Press, Cambridge), 213-233.
28. C. W. Misner (1969). The Mixmaster Universe. *Phys. Rev. Lett.* **22**, 1071-1074.
29. I. Oszvath and E. Schücking (1962). Finite Rotating Universe. *Nature* **193**, 1168-1169.

30. R. Penrose (1965). Gravitational collapse and space-time singularities. *Phys. Rev. Lett.* **14**, 57-59.
31. A. K. Raychaudhuri (1955). Relativistic cosmology. *Phys. Rev.* **98**, 1123-1126.
32. A. K. Raychaudhuri (1957). Relativistic and Newtonian cosmology. *Zs.f.Ap.* **43**, 161-164.
33. W. Rindler (1977). *Essential Relativity.* (Springer, Berlin), 243-244.
34. M. Ryan (1972). *Hamiltonian Cosmology.* (Springer, Berlin; Lecture Notes in Physics, Vol. 13).
35. E. Schücking (1957). *Naturwiss* **19**, 57.
36. P. A. Schilpp (Ed.) (1949). *Albert Einstein: Philosopher Scientist.* (2 Vols). (Open Court, La Salle).
37. J. M. Stewart and G. F. R. Ellis (1968). Solutions of Einstein's equations for a fluid which exhibits local rotational symmetry. *Journ. Math. Phys.* **9**, 1072-1082.
38. J. L. Synge (1937). Relativistic hydrodynamics. *Proc. Lond. Math. Soc.* **43**, 37.
39. A. H. Taub (1951). Empty Space-Times admitting a 3-parameter group of motions. *Ann. Math.* **53**, 472.
40. K. S. Thorne (1994). *Black Holes and Time Warps.* (Norton, New York).
41. F. J. Tipler, C. J. S. Clarke, and G. F. R. Ellis (1980). Singularities and Horizons: a review article. In *General Relativity and Gravitation: One Hundred years after the birth of Albert Einstein, Vol. 2*, Ed. A. Held (Plenum Press, New York), 97-206.
42. J. Wainwright and G. F. R. Ellis (Ed.) (1996). *The dynamical systems approach to cosmology.* In publication (Cambridge University Press, Cambridge).
43. E. C. Zeeman (1964). Causality implies the Lorentz group. *J. Math. Phys.* **5**, 490-493.

Kurt Gödel and the constructive Mathematics of A.A. Markov

Boris A. Kushner

Department of Mathematics
University of Pittsburgh at Johnstown
Johnstown, PA 15904, USA
e-mail borisvms.cis.pitt.edu

1.

I would like to dedicate this article to the memory of Dr. Oswald Demuth (12.9.1936 – 15.9.1988). Oswald was an excellent Mathematician and a dear friend of mine. I miss him so badly.

2.

The Russian School of constructive Mathematics was founded by A.A. Markov, Jr. (1903-1979) in the late 40-ies - early 50-ies [1]. In private conversations Markov used to state that he nurtured a type of constructive convictions for a very long time, long before the Second World War. This is an interesting fact if one considers that this was the time when Markov worked very actively in various areas of classical Mathematics and achieved first-rate results. Perhaps it is worth mentioning that Markov was a scientist with a very wide area of interest. In his freshman years he published works in Chemistry and he graduated from Leningrad University (1924) with a major in Physics. Besides Mathematics, he published works in theoretical Physics, Celestial Mechanics, Theory of Plasticity (cf., e.g., Markov and Nagorny [1988: introduction by Nagorny]; this monograph (originally in Russian, 1984) was completed and published by N.M. Nagorny after Markov's death). It is almost inevitable that a scientist of such universality arrives to philosophical and foundational issues. I believe that the explicitly "constructive" period of Markov's activities began with his work on Thue's Problem which had stood open since 1914. Thue's Problem was solved independently by A.A. Markov (Jr.) and E. Post in 1947. Markov began to develop his concept of so called *normal algorithms* as a tool to present his results on Thue Problem. Markov's publications on normal algorithms appeared as early as 1951 (Markov [1960; an English translation]). In 1954 Markov published his famous monograph

[1] This article is written for the Gödel '96 Proceedings. It was not and will not be published anywehere else in any form.

Theory of Algorithms (Markov [1961; an English Translation]). This monograph can be considered as probably the first systematic presentation of the general theory of algorithms together with related semiotical problems. In this monograph, in particular, one can find a mathematical theory of *words* as special types of sign complexes. As far as I know, such a theory was developed here for the first time. On the other hand, scrupulous proofs that such and such normal algorithm works on given words in a certain way can be considered as the first examples of program correctness verification. All in all, it seems that Markov's monograph is still underestimated by experts in Computer Science. At the same time Markov began to develop a mathematical worldview, and Mathematics in the framework of the above worldview, that was later to be known as "Markov's (or Russian) constructive Mathematics". In my opinion, Markov's constructive Mathematics (MCM; see, e.g.,Markov [1971], Kushner [1984], Kushner [1990], Kushner [1993 a,b])) is one of the three most important and coherent constructivist trends of our Century. The other two are Brouwer's Intuitionism (see, e.g., Heyting [1956], Troelstra [1977, 1990], Troelstra and van Dalen [1988], Kleene and Vesley [1965], Dragalin [1988], Beeson [1985]) and Bishop's constructive Mathematics (BCM) (see, e.g., Bishop [1967, 1970,1984], Bishop and Bridges [1985]) Chronologically, Markov stands between Brouwer and Bishop.

3.

Markov's constructive Mathematics (MCM) can be characterized by the following main features (cf., e.g., Markov [1971], Kushner [1984])

1. The objects of study are constructive processes and constructive objects arising as the results of these processes. The concept of constructive object is primitive. The main feature of constructive objects is that they are constructed according to definite rules from certain elementary objects, which are indecomposable in the process of these constructions. Hence we deal with objects of a completely combinatorial and finite nature. Practically, for developing MCM it is enough to consider a special type of constructive objects, namely words in one or another alphabet.

2. A special constructive logic is allowed to be used. This logic takes into account the specific nature of constructive objects and processes. In particular, tertium non datur principle and principle of double negation are not accepted as universal logical principles.

3. The abstraction of potential realizabality is accepted, but the abstraction of actual infinity is completely rejected.

4. The intuitive concept of effectiveness, computability etc is identified with one of the precise concept of algorithm (historically, with Markov's normal algorithms). This means that a version of Church's Thesis is accepted.

MCA differs, roughly speaking, with Intuitionism in features 1, 4 and with BCM, in feature 4. As is well-known, one of the principal achievements of Brouwer was his non-pointwise, Aristotelian-style theory of the real continuum. A very specific tool of choice sequences was developed and used to reach this goal. Choice sequences can be considered as developing, incomplete mathematical objects and their theory was, probably, the first sample of Mathematics of incomplete objects. In any case, they are not constructive objects in Markov's sense and therefore they are outside of MCM. On the other hand, Bishop refused to identify intuitive constructivity, effectiveness etc with ,say, recursiveness (see,e.g., Bishop [1967, 1984], Bishop and Bridges [1985]). It is worth noting that the main part of Brouwer's work on intuitionistic Mathematics was done before the concept of recursive function appeared on mathematical scene. Heyting once noted (Heyting [1962], van Dalen [1995]) that had it been the other way, Brouwer probably would not have introduced choice sequences and it would have been a pity. I can only agree with the last part of the statement - it would be a pity not to have today this marvelous concept. On the other hand, I am not sure about the strength of this "probably" above; in fact, Brouwer rejected the pointwise concept of continuum because of very deep philosophical reasons, hence, recursive functions would not have satisfied him in his task of creating a non-Cantorian theory of the continuum, anyway. The problem was not so much to grasp the volume of the intuitive notion of computability (that is all that recursive functions are about) but rather to present mathematically the Aristotelian idea of developing continuum. And this brings in mathematical objects that are incomplete in principle. It seems that Brouwer did not express in any written form his position with respect of Church's Thesis and, as far as I know, there no other evidences of his point of view in this respect. Nevertheless, it seems highly unlikely that Brouwer accepted Church's Thesis and, anyway, it was not used in the body of Intuitionistic Mathematics developed by him and his disciples. Be that as it may, both Brouwer and Bishop did not join the Church's Thesis Club, though Bishop considered the Thesis practically plausible. It is worth noting that Church's Thesis was taken for long time by the mathematical community almost for granted. In reality, this fundamental principle is not so evident, in particular philosophically. The attitude of such outstanding mathematical thinkers, as Brouwer and Bishop speaks for itself. And, as is known, Gödel was unconvinced by Church's Thesis (very impressive accounts of the early years of the theory of computability can be found in Feferman [1984] and Davis [1982] where the further bibliography can be found, as well) since Church failed to present conceptual analysis of the notion of finite algorithmic procedure. It was only after Turing's work with its analysis of the concept of mechanical procedure that Gödel was ready to accept the identification of intuitive and precise concepts of algorithm. E. Mendelson published recently an interesting work [Mendelson, 1990] on the subject. Along with Turing's conceptual analysis he considers the analysis of the nature of finitary processes that was undertaken by Kolmogorov and Uspensky (Kolmogorov-Uspensky [1958]), as another strong argument for accepting Church's Thesis. Mendelson goes so far as consider it as a legitimate proof of Church's Thesis.

An interesting feature of MCM is its pure syntactic mathematical universe. It is true that the same can be said with some reason about BCM. But Markov placed a special accent on this feature of his system. Consructive objects and constructive processes (algorithms) are the main (and essentially the only) Dramatis Personae on the scene. Thus constructive objects are considered to be initial data for algorithms and this point of view gives a very special touch to MCM which may be of interest for Computer Science. Let as consider an example. A classical real number can be defined as follows. Let α be a Cauchy sequence of rational numbers. This means that

$$\forall n \exists m \forall ij (i, j > m \supset |\alpha(i) - \alpha(j)| < 2^{-n}) \tag{3.1}$$

Clasically, real numbers are classes of equivalent Cauchy sequences. In MCM the arbitrary sequence above is to be replaced by an algorithm of the type $\mathbb{N} \to \mathbb{Q}$, where $\mathbb{N}$ is the set of natural numbers (and natural numbers are words of type $0, 0|, 0||...$) in the alphabet $\{0, |\}$ and $\mathbb{Q}$ is the set of rational numbers (which are words of a special type, as well). Of course, one can speak here about a recursive function with rational values. As for 3.1, the strictest constructive reading of it will be as follows. There is an algorithm β of type $\mathbb{N} \to \mathbb{N}$ such that

$$\forall nij (i, j > \beta(n) \supset |\alpha(i) - \alpha(j)| < 2^{-n}) \tag{3.2}$$

Such algorithm β we call a Cauchy modulus for α. The schemes of algorithms can be coded in some natural way by words (or by natural numbers). The code of α we denote by α^c. A constructive number is defined as a couple $\alpha^c * \beta^c$ where β is a Cauchy modulus for α. Therefore constructive real numbers are *words* in an alphabet. The set of constructive real numbers we denote by $\mathbb{D}$. $\mathbb{D}$ is an adequate continuum for MCM and usual Calculus can be readily developed over $\mathbb{D}$ using Markov's definition of constructive real functions (constructive function for the sake of shortness). A constructive function is an algorithm f of type $\mathbb{D} \to \mathbb{D}$ such that $f(x_1) = f(x_2)$ as soon as $x_1 = x_2$. (Equality of constructive real numbers can be introduced in an obvious way, see, e.g., Kushner [1984]). It should be noted that, as natural as it is, the concept of constructive real number is not completely evident. E.g., there is a temptation to consider computable systematic expansions. And the first definition of computable number published by Turing (Turing [1936/37]) was exactly of this type. A correction (Turing [1937]) followed immediately. Deficiencies of computable expansions are discussed in Kushner [1984]. The essence of those deficiencies was known already to Brouwer (see, e.g. Brouwer [1921]). It is enough to mention that, e.g., there is no algorithm for addition of computable systematic expansions.

A constructive real number as a syntactic object holds in itself information sufficient to find in an effective way rational approximations to the number with every desirable accuracy. Nevertheless, some interesting variations of this concept are possible. First of all it is possible to read 3.1 in a more liberal way, say as

$$\forall n \neg\neg \exists m \forall ij (i, j > m \supset |\alpha(i) - \alpha(j)| < 2^{-n}) \tag{3.3}$$

Formula 3.3 roughly speaking represents classical Cauchy property. We can consider now a new type of computable real numbers. The word α^c where α satisfies 3.3 we call a pseudonumber. The set of all pseudonumbers will be denoted by $\mathbb{P}$. $\mathbb{P}$ presents another model of constructive continuum. Reals from this continuum are computable in the sense that for every such number there is an algorithmic sequence of rational approximations that converges. But not only we do not have a recursive modulus of convergence included in the number-word, it can happen that such a modulus does not exist at all (this follows from a well-known result of Specker [1949], see, also, Kushner [1984]). Therefore there exist pseudonumbers that are not equal to any constructive real number. It is interesting that we can obtain two other variants of computable numbers in a rather syntactic way, by omitting information about a Cauchy modulus in the definition of constructive real numbers. We will call a word α^c an F-number (quasinumber) if there is (can not fail to be) a Cauchy modulus for α. Let $\mathbb{F}$ and $\mathbb{K}$ be the sets of all F-numbers and quasinumbers, respectfully. To compare the four models of constructive continuum let us imagine, for a moment, that they are placed on the classical real line. Every model singles out some computable numbers in the classical continuum. It is evident that $\mathbb{P}$ is wider in this sense than $\mathbb{D}$. But $\mathbb{D}, \mathbb{F}$ and $\mathbb{K}$ are from this point of view the same. They single out exactly the same points on the classical real line. The syntactic difference between constructive real numbers and F-numbers (quasinumbers) is evident and they do look as quite different initial data for algorithms. It is well-know that the information about a Cauchy modulus that is absent in F-numbers (quasinumbers) can not be restored in an effective way. Namely, there is no algorithm that finds for every F-number (quasinumber) p a constructive real number that is equal p (see, e.g., Kushner [1984]).

On the other hand, F-numbers and quasinumbers are words of the same type, every F-number is a quasinumber and there is no quasinumber p such that $p \neq q$ for every F-number q. The continua $\mathbb{F}$ and $\mathbb{K}$ are the same for a classical mathematician. One can not tell one from another from the classical point of view. But the difference is quite discernible constructively. There is a sequence of quasinumbers γ that is not a sequence of F-numbers. Really, in order to prove that γ is a sequence of F-numbers a constructivist should develop an algorithm that would give for every n a (code of) Cauchy modulus of $\gamma(n)$. For the sequence γ mentioned above such an algorithm does not exist (see, Kushner [1984]).

The difference between F-numbers and quasinumbers can be illustrated by a Brouwerian counterexample, as well. Consider an algorithm α such that

$$\alpha(n) = \begin{cases} 1, & : \text{ if there is a perfect} \\ & : \text{ number among the numbers } 2i+1 \text{ where } i \leq n \\ 0, & : \text{ otherwise} \end{cases}$$

It is evident that if there is no odd perfect number then

$$\forall n(\alpha(n) = 0)$$

and if $2i + 1$ is the least odd perfect number than

$$\alpha(n) = \begin{cases} 0, & \text{if} n < i \\ 1, & \text{if} n \geqslant i \end{cases}$$

It is evident that α can not fail to have a Cauchy modulus, hence α^c is a quasinumber. But nobody could present such a Cauchy modulus β today. Indeed, it is clear that an odd perfect integer exists iff $\alpha(\beta(1)) = 1$. Thus one can not state that the quasinumber α^c is an F-number.

All in all, we have four pretenders to bear the title of constructive continuum. It is interesting to notice that the completeness theorem holds for $\mathbb{D}$, $\mathbb{P}$, $\mathbb{F}$, but it does not hold for $\mathbb{K}$ (there is a (constructive) Cauchy sequence of quasinumbers that does not have the limit). Certainly, $\mathbb{D}$ looks like the most attractive constructive continuum. Since $\mathbb{D}$ is a countable set from the classical point of view (as are the other three models), the question arises whether our intuitive perception of continuity is grasped in $\mathbb{D}$. There is no immediate answer to this question. On one hand, $\mathbb{D}$ is not a countable set constructively, as Cantor's diagonal construction can be reproduced. Moreover, a constructive function f (which, as is known, is automatically constructively continuous) such that, say, $f(0) < 0$ and $f(1) > 0$, can not fail to have a root between 0 and 1. On the other hand, S.N. Manukyan [1976] constructed the following amazing counterexample

Theorem 3.1. *There are two constructive (and therefore continuous) planar curves $\bar{\varphi}_1$ and $\bar{\varphi}_2$ such that*

- $\bar{\varphi}_1(0) = (0,0), \bar{\varphi}_1(1) = (1,1)$;
- $\bar{\varphi}_2(0) = (0,1), \bar{\varphi}_2(1) = (1,0)$;
- *for every* $0 < t < 1$ *both* $\bar{\varphi}_1(t)$ *and* $\bar{\varphi}_2(t)$ *belong to the open unit square;*
- $\bar{\varphi}$ *and* $\bar{\varphi}_2$ *do not intersect.*

Thus the above continuous curves connect diametrically opposed vertices of the unit square, they do not leave this square and still they do not intersect!

One can consider this example as an argument that the immediate intuition of continuity is not grasped by $\mathbb{D}$. However, this argument can be met by reference to well-known topological examples, like Peano's curve, etc, that show that this very immediate intuition is not quite a reliable facility and sometimes even mislead us.

The splitting of a classical concept (the concept of Cauchy sequence above) into several constructive concepts is a common thing in any constructive mathematics. But the syntactic splitting (e.g., constructive real numbers versus F-numbers or quasinumbers) is very characteristic for MCM with its syntactic mathematical universe. The same can be said about the subtle difference between quasinumbers and F-numbers.

4.

Both MCM and BCM took a lot from Brouwer, especially in their critical approach to the set-theoretical (classical) Mathematics and in their understanding of constructive logical operators. I have tried to describe the mathematical and human relations between Markov and Bishop, between MCM and BCM in my essay [1993a] (there exists a Russian version of this essay Kushner [1992]). It seems in general that Bishop's approach to constructive mathematics was mostly pragmatic and foundational problems did not attract him - at least, they were not supposed to be brought in at the expense of concrete mathematical activities. Nevertheless, Bishop still did not avoid the eternal problem of interpretation of implication, this host of Hamlet's father of any constructive mathematics (see, Bishop [1970]). It is worth mentioning that Markov spent the last years of his life struggling to develop a large semantical system to achieve, above all, a satisfactory theory of implication (see, e.g., Markov [1971,1976]). BCM can be considered as neutral in the sense that its results are acceptable both intuitionistically and in MCM. So the remarkable body of mathematical analysis that was built in BCM by Bishop himself and his disciples contributes to the claims of Intuitionism and MCM.

Very interesting remarks revealing Markov's position with respect of Intuitionism can be found in his Editor's Comments to the Russian translation of Heyting's book [1956] (Markov [1965]). As is known, Heyting's book is written artistically in the form of a discussion between several dramatis personae, like Clas, Int, Form etc. In his remarks Markov introduced a new person Con (a Constructivist). Perhaps, it would be interesting to see an edition of Heyting's books with Markov's remark translated and incorporated into the text.

5.

60-ies - 70-ies were the best years for MCM. It had active centers in Moscow (headed by A.A. Markov), in Leningrad (headed by N.A. Shanin), in Erevan (headed by I.D. Zaslavsky) and in Prague (headed by O. Demuth). Numerous impressive results were obtained in constructive Analysis, constructive

Logic, Theory of Algorithms, Theory of Complexity of Algorithms and Calculations, and Philosophy of Mathematics. Each of the above centers had a particular face. E.g., pioneering works on automatic theorem proving were done in Leningrad under Shanin's leadership. A body of work on constructive functional analysis was created in Prague by Oswald Demuth and his disciples (see, e.g., Demuth [1965, 1978, 1980], Kushner [1984], Kushner [1990]). Oswald was Markov's PhD student in Moscow University in the same years approximately that I worked on my Thesis under Markov, too. I remember vividly his talks in seminars, his very specific and charming Russian language and our discussions about his approach to the constructive integral of Lebesgue. In the tragic days of August, 1968 he was in Leningrad. He returned to his Mother-Land immediately and took all the consequences of his dignified rejection of Soviet occupation and the events that followed.

The bold attempt to develop Mathematics in the framework of a coherent constructive mathematical worldoutlook that was undertaken by A.A. Markov (Jr.) will beyond doubt be remembered as an exciting chapter in the History of the Mathematics of our century.

6.

The influence of Gödel's works on Markov's program was not, probably, direct but it was an essential one. Exactly as one who listens to the grandiose symphonies of Shostakovich can feel that this great Master has heard the gigantic symphonic works of Mahler, one feels, while reading Markov's fundamental papers on constructive Mathematics, that he was deeply familiar with the works and ideas of Hilbert, Brouwer and Gödel. It is very characteristic of Markov that he translated into Russian, edited and published in 1948 in *Uspekhi Mathematicheskikh Nauk* Gödel's work [1940] on the consistency of the Continuum hypothesis and in the same year published his own short paper about the dependence of axioms in the original Bernays-Gödel's system (Markov [1948]) (evidently, this paper originated in Markov's work on translating Gödel). And this happened in the year when Markov embarked on his revolutionary activities on developing his own constructive mathematics! I was too young then, but later in the early 60-ies I heard many times Markov's declarations like "I do not understand it. It is something classical, no, no...Do not even tell it to me..." I always felt a mephistophelian sarcasm behind such public statements. I do not know if Markov ever met Gödel. I doubt that such a meeting ever took place. Generally, in the political climate of those years personal contacts between Russian and Western mathematicians were very scarce. Anyway, I do not think that two men would get along if they met. Markov, like the other two great constructive leaders Brouwer and Bishop, was a quite outspoken person. He just loved to express opinions of paradoxical nature, to amaze and to shock colleagues at any convenient (or non-convenient) occasion by declarations that amounted to mathematical sacrilege. It was sometimes as though somebody would deny the Bible in a Church. On the other hand, as I got to know from an excellent account of S. Feferman [1984], Gödel was quite a different person. It looks that he preferred to be left alone with his great ideas and shied away of every type of publicity. He exercised an extreme caution and did not express openly his strong Cantorian-like platonistic convictions. And it seems that he formulated his incompleteness results in terms of provability, rather than truth, just to avoid discussions (inevitable in those years) on "what the truth is". It somehow reminds me Gauss's reservation with respect of his discovery of non-Euclid Geometry.

Be that as it may, the presence of the titanic personality of Gödel was always felt in our discussion and seminars.

I believe that MCM was influenced the most by three of Gödel's results and ideas: 1) the definition of recursive functions; 2) the incompleteness theorems; 3) the concept of a computable function of a finite type.

Though Markov had introduced his own precise concept of algorithm (normal algorithms of A.A. Markov) , especially designed for his constructive program, and developed an original and deep theory of normal algorithms, one should admit that this work was inspired and influenced by ideas and techniques of the earlier concepts of recursive functions and Turing/Post machines. As is known, Gödel was one of the pioneers in theory of recursive functions. It was he who formulated the first definition of recursive functions that is known today as the Herbrand-Gödel definition. The dramatic history of the first years, better to say months, of the theory of computability is presented in Davis [1982]. It would be simply impossible even to formulate Markov's constructive program without the pre-war achievements in the theory of computability.

As is well-known, the main idea of Hilbert's foundational program was a justification of classical mathematics by a finite proof of its consistency. Sure enough, neither Brouwer nor Markov nor Bishop would buy such a proof as a convincing argument toward legitimation of classical mathematics, as such. The constructive tendency grew not so much from paradoxes, as from intellectual doubts about the main philosophical concepts of classical mathematics, especially about actual infinity and the universality of the tertium non datur principle. (Incidentally, the identification of mathematical existence and consistency was challenged already by H. Poincaré). The most convincing proof of consistency would not make those concepts more feasible. Constructivism, as, probably, every "ism" is about principles, not paradoxes. Moreover, the well-known opinion that Gödel's incompleteness results dealt a death blow to Hilbert's hopes seems exaggerated. The point is that Hilbert's finitism put extremely strong, evidently too strong restrictions on possible proofs of consistency. This was exactly Hilbert's reaction to Gödel's results (see his preface to the first edition (1934) of Hilbert and Bernays' monograph [1968]). I'll cite from Feferman [1993] an excellent description that Gödel gave to Hilbert's finitism in the lecture "The present situation in the foundations of mathematics" delivered in December 1933 to a joint Meeting of The Mathematical Association of America and American Mathematical Society (Cambridge, Massachusetts):

1. The applications of the notion of "all" or "any" is to be restricted to those infinite totalities for which we can give a finite procedure for generating all their elements [such as integers]...

2. Negation must not be applied to propositions stating that something holds for all elements, because this would give existence propositions...[these] are to have a meaning in our system only in the sense that we have found an example but, for the sake of brevity, do not state it explicitly...

3. And finally we require that we should introduce only such notions as are decidable for any particular element and only such functions as can be calculated for any particular element.

It is evident that constructivists are usually far more liberal in their restrictions. Hence, there could be consistency proofs that would satisfy them. In fact, I think that Gentzen's proof is one of them. Various other proofs in this or that extensions of Hilbert's finitism were published subsequently. I would mention recent work of N.M. Nagorny [1995]: the author, a known constructivist of Markov's school, states that his proof of consistency of classical (formal) arithmetic is neutral, i.e. it would be accepted by both intuitionists and constructivists of Markov's school (I think that the same holds for BCM, as well).

Nevertheless, one should not underestimate the significance of Gödel's incompleteness results for constructive mathematics. One of the philosophical consequences of those results was the understanding that Hilbert's goal can not be reached, at least in full. Therefore the conceptual problems that classical mathematics was faced with were more deep and disturbing than this great mathematician believed. It goes without saying that this understanding created more a favorable psychological climate for constructive mathematics and helped to recruit new champions for it. On the other hand, it is worth noting that the conceptual and technical apparatus developed in the framework of Hilbert's program and Gödel's works turned out to be indispensable in building of Markov's constructive mathematics, which is based on a precise concept of an algorithm.

In his work of 1958 Gödel suggested a new interpretation of intuitionistic arithmetic (so called Dialectica-interpretation) by means of computable functions of finite types (the last concept was introduced in the same paper). Hence, the task of developing of an universe of computable functions of finite types arises. Two way of approaching the problem are evident. One is to enumerate objects of lower types and use such Gödel numbers as initial data for functions of higher types. Another is to use approximations to function-arguments, i.e. some topological structure for functions of lower types. In the first case we would speak of Markov operators, in the second of Kleene operators. An interesting theory of computable functions of finite types was later developed topologically in the frameworks of Markov's constructive mathematics by Chernov [1972 a-c]. It is worth noting that everywhere defined constructive functions (the counterpart to the classical concept of everywhere defined real function) can be considered in the spirit of Gödel's approach as computable functions of type ((0,0),(0,0)) with some restrictions on the domain and closure conditions on the range. As was mentioned above, two main ways to operate constructively with (0,0) objects are known: using approximations (e.g. in Bair space) or Gödel codes (numbers) of (0,0) objects. In the first case we arrive to Kleene's partial-recursive operators, in the second case we deal with Markov's constructive functions. It seems that more information about the argument-function is available for a Markov operator than for an operator of Kleene which uses only "beginnings" of the argument-functions. This effect can be really felt in the case of not everywhere defined operators (Muchnik-Friedberg counterexample, see, e.g., Kushner [1984]). But the Kreisel-Lacombe-Schoenfield-Tseitin Continuity Theorem (see, e.g., Kushner [1984]) states that the two above approaches are equivalent for constructive functions that are everywhere defined on Markov's constructive continuum. On the other hand, some results of the author (Kushner [1982]) show that the "Gödel numbers" approach gives a wider class of computable functions than partial-recursive operators if one considers functions everywhere defined on a more liberal version of Markov's constructive continuum, namely $\mathbb{P}$. These results turn out to be closely related to the problem of the compactification of constructive continuum and to uniform continuity of constructive functions. We mention the following theorem (technical detail can be found in Kushner [1982, 1984]).

Theorem 6.1. *1. If a constructive function f is everywhere defined and a Kleene operator that computes f on the closed constructive unit interval is defined for all pseudonumbers of this interval, then f is constructively uniformly continuous on this interval.*

2. There is an everywhere defined constructive function g such that there is an algorithm G of type $\mathbb{P} \to \mathbb{P}$ that extends g and, nevertheless, g is effectively non-uniformly continous on the closed unit interval.

Acknowledgement. I am gratefull to Dr. A. Wilce for many useful discussions.

References

1. M. J. Beeson. *Foundations of Constructive Mathematics.* Springer, Berlin, 1985.
2. E. Bishop. *Foundations of Constructive Analysis.* MvGraw- Hill, New York, 1967.

3. E. Bishop. Mathematics as a numerical language. *Intuitionism and Proof Theory (Proc.Conf, Buffalo, N.Y.)*, North-Holland, Amsterdam, 53-71, 1970.
4. E. Bishop. Schizophrenia in Contemporary Mathematics. *Erret Bishop: Reflections on Him and His Research* (San Diego California, 1983) (M. Rosenblatt, Editor), AMS, *Contemp. Math.*, vol.36, Providence, Rhode Island, 1984, 1-32.
5. E. Bishop and D. Bridges. *Constructive Analysis*. Springer-Verlag, Berlin-heidelberg-New York-Tokyo, 1985.
6. L. E. I. Brouwer. Bezitzt jede reelle Zahl eine Dezimalbruchentwicklung? *Nederl.Akad. Wetensch. Vershlagen* , 29, 803-812. Also in *Math. Ann* 83,201-210, 1921, 1921.
7. V. P. Chernov. Topologicheskie varianty teoremy o nepreryvnosti otobrazheniy i rodstvennye teoremy. Russian, *Zapiski Nauchnykh Seminarov LOMI*, 32, 1972a, 129-139.
8. V. P. Chernov. O konstruktivnykh operatorakh konechnykh tipov. Russian, *Zapiski Nauchnykh Seminarov LOMI* , 32, 1972b, 140-147.
9. V. P. Chernov. Klassifikaciya prostranstv operatorov konechnykh tipov. Russian, *Zapiski Nauchnykh Seminarov LOMI* , 32, 1972c, 148-152.
10. D. van Dalen. Why Constructive Mathematics? *The Foundational Debate. Complexity and Constructivity in Mathematics and Physics*, W. Depauli-Schimanovich at al, Eds. Vienna Circle Yearbook, Kluwer, Dordrecht, 141-158, 1995.
11. M. Davis. Why Gödel Didn't Have Church's Thesis. *Information and Control*, 54, 3-24, 1982.
12. O. Demuth. Lebesque integration in constructive analysis Russian, *Dokl. Akad. Nauk SSSR* (Engl. transl. in *Soviet Math. Dokl.* 6 (1965), 160, 1965, 1239-1241.
13. O. Demuth. Nekotorye voprosy teorii konstruktivnykh funktsiy deystvitel'noy peremennoy. Russian, *Acta Universitatis Carolinae, Math. et Phys.*, 19, No1, 61-69, 1978.
14. O. Demuth. O konstruktivnom integrale Perrona. Russian, *Acta Universitatis Carolinae, Math. et Phys.*, 21, No1, 3-57, 1980.
15. A. G. Dragalin. *Mathematical Intuitionism: Introduction to proof theory*. AMS (Translation from the Russian; Russian original 1979) , Providence, Rhode Island , 1988.
16. S. Feferman. Kurt Gödel:Conviction and Caution. *Philosophia Naturalis*, 21 (2-4), 546-562, 1984.
17. S. Feferman. Gödel's *Dialectica* interpretation and its two-way stretch. *Computational Logic and Proof Theory, Proc. of the Third Kurt Gödel Colloquium*, G. Gottlob et al, Eds, Lecture Notes in Computer Science, 713, 1993, 23-40.
18. Gödel. *The consistency of the Continuum Hypothesis*. Princeton University Press, Princeton, New Jersey, 1940.
19. Gödel. Über eine bisher noch nicht benütze Erweiterung des finititen Standpunktes. *Dialectica*, 12, 240-251, 1958.
20. A. Heyting. *Intuitionism. An Introduction*. North-Holland, Amsterdam, 1956.
21. A. Heyting. After thirty years. *Logic, Methodology and Philosophy of Science, Proceedings of the 1960 International Congress* , E. Nagel et all, Eds, Stanford University Press, Stanford, California , 194-197, 1962.
22. D. Hilbert and P. Bernays. *The Grundlagen der Mathematik, 1*. Springer-Verlag, Berlin-Heidelberg-New York, 1968.
23. S. C. Kleene and R. E. Vesley. *The Foundations of Intuitionistic Mathematics. Especially in Relation to Recursive functions*. North-Holland, Amsterdam, 1965.
24. A. N. Kolmogorov and V. A. Uspensky. K opredeleniyu algoritma. Russian, *Uspekhi Mathematicheskikh Nauk*, 13, vyp. 4(82), 3-28, 1958.

25. B. A. Kushner. Some extensions of Markov's constructive continuum and their applications to the theory of constructive functions. *The L.E.J. Brouwer Centenary Symposium*, A. S. Troelstra and D. van Dalen, Eds, North-Holland Publishing Company, Amsterdam-New York-Oxford, 1982 , 261-273.
26. B. A. Kushner. *Lectures on Constructive Mathematical Analysis.* (Translation from the Russian; Russian original 1973) , AMS, Providence, Rhode Island, 1984.
27. B. A. Kushner. Printsip bar-induktsii i teoriya kontinuuma u Brauera. Russian, *Zakonomernosti razvitiya sovremennoy matematiki* , Nauka, Moscow, 1987, 230-250.
28. B. A. Kushner. Ob odnom predstavlenii Bar-Induktsii. Russian, *Voprosy Matematicheskoy Logiki i Teorii Algoritmov*, Vychislitel'ny Tsentr AN SSSR, Moscow, 1988, 11-18.
29. B. A. Kushner. Markov's Constructive Mathematical Analysis: the Expectations and Results. *Mathematical Logic*, P. Petkov, Ed. , 53-58, 1990, Plenum Press, New York-London.
30. B. A. Kushner. A Version of Bar-Induction. Abstract, 1992, 57,No1 *The Journal of Symbolic Logic.*
31. B. A. Kushner. Markov i Bishop. Russian, *Voprosy Istorii Estestvoznaniya i Tekhniki*, 1, 1992, 70-81.
32. B. A. Kushner. Markov and Bishop. *Golden Years of Moscow Mathematics*, S. Zdravkovska, P. Duren, Eds, 179-197, 1993a , AMS-LMS, Providence, Rhode Island.
33. B. A. Kushner. Konstruktivnaya matematika A.A. Markova: nekotorye razmyshleniya. Russian, *Modern Logic*, 3, No2, 119-144, 1993b.
34. S. N. Manukyan. O nekotorykh topologicheskikh osobennostyakh konstruktivnykh prostykh dug. Russian, *Issledovaniya po teorii algorifmov i matematicheskoy logike*, B. A. Kushner and A. A. Markov, Eds , Vychislitel'ny Tsentr AN SSSR , Moscow, 1976, 122-129 .
35. A. A. Markov. O zavisimosti aksiomy B6 ot drugikh aksiom sistemy Bernaysa-Gedelya. Russian, *Izvestiya Akademii Nauk SSSR, ser. matem.* , 12, 1948, 569-570.
36. A. A. Markov. The theory of algorithms. *Amer. Math. Soc. Transl.*, (2) 15, 1960, AMS (Translation from the Russian, Trudy Instituta im. Steklova 38 (1951), 176-189) , Providence, Rhode Island.
37. A. A. Markov. *The theory of algorithms.* Israel Programm Sci. Transl. (Translation from the Russian, Trudy Instituta im. Steklova 42 (1954)) , Jerusalem, 1961.
38. A. A. Markov. On constructive mathematics. *Amer.Math.Soc. Transl.*, (2) 98, 1971, AMS (Translation from the Russian, Trudy Instituta im. Steklova, 67, 8-14, (1964)) , Providence, Rhode Island.
39. A. A. Markov. Comments of the Editor of the Russian translation of Heyting's book *Intuitionism.* Russian, A. Geyting, *Intuitsionizm.* Russian, Moskva, Mir, 1965, 161-193.
40. A. A. Markov. Essai de construction d'une logique de la mathematique constructive. *Revue intern. de Philosophie*, 1971, 98, Fasc.4.
41. A. A. Markov. Popytka postroeniya logiki konstruktivnoy matematiki. Russian, *Issledovaniya po teorii algorifmov i matematicheskoy logike*, B. A. Kushner and A. A. Markov, Eds , Vychislitel'ny Tsentr AN SSSR , Moscow, 1976, 3-31, A Russian version of Markov 1971.
42. A. A. Markov and N. M. Nagorny *The Theory of Algorithms.* Kluwer Academic Publishers (Translation from the Russian; Russian original 1984), Dordrecht-Boston-London, 1988.

43. E. Mendelson. Second thought about Church's thesis and mathematical proofs. *The Journal of Philosophy*, 87, 225-233 , 1990.
44. N. M. Nagorny. *K voprosu o neprotivorechivosti klassicheskoy formal'noy arifmetiki.* Russian, Computing Center RAN, Moscow , 1995 .
45. H. Rogers, Jr. *Theory of Recursive Functions and Effective Computability.* McGraw-Hill Book Company, New York, 1967.
46. E. Specker. Nicht konstruktiv beweisbare Sätze der Analysis. *The Journal of Symbolic Logic*, 14, 145-158, 1949.
47. A. S. Troelstra. On the early history of intuitionistic logic. *Mathematical Logic*, P. P. Petkov, Ed., Plenum Press , New York-London, 1990, 3-17.
48. A. S. Troelstra. *Choice Sequences. A Chapter of Intuitionistic Mathematics.* Clarendon Press, Oxford, 1977.
49. A. S. Troelstra and D. van Dalen. *Constructivism in Mathematics. An Introduction. Vol.1-2*, North-Holland , Amsterdam-New York-Oxford-Tokyo, 1988.
50. A. M. Turing. On computable numbers, with an application to the Entscheidungsproblem. 1936-37, 42 (2), *Proc. London Math. Soc.*, 230-265.
51. A. M. Turing. Correction. 1937, 43 (2), *Proc. London Math. Soc.*, 544-546.

Hao Wang as Philosopher

Charles Parsons

Harvard University
Department of Philosophy

In this paper I attempt to convey an idea of Hao Wang's style as a philosopher and to identify some of his contributions to philosophy. Wang was a prolific writer, and the body of text that should be considered in such a task is rather large, even if one separates off his work in mathematical logic, much of which had a philosophical motivation and some of which, such as his work on predicativity, contributed to the philosophy of mathematics. In a short paper one has to be selective. I will concentrate on his book *From Mathematics to Philosophy* [1974][1], since he considered it his principal statement, at least in the philosophy of mathematics. It also has the advantage of reflecting his remarkable relationship with Kurt Gödel (which began with correspondence in 1967[2]) while belonging to a project that was well under way when his extended conversations with Gödel took place. Although Gödel's influence is visible, and in some places he is documenting Gödel's views by arrangement with their author, the main purpose of the book is to expound Wang's philosophy. Although I will comment on the Wang-Gödel relation, a discussion of Wang's work as source for and interpretation of Gödel, work which included two books written after Gödel's death ([1987] and [199?]) will have to be deferred until another occasion.

1. Style, convictions, and method

Wang's writings pose difficulties for someone who wishes to sort out his philosophical views and contributions, because there is something in his style that makes them elusive. FMP, like other writings of Wang, devotes a lot of space to exposition of relevant logic and sometimes mathematics, and of the work and views of others. Sometimes the purpose of the latter is to set the views in question against some of his own (as with Carnap, 381-384); in other cases the view presented seems to be just an exhibit of a view on problems of the general sort considered (as with Aristotle on logic, 131-142). The presence of expository sections might just make Wang's own philosophizing a little harder to find, but it is not the most serious difficulty his reader faces. That

[1] This work is referred to as FMP and cited merely by page number.

[2] In fact Wang first wrote to Gödel in 1949, and they had a few isolated meetings before 1967. But the closer relationship originated with an inquiry of Wang with Gödel in September 1967 about the relation of his completeness theorem to Skolem's work. Gödel's reply is the first of the two letters published in FMP, pp. 8-11.

comes from a typical way Wang adopts of discussing a philosophical issue: to raise questions and to mention a number of considerations and views but with a certain distance from all of them. This makes some of his discussions very frustrating. An example is his discussion (FMP, ch. viii) of necessity, apriority, and the analytic-synthetic distinction. Wang is sensitive to the various considerations on both sides of the controversy about the latter distinction and considers a greater variety of examples than most writers on the subject. But something is lacking, perhaps a theoretical commitment of Wang's own, that would make this collection of expositions and considerations into an *argument*, even to make a definite critical point in a controversy structured by the views of others.[3]

This manner of treating an issue seems to reflect a difference in philosophical aspiration both from older systematic philosophy and from most analytic philosophy. In the Preface to FMP Wang writes:

> This book certainly makes no claim to a philosophical theory or a system of philosophy. In fact, for those who are convinced that philosophy should yield a theory, they may find here merely data for philosophy. However, I believe, in spite of my reservations about the possibility of philosophy as a rigorous science, that philosophy can be relevant, serious, and stable. Philosophy should try to achieve some reasonable overview. There is more philosophical value in placing things in their right perspective than in solving specific problems (x).

Nonetheless one can identify certain convictions with which Wang undertook the discussions in FMP and other works. He is very explicit about one aspect of his general point of view, which he calls "substantial factualism." This is that philosophy should respect existing knowledge, which has "overwhelming importance" for philosophy. "We know more about what we know than how we know what we know" (1). Wang has primarily in mind mathematical and scientific knowledge. Thus he will have no patience with a proposed "first philosophy" that implies that what is accepted as knowledge in the scientific fields themselves does not pass muster on epistemological or metaphysical grounds, so that the sciences have to be revised or reinterpreted in some fundamental way. He would argue that no philosophical argument for modifying some principle that is well established in mathematical and scientific practice could possibly be as well-grounded as the practice itself.[4]

Factualism as thus stated should remind us of views often called naturalism. In rejecting first philosophy, Wang is in agreement with W. V. Quine, as he seems to recognize (3), and yet his discussions of Quine's philosophy

[3] But see the remarks below on Wang's discussion of "analytic empiricism".

[4] All views of this kind have to recognize the fact that scientific practice itself undergoes changes, sometimes involving rejection of previously held principles. There is a fine line between altering a principle for reasons internal to science and doing so because of a prior philosophy.

emphasize their disagreements.[5] Wang's factualism differs from a version of naturalism like Quine's in two respects. First, natural science has no especially privileged role in the knowledge that is to be respected. "We are also interested in less exact knowledge and less clearly separated out gross facts" (2). Even in the exact sphere, Wang's method gives to mathematics an autonomy that Quine's empiricism tends to undermine. Second, in keeping with the remark from the Preface quoted above, Wang has in mind an essentially descriptive method. Quine's project of a naturalistic epistemology that would construct a comprehensive *theory* to explain how the human species constructs science given the stimulations individuals are subjected to is quite alien to Wang. In one place where Wang criticizes epistemology, his target is not only "foundationalism"; he says his point of view "implies a dissatisfaction with epistemology as it is commonly pursued on the ground that it is too abstract and too detached from actual knowledge" (19). That comment could as well have been aimed at Quine's project of naturalistic epistemology as at "traditional" epistemology.[6] Wang proposes to replace epistemology with "epistemography which, roughly speaking, is supposed to treat of actual knowledge as phenomenology proposes to deal with actual phenomena" (ibid.). But he does not make that a formal program; I'm not sure that the term "epistemography" even appears again in his writings. One way of realizing such an aspiration is by concrete, historical studies, and there is some of this in Wang's writing.[7]

There were, I think, convictions related to his "factualism" that are at work in Wang's work from early on. One I find difficult to describe in the form of a thesis; one might call it a "continental" approach to the foundations of mathematics, where both logicism after Frege and Russell and the Vienna Circle's view of mathematics and logic have less prominence, and problems arising from the rise of set theory and infinitary methods in mathematics, and their working out in intuitionism and the Hilbert school, have more. In an early short critical essay on Nelson Goodman's nominalism he wrote:

> ... there is ... ground to suppose that Quine's general criterion of using the values of variables to decide the "ontological commitment" of a theory is not as fruitful as, for instance, the more traditional ways of distinguishing systems according to whether they admit of infinitely many things, or whether impredicative definitions are allowed, and so on.[8]

Wang's sense of what is important in foundations probably reflects the influence of Paul Bernays, under whose auspices he spent much of the academic

[5] See especially [Wang 1985] and [1986].

[6] In a brief comment on Quine's project ([1985], p. 170), Wang expresses a similar philosophical reservation but also asks whether the time is ripe for such a program to achieve *scientific* results.

[7] Good examples are [1957] and the chapter on Russell's logic in FMP.

[8] [1953], pp.416-417 of the reprint in FMP.

year 1950-51 in Zürich. Wang's survey paper [1958] is revealing. There he classifies positions in the foundations of mathematics according to a scheme he attributes to Bernays (and which indeed can be extracted from [Bernays 1935]), so that the article ascends through strict finitism (which he prefers to call "anthropologism"), finitism, intuitionism, predicativism, and platonism. He shared the view already expressed by Bernays that one does not need to make a choice between these viewpoints and that a major task of foundational research is to formulate them precisely and analyze their relations. His own work on predicativity, most of which was done before 1958, was in that spirit. Although he was familiar with the Hilbert school's work in proof theory and already in the mid-1950's collaborated with Georg Kreisel, his own work on problems about the relative logical strength of axiom systems made more use of notions of translation and relative interpretation than of proof-theretic reductions.[9]

Another conviction one can attribute to the early Wang, more tenuously connected with factualism, is of the theoretical importance of computers. Interest in computability would have come naturally to any young logician of the time; recursion theory and decision problems were at the center of interest. Wang concerned himself with actual computers and spent some time working for computer firms. His work in automatic theorem proving is well known. A whole section of the collection [1962] consists of papers that would now be classified as belonging to computer science. Issues about computational feasibility had some concrete reality for him.

One can see Wang's familiarity with computers at work in the essay [1961], partly incorporated into chapter vii of FMP. This essay is one of the most attractive examples of Wang's style. Many of the issues taken up seem to arise from Wittgenstein's *Remarks on the Foundations of Mathematics* [1956], but Wittgenstein's name is not mentioned; in particular he does not venture to argue for or against Wittgenstein's general point of view as he might interpret it. But the notion of perspicuous proof, the question whether a mathematical statement changes its meaning when a proof of it is found, the question

[9] Here I might make a comment about my own brief experience as Wang's student. In the fall of 1955, as a first-year graduate student at Harvard, I took a seminar with him on the foundations of mathematics. I had begun the previous spring to study intuitionism, but without much context in foundational research. Wang supplied some of the context. In particular he lectured on the consistency proof of [Ackermann 1940]. I knew of the existence of some of Kreisel's work (at least [Kreisel 1951]) but was, before Wang's instruction, unequipped to understand it. In the spring semester, in a reading course, he guided me through [Kleene 1952]. Unfortunately for me he left Harvard at the end of that semester, but his instruction was decisive in guiding me toward proof theory and giving me a sense of its importance.

Wang's seminar was memorable for another reason. Its students included two undergraduates, David Mumford and Richard Friedberg. Friedberg gave a presentation on problems about degrees growing out of Post's work; I recall his mentioning Post's problem and perhaps indicating something of an approach to it. It was just a few weeks after the seminar ended that he obtained his solution.

whether contradictions in a formalization are a serious matter for mathematical practice and applications, and a Wittgensteinian line of criticism of logicist reductions of statements about numbers are all to be found in Wang's essay. But it could only have been written by a logician with experience with computers; for example it is by a comparison with what "mechanical mathematics" might produce that Wang discusses the theme of perspicuity. Computers and Wittgenstein combine to enable Wang to present problems about logic in a more concrete way than is typical in logical literature, then or now. But except perhaps for the criticism of Frege and Russell on '$7+5=12$', one always wishes for the argument to be pressed further.

In the remainder of the paper I shall discuss three themes in Wang's philosophical writing where it seems to me he makes undoubted contributions that go beyond the limits of his descriptive method. The first is the concept of set, the subject of chapter vi of FMP. The second is the range of questions concerning minds and machines, the subject of chapter x and returned to in writings on Gödel and the late article [1993]. The third is the discussion of "analytic empiricism", a term which he uses to describe and criticize the positions of Rudolf Carnap and W. V. Quine. This only becomes explicit in [1985] and [1985a]. Each of these discussions reflects the influence of Gödel, but it is only of the second that one could plausibly say that Wang's contribution consists mostly in the exposition and analysis of Gödel's ideas.

2. The concept of set

Chapter vi of FMP is one of the finest examples of Wang's descriptive method. It combines discussion of the question how an intuitive concept of set motivates the accepted axioms of set theory with a wide-ranging exploration of issues about set theory and its history, such as the question of the status of the continuum hypothesis after Cohen's independence proof. As an overview and as a criticsm of some initially plausible ideas,[10] it deserves to be the first piece of writing that anyone turns to once he is ready to seek a sophisticated philosophical understanding of the subject. Not of course the last; in particular, even given the state of research when it was published, one might wish for more discussion of large cardinal axioms,[11] and some of the historical picture would be altered by the later work of Gregory Moore and Michael Hallett.

[10] For example, that independence from ZF itself establishes that a set-theoretic proposition lacks a truth-value (194-196). This discussion has a curious omission, of the obvious point that if ZF is consistent, then on the view in question the statement of its consistency lacks a truth-value. Possibly Wang thought the holder of this view might bite this particular bullet; more likely the view is meant to apply only to properly set-theretic propositions, so that the theorems of a progression of theories generated by adding consistency statements would be conceded to have a truth-value.

[11] Wang himself seems to have come to this conclusion; see [1977].

The inconclusiveness that a reader often complains of in Wang's philosophical writing seems in the case of some of the issues discussed here to be quite appropriate to the actual state of knowledge and to respond to the fact that philosophical reflection by itself is not likely to make a more conclusive position possible, not only on a specific issue such as whether CH has a definite truth-value but even on a more philosophically formulated question such as whether "set-theoretical concepts and theorems describe some well-determined reality" (199). (Incidentally, Wang is here suspending judgment about a claim of Gödel.)

The most distinctive aspect of the chapter, in my view, is his presentation of the iterative conception of set and "intuitive" justifications of axioms of set theory. It is very natural to think of the iterative conception of set in a genetic way: Sets are formed from their elements in successive stages; since sets consist of "already" given objects, the elements of a set must be available (if sets, already formed) at the stage at which the set is formed. Wang's treatment is the most philosophically developed presentation of this idea. A notion of "multitude" (Cantor's *Vielheit* or *Vieles*) is treated as primitive; in practice we could cash this in by plural or second-order quantification. Exactly how this is to be understood, and what it commits us to, is a problem for Wang's account, but not more so than for others, and for most purposes we can regard the term "multitude" as a place-holder for any one of a number of conceptions.

Wang says, "We can form a set from a multitude only in case the range of variability of this multitude is in some sense intuitive" (182). One way in which this condition is satisfied is if we have what Wang calls an intuitive concept that "enables us to overview (or look through or run through or collect together), in an *idealized* sense, all the objects in the multitude which make up the extension of the concept" (ibid.). Thus he entertains an idealized concept of an infinite intuition, apparently intuition of objects. An application of this idea is his justification of the axiom of separation, stated "If a multitude A is included in a set x, then A is a set" (184):

> Since x is a given set, we can run through all members of x, and, therefore, we can do so with arbitrary omissions. In particular, we can in an idealized sense check against A and delete only those members of x which are not in A. In this way, we obtain an overview of all the objects in A and recognize A as a set.

Some years ago I argued that the attempt to use intuitiveness or intuitability as a criterion for a multitude to be a set is not successful. The idealization that he admits is involved in his concept of intuition cuts it too much loose from intuition as a human cognitive faculty.[12] For example, the set x can be very large, so that "running through" its elements would require

[12] [Parsons 1977], pp. 275-279, a paper first presented in a symposium with Wang. (His paper is [1977].)

something more even than immortality: a structure to play the role of time that can be of as large a cardinality as we like. Moreover, it seems we need to be omniscient with regard to A, in order to omit just those elements of x that are not in A. It is not obvious that intuitiveness is doing any work that is not already done by the basic idea that sets are formed from given objects.

I am, however, not sure that my earlier critical discussion captured Wang's underlying intention. Two things raise doubts. First, Wang lists five principles suggested by Gödel "which have actually been used for setting up axioms." The first is "Existence of sets representing intuitive ranges of variability, i. e. multitudes which, in some sense, can be 'overviewed'" (189). This suggests that Gödel gave some level of endorsement to Wang's conception.[13] Second, Wang evidently saw some justice in my criticisms, at least of some of his formulations (see [1977], p. 327) but still holds that the just quoted principle "is sufficient to yield enough of set theory as a foundation of classical mathematics and has in fact been applied ... to justify all the axioms of ZF ([1977], p. 313).

Wang's use of the term "intuition" in chapter vi of FMP is confusing. I don't think it is entirely consistent or fits well either the Kantian paradigm or the common conception of intuition as a more or less reliable inclination to believe. The way the term "intuitive range of variability" is used also departs from Gödel's use of "intuition" in his own writings, although it could be an extension of it rather than inconsistent with it.

One possibly promising line of attack is to think of what Wang calls "overview" as conceptual. In going over some of the same ground in note 4 of [1977], Wang encourages this reading; for example he seems to identify being capable of being overviewed with possessing unity. A reason for being confident that the natural numbers are a set is that the concept of them as what is obtained by iterating the successor operation beginning with 0 gives us, not only a clear concept of natural number, but some sort of clarity about the *extension* of natural number, what will count as a natural number. It is this that makes the natural numbers a "many that can be thought of as one."[14] For well-known reasons we cannot obtain that kind of clarity about all sets or all ordinals. The difficult cases are the situations envisaged in the power set axiom and the axiom of replacement. In the footnote mentioned, Wang discusses the set of natural numbers in a way consistent with this

[13] However, it appears from remarks quoted in [Wang 199?] that Gödel's understanding of "overview" included some of what I was critizing. Gödel is reported to say, "The idealized time concept in the concept of overview has something to do with Kantian intuition" (in remark 71.17 (in ch. 7)). In remark 71.18 he speaks of infinite intuition, and of "the process of selecting integers as given in intution." Speaking of idealizations, he says, "What this idealization ... means is that we conceive and realize the possibility of a mind which can do it" (71.19). It should be clear that when Gödel talks of idealized or infinite intution in these remarks, he is not attributing such a faculty to humans; cf. also 71.15.

[14] Cantor in 1883 famously characterizes a set as "jedes Viele, welches sich als Eines denken läßt" ([1932], p. 204).

approach and then reformulates his treatment of the power set axiom as follows:

> Given a set b, it seems possible to think of all possible ways of deleting certain elements from b; certainly each result of deletion remains a set. The assumption of the formation of the power set of b then says that all these results taken together again make up a set ([1977], p. 327).

The difference between thinking of all possible ways of deleting elements from a given set and thinking of all possible ways of generating sets in the iterative conception is real, but not so tangible as one would like, as Wang concedes by talking of the *assumption* of the formation of the power set. There is a way of looking at the matter that Wang does not use, although other writers on the subject do.[15] In keeping with Wang's own idea that it is the "maximum" iterative conception that is being developed, any multitude of given objects can constitute a set. Suppose now that a set x *is* formed at stage α. Then, since its elements must have been given or available at that stage, any subset of x could have been formed at stage α. But we would like to say that *all* the subsets of x are formed at stage α, so that $\mathcal{P}(x)$ can be formed at stage $\alpha + 1$. This amounts to assuming that if a set could have been formed at α, then it is formed at α. This is a sort of principle of plenitude; it could be regarded as part of the maximality of the maximum iterative conception. But it is certainly not self-evident.[16]

Wang's discussion of the axioms has the signal merit that he works out what one might be committed to by taking seriously the idea that sets are *formed* in successive stages. It thus stimulates one to attempt a formulation of the iterative conception that does not require that metaphor to be taken literally. The task is not easy. An attempt of my own (in [Parsons 1977]) relied on modality in a way that others might not accept and might be objected to on other grounds. Moreover, much of his discussion, for example of the axiom of replacement, can be reformulated so as not to rely on highly idealized intuitability.

I have left out of this discussion the a posteriori aspect of the justification of the axioms of set theory. Following Gödel, Wang does not neglect it, although in the case of the axioms of ZF itself, I think he gives it less weight than I would.

[15] For example [Boolos 1971], p. 494 of reprint; cf. [Parsons 1977], p. 274.

[16] Cf. [Parsons 1995], pp. 86-87. But there I asked why we should accept the appeal to plenitude in the case of subsets of a set but not in the case of sets in general or ordinals. This question was thoroughly confused. If by "sets in general" is meant all sets, then there is no stage at which they *could* have been formed, and hence no application for plenitude; similarly for ordinals. If one means *any* set, then in the context of the iterative conception, the version of plenitude leading to the power set axiom already implies that when a set could have been formed (i. e. when its elements are available) it is formed.

3. Minds and machines

Clearly Wang was prepared by experience to engage himself seriously with Gödel's thought about the concept of computability and about the question whether the human power of mathematical thought surpasses that of machines. Chapter x of FMP contains the first informed presentation of Gödel's views on the subject, and Wang returned to the question several times later.

Most of the chapter is a judicious survey of issues about physicalism, mechanism, computer simulation as a tool of psychological research, and artificial intelligence. The general tone is rather skeptical of the claims both of the mechanist and the anti-mechanist sides of the debates on these subjects. Section 6 turns to arguments in this area based on the existence of recursively unsolvable problems or on the incompleteness theorems; the difficulties of establishing conclusions about human powers in mathematics by means of these theorems are brought out, but as they had already been in the debate prompted by J. R. Lucas's claim that the second incompleteness theorem shows that no Turing machine can model human mathematical competence ([Lucas 1961]).

Only in the last section of the chapter to Gödel's views enter, and Wang confines himself to reporting. On the basis of the then unpublished 1951 Gibbs Lecture, the "two most interesting rigorously proved results about minds and machines" are said to be (1) that the human mind is incapable of formulating (or mechanizing) all its mathematical intuitions; if it has formulated some of them, this fact yields new ones, e.g. the consistency of the formalism. (2) "Either the human mind surpasses all machines (to be more precise: it can decide more number theoretical questions than any machine) or else there exist number theoretical questions undecidable for the human mind" (324).[17] The disjunction (2) is now well known. A statement follows of Gödel's reasons for rejecting the second disjunct, expressing a point of view that Wang calls "rationalistic optimism".[18] As an expression of the view that "attempted proofs of the equivalence of mind and machines are fallacious" there follows a one-paragraph essay by Gödel criticizing Turing (325-326).[19] Wang reports Gödel's view that the argument attributed to Turing would be valid if one added the premisses (3) "there is no mind separate from matter" and (4) "the brain functions basically like a digital computer" (326).[20] He also reports

[17] Cf. [Gödel *1951], pp. 307-308, 310.

[18] Wang subsequently stated ([1993], p. 119) that the paragraph consisting of this statement (324-325) was written by Gödel. He also described the formulations of (1) and (2) as "published with Gödel's approval" ([1993], p. 118 n. 12).

[19] This essay is an alternate version of Remark 3 of [Gödel 1972a]. Wang states ([1993], p. 123), presumably on Gödel's authority, that the version he published is a revision of what subsequently appeared in the *Collected Works*.

[20] These theses are numbered (1) and (2) in FMP; I have renumbered them to avoid confusion with the theses of Gödel already numbered (1) and (2).

Gödel's conviction that (1) will be disproved, as will mechanism in biology generally.

Wang dropped a kind of bombshell by this reporting of Gödel's views, with very little explanation and only the background of his own discussion of the issues. But he did not leave the matter there. It is taken up in [1987] (pp. 196-198), but only to give a little more explanation of Gödel's theses. The questions are pursued in much greater depth in [1993]. Much of this essay consists of commentary on Gödel, either on the views just mentioned reported in FMP or on other remarks made in their conversations or in other documents such as the 1956 letter to John von Neumann containing a conjecture implying $P = NP$.[21]

By "algorithmism" about a range of processes Wang means that the thesis that they can be captured or adequately modeled by an algorithm (and so simulated by a Turing machine). Physicalism and algorithmism about mental processes have to be distinguished, because one can't take for granted that *physical* processes can be so modeled.

General physical grounds will support algorithmism about the *brain* (some version of Gödel's (4)) only if physical processes are in some appropriate sense computable. Wang thought that it was such grounds that one should look for in order to decide whether to accept (4).[22] Gödel thought it practically certain that physical laws, in their observable consequences, have a finite limit of precision. Wang concludes that numbers obtained by observations can be approximated as well as makes observational sense by computable numbers, and therefore the best approach to the question of algorithmism about the physical is to ask whether physical theories yield computable predictions on the basis of computable initial data. Wang was skeptical about the physical relevance of the well-known negative results of Pour-El and Richards and mentions a conjecture of Wayne Myrvold that "noncomputable consequences cannot be generated from computable initial data within quantum mechanics" ([1993], p. 111).[23] Gödel is reported to have said that physicalism amounts to algorithmism, but Wang expresses doubts because the conclusion is based on arguments that tend to show that computable theories will agree with observation, but that is not the only requirement on a physical theory. Although he is not able to formulate the point in a way he finds clear, he is also given pause by the nonlocality of quantum mechanics.

Thus it is not so clear one way or the other whether physical processes are algorithmic. Wang does not go far into the question whether mental processes, considered apart from any physicalist or anti-physicalist thesis, are

[21] The letter is now published, with an English translation, in [Clote and Krajíček 1993], p. vii-ix.

[22] Wang evidently regards it as most prudent to consider the brain as a physical system, leaving open the question whether characteristic properties of the mind can be attributed to the brain.

[23] Cf. [Myrvold 1995], which contains results and discussion relevant to this question, not limited to quantum mechanics.

algrorithmic, although he gives a useful commentary on Gödel's remarks on Turing's alleged argument for mechanism about the mind (FMP p. 326 and [Gödel 1972a]).[24] The question about the mental has been much discussed by philosophers of mind, often in a highly polemical way. Wang may have thought it not fruitful to engage himself directly in that debate. He follows Gödel in confining himself to what can be said on the basis of rather abstract considerations and the structure of mathematics. There the upshot of his remarks is that is very difficult to make Gödel's considerations about the inexhaustibility of mathematics into a convincing case for anti-mechanism.

Considered as a whole, Wang's discussions of these issues consist of a mixture of his characteristic descriptive method with commentary on Gödel's rather cryptically expressed views. The latter does lead him into an analytical investigation of his own, particularly into whether processes in nature are in some sense computable, and if so what that sense is. Philosophers of mind, when they discuss the question of mechanism, tend to take the "machine" side of the equation for granted. The value of Wang's exploration of Gödel's thoughts is in bringing out that one cannot do that.

4. Wang on "analytic empiricism"

Underlying FMP is undoubtedly a rejection of empiricism as unable to give an adequate account of mathematical knowledge. This had long been recognized as a problem for empiricism. The Vienna Circle thought it saw the way to a solution, based on the reduction of mathematics to logic and Wittgenstein's conception of the propositions of logic as tautologies, as "necessarily true" because they say nothing. The most sophisticated philosophy of logic and mathematics of the Vienna Circle is that of Carnap. Quine rejected Carnap's views on these matters but maintained his own version of empiricism. In [1985a], Wang undertakes to describe a position common to both and to criticize it as not giving an adequate account of mathematics. A more general treatment of Carnap and Quine, in the context of a development beginning with Russell, is given in [1985].

Wang states the "two commandments of analytic empiricism" as follows ([1985a], p. 451):

(a) Empricism is the whole of philosophy, and there can be nothing (fundamental) that can be properly called conceptual experience or conceptual intuition.
(b) Logic is all-important for philosophy, but analyticity (even necessity) can only mean truth by convention.

[24] Wang also raises the corresponding question about the evolutionary process in biology but does not pursue it far, although his quotations from [Edelman 1992] are very provocative.

(b) implies that the Vienna Circle's solution to the problem posed by mathematics for empiricism will be to view it as true by convention. Wang attributes this view to Carnap and criticizes it on lines close to those of Gödel's remarks on the subject ([*1951] and [*1953/9]). Apart from sorting out some different elements in the position being criticized and considering options as to changing one or another of them, Wang's main addition to Gödel's discussion is to include the views of Quine, on whose philosophical writings Gödel nowhere comments (even in the remarks from conversations in [Wang 199?]). Wang brought out the interest of Gödel's views for the Carnap-Quine debate, by pointing to the fact that the stronger sense of analyticity that Gödel appeals to allows the thesis that mathematics is analytic to be separated from the claim that mathematics is true by convention or by virtue of linguistic usage, so that Gödel's view is a third option. I argue elsewhere that Gödel, in his reliance on his notion of concept, does not really have an answer to the deeper Quinian criticism of the ideas about meaning that underlie the analytic-synthetic distinction.[25] But that is not the end of the matter.

Concerning Carnap, Wang assumes, with Gödel, that Carnap aspires to answer a somewhat traditionally posed question about the nature of mathematical truth. His discussion is vulnerable to the objection to Gödel posed by Warren Goldfarb,[26] who points out reasons for questioning that assumption. It is hard to be comparably clear about how Wang intends to criticize Quine on these issues. In [1985a] he makes rather generalized complaints, for example against Quine's tendency to obliterate distinctions. The extended discussion of Quine's philosophy in [1985] is not very helpful; it is rather rambling and does not really engage Quine's arguments.[27]

I think one has to reconstruct Wang's reasons for the claim that Quine's version of empiricism does not give a more satisfactory account of mathematics than that given by logical positivism. An indication of the nature of the disagreement is the difference already noted between Wang's factualism and Quine's naturalism. Wang grants an autonomy to mathematics that Quine seems not to; mathematical practice is answerable largely to internal considerations and at most secondarily to its application in science. By contrast, for Quine mathematics forms a whole of knowledge with science, which is then answerable to observation. At times he views mathematics instrumentally, as serving the purposes of empirical science. This leads Quine to some reserve toward higher set theory, since one can have a more economical scientific theory without it.

[25] [Parsons 1995a], written in 1990. This paper was influenced and in some respects inspired by [Wang 1985a].

[26] See Goldfarb's introductory note to [*1953/9], [Gödel 1995], pp. 329-330.

[27] The reader gets the impression that some rather basic features of Quine's outlook repel Wang, much as he admires Quine's intellectual virtuosity and persistence. The rather limited focus of the present paper means that it does not attempt to do justice to [Wang 1985].

Wang's basic objection, it seems to me, is that Quine does not have a descriptively adequate account of mathematics, because he simply does not deal with such matters as the more direct considerations motivating the axioms of set theory, the phenomena described by Gödel as the inexhaustibility of mathematics, and the essential uniqueness of certain results of analysis of mathematical concepts such as that of natural number (Dedekind) or mechanical computation procedure (Turing). Thus he writes:

> In giving up the first commandment of analytic empiricism, one is in a position to view the wealth of the less concrete mathematical facts and intuitions as a welcome source of material to enrich philosophy, instead of an irritating mystery to be explained away ([1985a], p. 459).

His primary argument against Quine, then, would consist of descriptions and analyses of various kinds, such as those discussed above concerning the concept of set and the axioms of set theory.

Wang goes further in describing mathematics as "conceptual knowledge", where he evidently means more than to give a label to descriptive differences between the mathematician's means of acquiring and justifying claims to knowledge and the empirical scientist's. He is evidently sympathetic to Gödel's way of regarding mathematics as analytic and thus true *by virtue* of the relations of the concepts expressed in mathematical statements. If that is actually Wang's view, it is then troubling that one does not find in his writings any real response to the very powerful objections made by Quine against that view or the availability of a notion of concept that would underwrite it.

Closer examination shows, however, that Wang does not commit himself to Gödel's view. In a later paper, Wang returns to the theme of conceptual knowledge. He states what he calls the Thesis of Conceptualism:

> Given our mathematical experience, the hypothesis stating that concepts give shape to the subject-matter (or universe of discourse) of mathematics is the most natural and philosophically, the most economical.[28]

Wang's discussion of this thesis is too brief to give us a very clear idea of its meaning. One theme that emerges is the importance of a general overview or understanding of mathematical situations and also of organizing mathematical knowledge in terms of certain central concepts; he cites Bourbaki as a way of arguing for some version of the thesis, presumably because of the idea underlying their treatise of organizing mathematics around certain fundamental structures. Wang recognizes that he has not given an explanation

[28] [1991], p. 263. Wang is at pains to distinguish this thesis from realism, which is stated as a separate thesis about which he is more reserved though not unsympathetic.

of the notion of concept.[29] But even though this discussion is undeveloped, it should be clear that he is less committed to the philosophical notion of conceptual truth, of which analyticity as conceived by Gödel is an instance, than appears at first sight. In this passage, what Wang contrasts conceptual knowledge with is not empirical knowledge but technical knowledge or skill.

5. Conclusion

In his discussions of Carnap and Quine (especially in [1985]), Wang expresses general dissatisfaction with the philosophy of his own time. That is not only context in which he does so. Neither time nor space would permit a full discussion of his reasons. By way of conclusion I want to mention a respect in which Wang's philosophical aspirations differed importantly from those of most of his philosophical contemporaries, of different philosophical tendencies and differing levels of excellence. The major philosophers of the past have left us systematic constructions that are or at least aspire to be rather tightly organized logically, and many of the problems of interpreting them derive from the difficulty of maintaining such a consistent structure while holding a comprehensive range of views. Systematic constructions have been attempted by some more conservative figures of recent times as well as some of the important "continental" philosophers, in particular Husserl. But it has also been the aim of some figures in the analytic tradition whom Wang knew well, such as Quine and Dummett. Those who have not sought such systematicity have generally approached problems in a piecemeal manner or relied heavily on a technical apparatus.

As I see it, Wang's aspiration was to be synoptic, in a way that the more specialized analytic philosophers do not rise to, but without being systematic, which I think he saw as incompatible with descriptive accuracy and a kind of concreteness that he sought. Moreover, part of what he sought to be faithful to was philosophical thought itself as it had developed, so that he was led into an eclectic tendency. All but the last of these aspirations might be thought to be shared with a figure who greatly interested Wang but whom I have mentioned only in passing so far: Ludwig Wittgenstein. Yet Wang was no Wittgensteinian. I am probably not the person to explain why. But clearly there were deep differences in their approaches to mathematics and logic, and Wang could, I think, never accept the idea that the philosophical problems should *disappear* if one does one's descriptive work right.

In Gödel, Wang encountered someone who shared many of his views and aspirations but who did have the aspiration to system that he himself re-

[29] [1991], p. 264. In 1959 Gödel made a similar admission in a letter to Paul Arthur Schilpp explaining why he had not finished [*1953/9], quoted in [Parsons 1995a], p. 307. Sidney Morgenbesser informs me that in discussions he and Wang had in the last years of Wang's life the notion of concept figured prominently, but it appears that Wang did not arrive at a settled position.

garded skeptically. I think this was the most significant philosophical difference between them.[30] Because of its rather general nature, it is most often manifested by Wang's reporting and working out rather definite claims of Gödel about which he himself suspends judgment. On some matters, such as the diagnosis of his problems with analytic empiricism, Gödel's influence made it possible for Wang to express his own views in a more definite way; one might say that the ideal of systematicity imposed a discipline on Wang's own philosophizing even if he did not in general share it. I would conjecture that the same was true of Wang' s treatment of the concept of set, although in that case it is a difficult to pin down Gödel's influence very exactly.

Wang has written that Gödel did not believe he had fulfilled his own aspirations in systematic philosophy.[31] Very often he also gives the impression that he did not find what he himself was seeking in philosophical work. Both Gödel and Wang struggled with the tension inherent to the enterprise of general philosophizing taking off from a background of much more specialized research. I would not claim for Wang that he resolved this tension satisfactorily, and often, even when he is most instructive, what one learns from him is less tangible than what one derives from the arguments of more typical philosophers, even when one disagrees. I hope I have made clear, however, that Wang's "logical journey" gave rise to philosophical work that is of general interest and to contributions that are important from standpoints quite different from his own.

References

Cited works of Hao Wang

[1953] What is an individual? *Philosophical Review* 62, 413-420. Reprinted in [1974], Appendix.

[1957] The axiomatization of arithmetic. *The Journal of Symbolic Logic* 22, 145-158. Reprinted in [1962].

[1958] Eighty years of foundational studies. *Dialectica* 12, 466-497. Reprinted in [1962].

[1961] Process and existence in mathematics. In *Essays on the Foundations of Mathematics, dedicated to Prof. A. H. Fraenkel on his 70th birthday*, pp. 328-351. Jerusalem: Magnes Press, The Hebrew University of Jerusalem.

[30] On some issues outside the scope of the present discussion, there are disagreements related to cultural differences. Where Gödel turns to general questions of metaphysics and *Weltanscauung*, he regards the rational theism of pre-Kantian rationalism seriously and manifests a Leibnizian optimism. In his response to these views (which he says Gödel did not discuss much with him), Wang seems to me to show his Chinese cultural background.

[31] [1987], pp. 192, 221.

[1962] *A Survey of Mathematical Logic.* Peking: Science Press. Also Amsterdam: North-Holland, 1963. Reprinted as *Logic, Computers, and Sets.* New York: Chelsea, 1970.

[1974] *From Mathematics to Philosophy.* London: Routledge and Kegan Paul. Also New York: Humanities Press.

[1977] Large sets. In [Butts and Hintikka 1977], pp. 309-334.

[1985] *Beyond Analytic Philosophy. Doing Justice to What We Know.* Cambridge, Mass.: MIT Press.

[1985a] Two commandments of analytic empiricism. *The Journal of Philosophy* 82, 449-462.

[1986] Quine's logical ideas in historical perspective. In Lewis E. Hahn and Paul Arthur Schilpp (eds.), *The Philosophy of W. V. Quine*, pp. 623-643. La Salle, Ill.: Open Court.

[1987] *Reflections on Kurt Gödel.* Cambridge, Mass.: MIT Press.

[1991] To and from philosophy – Discussions with Gödel and Wittgenstein. *Synthese* 88, 229-277.

[1993] On physicalism and algorithmism: Can machines think? *Philosophia Mathematica* Series III, 1, 97-138.

[199?] *A Logical Journey: From Gödel to Philosophy.* Cambridge, Mass.: MIT Press. (forthcoming)

Other writings

Ackermann, Wilhelm, [1940] Zur Widerspruchsfreiheit der Zahlentheorie. *Mathematische Annalen* 117 (1940), 162-194.

Benacerraf, Paul, and Hilary Putnam (eds.), [1983] *Philosophy of Mathematics: Selected Readings.* 2d ed. Cambridge University Press.

Bernays, Paul, [1935] Sur le platonisme dans les mathématiques. *L'ensiegnement mathématique* 34, 52-69. English translation in [Benacerraf and Putnam 1983].

Boolos, George, [1971] The iterative conception of set. *The Journal of Philosophy* 68, 215-231. Reprinted in [Benacerraf and Putnam 1983].

Butts, Robert E., and Jaakko Hintikka (eds.), [1977] *Logic, Foundations of Mathematics, and Computability Theory.* Dordrecht: Reidel.

Cantor, Georg, [1932] *Gesammelte Abhandlungen mathematischen und philosophischen Inhalts.* Edited by Ernst Zermelo. Berlin: Springer.

Clote, Peter, and Jan Krajíček (eds.), [1993] *Arithmetic, Proof Theory and Computational Complexity.* Oxford: Clarendon Press.

Edelman, Gerald M., [1992] *Bright Air, Brilliant Fire.* New York: Basic Books.

Gödel, Kurt, [*1951] Some basic theorems of the foundations of mathematics and their applications. 25th Josiah Willard Gibbs Lecture, American Mathematical Society. In [Gödel 1995], pp. 304-323.

Gödel, Kurt, [*1953/9] Is mathematics syntax of language? Versions III and V in [Gödel 1995], pp. 334-362.

Gödel, Kurt, [1972a] Some remarks on the undecidability results. In [Gödel 1990], pp. 305-306.

Gödel, Kurt, [1990] *Collected Works, volume II: Publications 1938-74*. Edited by Solomon Feferman, John W. Dawson, Jr., Gregory H. Moore, Robert M. Solovay, and Jean van Heijenoort. New York and Oxford: Oxford University Press.

Gödel, Kurt, [1995] *Collected Works, volume III: Unpublished Essays and Lectures*. Edited by Solomon Feferman, John W. Dawson, Jr., Warren Goldfarb, Charles Parsons, and Robert M. Solovay. New York and Oxford: Oxford University Press.

Kleene, S. C., [1952] *Introduction to Metamathematics*. New York: Van Nostrand. Also Amsterdam: North-Holland and Groningen: Noordhoff.

Kreisel, G., [1951] On the interpretation of non-finitist proofs, part I. *The Journal of Symbolic Logic* 16, 241-267.

Lucas, J. R., [1961] Minds, machines, and Gödel. *Philosophy* 36, 112-127.

Myrvold, Wayne C., [1995] Computability in quantum mechanics. In W. DePauli-Schimanovich et al. (eds.), *The Foundational Debate*, pp. 33-46. Dordrecht: Kluwer.

Parsons, Charles, [1977] What is the iterative conception of set? In [Butts and Hintikka 1977], pp. 335-367. Reprinted in [Parsons 1983] and [Benacerraf and Putnam 1983]. Cited according to the reprint in [Parsons 1983].

Parsons, Charles, [1983] *Mathematics in Philosophy: Selected Essays*. Ithaca, N. Y.: Cornell University Press.

Parsons, Charles, [1995] Structuralism and the concept of set. In Walter Sinnott-Armstrong et al. (eds.), *Modality, Morality, and Belief: Essays in honor of Ruth Barcan Marcus*, pp. 74-92. Cambridge University Press.

Parsons, Charles, [1995a] Quine and Gödel on analyticity. In Paolo Leonardi and Marco Santambrogio (eds.), *On Quine: New Essays*, pp. 297-313. Cambridge University Press.

Wittgenstein, Ludwig, [1956] *Remarks on the Foundations of Mathematics*. Edited by G. E. M. Anscombe, R. Rhees, and G. H. von Wright. With a translation by G. E. M. Anscombe. Oxford: Blackwell.

A bottom-up approach to foundations of mathematics *

Pavel Pudlák**

Mathematical Institute
Academy of Sciences
Prague, Czech Republic

1. Introduction

There are two basic properties of a logical system: consistency and completeness. These two properties are important for systems used for particular special purposes (say, in artificial intelligence), as well as for systems proposed as foundations of the all of mathematics. Therefore, any systematic study of the foundations of mathematics should address questions of consistency and completeness. Gödel's theorems provide negative information about both: any reasonable sufficiently strong theory is unable to demonstrate its own consistency and any such system is incomplete.

There has been impressive success in proving independence results for set theory. Cohen's forcing method and its boolean valued version was applied to solve the continuum problem and to prove a lot of other sentences to be independent of set theory. This may give the impression that logic is doing very well in studying the independence phenomenon which is only partly true. The known independent sentences in set theory express statements about infinite sets. The type of the infiniteness in these independence results is, in a sense, of higher order than in classical mathematics. For instance, the continuum hypothesis talks about the cardinality of the set of real numbers, which is in any case uncountable. While most of classical mathematics is about real numbers, almost everything there can be encoded using only a countable number of elements, hence can be expressed as a statement about natural numbers. As an example, consider a continuous real function. Clearly, such a function is described by its values on rationals, thus statements about continuous functions can be written as arithmetical formulas. For arithmetical formulas, current methods of logic almost completely fail. An exception are Paris-Harrington type independence results [28]. Still it is true that no problem from classical mathematics (i.e. number theory, algebra, calculus) was proved independent from a logical theory before it had actually been solved. Furthermore, it seems that even the most difficult classical results in number

* This paper is in its final form and no similar paper has been or is being submitted elsewhere.

** Partially supported by grant No. A1019602 of AVČR and US-Czechoslovak Science and Technology Program grant No. 93025

theory and finite combinatorics can be proved in Peano arithmetic, a theory which is much weaker than Zermelo-Fraenkel set theory.

Therefore one of the principal goals of mathematical logic should be development of new techniques for proving independence. Naturally we should start with weak theories and extend our techniques to stronger ones. This is what I referred to in the title as *the bottom-up approach.* We can view the independence result of Paris and Harrington for Peano arithmetic as the first result in this direction. Later most of the research concentrated on weaker systems called bounded arithmetic. The reason for studying bounded arithmetic is not only the fact that it is much weaker. Bounded arithmetic is interesting mainly because of its connection to computational complexity. There seems to be a correspondence between various theories from the class of bounded arithmetic and complexity classes. This correspondence manifests itself in many ways; e.g. to separate two theories looks to be almost as hard as to separate the corresponding complexity classes.

It turned out that bounded arithmetic is not the very bottom level from which one should start. Independence and separation problems for bounded arithmetic can be further reduced to combinatorial problems about propositional calculus, namely, questions about the *lengths* of proofs. Here we come even closer to computational complexity, as proofs in propositional calculus are essentially nondeterministic computations.

Concerning the consistency question there has been practically no progress whatsoever. Maybe, it is because this is not a mathematical question and we should discard it for this reason. However the sentences expressing consistency of a theory are a very important tool in proof theory. They can be used to separate a weak theory from a strong one, but not in the case of bounded arithmetic, as we shall see below.

There is a lot of activity around bounded arithmetic, the complexity of propositional calculus and many questions studied in theoretical computer science are related to it too. The purpose of this paper is to survey some results which should give an idea to an outsider of what is going on in this field and explain motivations for the studied problems. We recommend [3, 5, 15, 11, 34] to those who want to learn more about this subject.

2. Basic concepts

2.1

The basic theory used in logic for studying natural numbers is the so called *Robinson's arithmetic,* or Q, which is axiomatized by the following axioms:

$$S(x) \neq 0,\ S(x) = S(y) \rightarrow x = y,\ x \neq 0 \rightarrow \exists y(x = S(y)),$$

$$x + 0 = x,\ x + S(y) = S(x + y),\ x \cdot 0 = 0,\ x \cdot S(y) = x \cdot y + x,$$

$$x \leq y \equiv \exists z(z + x = y).$$

These axioms express the inductive properties of the operations of the successor S, addition and multiplication, the last axiom is just a definition of $\leq$. The theory is useless for practical purposes as it does not prove even such basic statements as is the associativity law for addition.

Another basic system, called *Peano arithmetic* or PA, consists of Robinson's arithmetic and induction axioms

$$(\varphi(0) \wedge \forall x(\varphi(x) \rightarrow \varphi(S(x)))) \rightarrow \forall x \varphi(x), \tag{2.1}$$

for all formulas in the language $\{0, S, +, \cdot, \leq\}$.

A bounded quantifier is a quantifier of the form $\forall x \leq \tau$ or $\exists x \leq \tau$ where τ is a term not containing the variable x. These constructs are, of course, not present in the usual first order language, so we treat them as abbreviations. (Alternatively, they can be used as part of the language, if we add appropriate logical rules.) A formula where all quantifiers are bounded is called *a bounded formula* or Δ_0 or Σ_0. Formulas of the form $\forall x_1 \dots \forall x_n \varphi$, $\exists x_1 \dots \exists x_n \varphi$, $\forall x_1 \dots \forall x_n \exists y_1 \dots \exists y_m \varphi$, $\exists x_1 \dots \exists x_n \forall y_1 \dots \forall y_m \varphi$, etc. where φ is a bounded formula are called, respectively, Π_1, Σ_1, Π_2, Σ_2 etc. formulas.

In this hierarchy the simplest sentences which are unprovable in arithmetical theories are Π_1. This is because a true Σ_1 sentence is already provable in Q (take the numbers which witness the existential quantifiers and check the Δ_0 part using the axioms of Q). Also because of their simplicity Π_1 sentences are the most interesting ones from the point of view of provability. Many famous problems can be stated as Π_1 sentences: Fermat's last theorem, the four color theorem (both being eventually proved). The famous Riemann hypothesis can be also stated as a Π_1 sentence see [22, 24], though it is not so easy to prove. Several conjectures about the distribution of primes are Π_1 sentences.

The most important problem in theoretical computer science is the question if $P = NP$. The statement $P \neq NP$, which most people believe is true, can be expressed as a Π_2 sentence (this follows from the existence of an NP complete set). Thus we may hope that we can prove its independence in the same way as the independence of Paris-Harrington type sentences, but it is very unlikely. Let $n(k)$ be the least n such that satisfiability of formulas of length n cannot be computed by a circuit of size n^k. A Paris-Harrington type independence proof would require to prove that $n(k)$, as a function of k, grows extremely fast. This is the same as to say that the circuit complexity of satisfiability is $n^{f(k)}$ where $f(k)$ grows extremely slowly. Quite on the contrary, the general opinion is that $f(k)$ is probably of the order n^c, for some constant c, and definitely at least, say, $\varepsilon \log n$. The statement that satisfiability does not have circuits of size $n^{\varepsilon \log n}$ is a Π_1 sentence. The same can be said about other conjectures in complexity theory: they are not Π_1 sentences, but we believe that they will follow from stronger Π_1 sentences.

Let us note that we are interested only in theories one of whose models is the *standard model* which is the set of natural numbers with the usual operations (subtheories of the *true arithmetic*, in particular they are ω-*consistent*). For such a theory T, if we prove that a Π_1 sentence φ is consistent with T, then φ is true. The standard model is isomorphic to an initial segment of any model of an arithmetical theory, hence any consistent Π_1 sentence is true. (By an *initial segment* we mean a nonempty subset which contains with each number all smaller numbers.)

The fact that so many important problems are Π_1 sentences is an important reason for looking for techniques by which one can prove independence of such sentences. In fact, as shown above on the example of $P = NP$?, we use sentences of higher quantifier complexity only because we are not quite sure what is the right conjecture. For more information about this subject we recommend to consult Kreisel's paper [22].

2.2

It is necessary to have some background in complexity theory in order to appreciate the importance of the bottom-up approach. We shall assume that the reader knows some basics, in particular knows the definitions of the classes P and NP. For a logician it is easy to define a hierarchy of classes extending NP where $NP = \Sigma_1^p$, Π_1^p are complements of the NP classes (also denoted by $coNP$), Σ_2^p are classes defined by polynomially bounded existential quantifiers followed by universal polynomially bounded quantifiers bounding variables in a predicate from P etc. The union of these classes is called *the Polynomial Hierarchy* and denoted by PH. If we use *linearly bounded computations* instead of polynomially bounded ones, we obtain *the Linear Hierarchy*. The Linear Hierarchy when restricted to sets of natural numbers coincides with sets definable by Δ_0 formulas.

3. Independent sentences

In this section we shall explain why we are not satisfied with the current methods of proving independence results. The main reason is that, except for Gödel's theorem which gives only some special formulas, no general method is known for proving independence of Π_1 sentences.

Cohen's forcing method can be roughly described using model theory as follows. We start with a model M of ZF and extend it to M' by adding new sets. When doing that we must follow some rules in order to get a model of ZF again. The details are not important here, what we need is only the fact that the ordinal numbers are used as a sort of skeleton for building the new universe. As a result the ordinal numbers of the new model M' are the same as the ones of M. Furthermore, the concept of being finite is also preserved by the extension. The natural numbers are the finite ordinals hence, in particular, they are the same in both models. Consequently the arithmetical sentences which are true in M' are the same as those in M. Therefore this method does not give any independent arithmetical sentence.

Modifications of the forcing method have been used in various situations, so it is not completely excluded that some variant can give arithmetical independent sentences. However in order to get something interesting one would have to come up with a substantial change. To see what kind of problems it is necessary to overcome let us mention at least one. Suppose M is just a model of the natural numbers, say a model of Peano arithmetic. Trying to follow the original idea of forcing we attempt to add new numbers *between* the old ones, i.e. for some $a \in M$ there will be some new $b \in M'$ $b < a$. However there is a number $c \in M$ which codes (say using Gödel β-function) all the numbers of M below a. This c will have the same property in M', so we cannot add such a b!

Let us turn now to Paris-Harrington type independence results. Again we shall not talk about details and concentrate only on the overall structure of the argument. Let M be a nonstandard model of PA, say, a countable one. Instead of enlarging the model we look for an initial part M' of M which is also a model of PA. It turns out that M' is a model of PA if it is closed with respect to certain functions definable in M. The point of the argument is that it is possible to characterize, using some natural combinatorial sentences, where such segments can be found. Various combinatorial characterizations give various independent combinatorial sentences. The fact that we use initial segments of a given model is a serious limitation of the method. This method can prove independence of sentences which are Π_2 and, moreover, directly or indirectly refer to fast growing functions. Also it is known that these sentences are equivalent in PA to the Σ_1 reflection principle. This is not necessarily a drawback, as we consider equivalence with respect to a strong theory, but still it means that the class of such sentences is restricted.

It is obvious that we cannot prove independence of a Π_1 sentence in this way: any such sentence is preserved on initial segments. The only way that we know of to prove independence of a Π_1 sentence is Gödel's theorem. Gödel showed that a suitable diagonal sentence (*"I am not provable in PA"*) is not provable in PA. Then he proved that the sentence is equivalent to a sentence Con_{PA} asserting the consistency of PA. The same argument works for any sufficiently strong theory; with some modification it can be applied even to Robinson's Q. The above remark that the Paris-Harrington sentence is equivalent to the Σ_1 reflection principle suggests that one might prove the independence of a concrete combinatorial statement by showing that it implies the consistency of PA or another theory, but so far there are no such results.

By "concrete combinatorial" we mean a sentence expressing some principle from finite combinatorics or number theory. If one takes into account infinite principles, then such reductions to consistency statements are quite common. In this way one proves e.g. that some large cardinal axiom is stronger than another one.

4. Bounded arithmetic

4.1

By these words we denote a class of arithmetical theories which are axiomatized by induction axioms restricted to bounded formulas in various languages and with various modifications, restrictions, etc. We should stress again that the reason for restricting to weak theories is not that we accept some philosophy of "feasible mathematics" where only certain constructive reasoning is allowed. We want to study metamathematical properties of such theories *using any mathematical means, however nonconstructive.* The true reason why we are interested in these systems is that they seem to be more amenable to logical analysis and for their close relation to complexity theory.

The oldest considered system is $I\Delta_0$ which is Q plus induction axioms (2.1) for Δ_0 formulas, introduced in [27]. If M is any model of $I\Delta_0$, then any infinite initial segment closed under multiplication is a model of $I\Delta_0$ too. This implies that if $I\Delta_0$ proves that a function $f(x)$ is defined for all numbers, then the function is bounded by a polynomial in x. In particular it is not provable in $I\Delta_0$ that $x^{\lfloor \log_2(x+1) \rfloor}$, 2^x etc. are defined for all numbers.

We can view a number x as encoding a binary string of length $\approx \log x$. The strings of quadratic length, i.e. of length $(\log x)^2$, can be encoded by numbers of size $\approx x^{\lfloor \log_2(x+1) \rfloor}$, the length $(\log x)^4$ requires the function $x^{\lfloor \log_2(x+1) \rfloor}$ iterated twice and so on. Hence the function $x^{\lfloor \log_2(x+1) \rfloor}$ gives us the right growth rate if we want to extend the lengths of strings polynomially, e.g. using polynomial time computations. This is the reason for extending $I\Delta_0$ by an axiom saying that is a total function. This theory is usually denoted by $I\Delta_0 + \forall x \exists y \; (y = x^{\lfloor \log_2(x+1) \rfloor})$ or shortly $I\Delta_0 + \Omega_1$. (The last formula is just an abbreviation of a rather complicated formula, since we do not have x^y in our language.) This theory is capable of formalizing properly polynomial time computations, not only deterministic, but in fact with finitely many alternations - this is what is needed for machine based definitions of the Polynomial Hierarchy.

4.2

Having these two theories we can give an example of a basic problem in bounded arithmetic which seems to be closely connected with a problem in complexity theory.

Problem 4.1. Is every Π_1 sentence provable in $I\Delta_0 + \Omega_1$ provable in $I\Delta_0$?

The corresponding problem in complexity theory is:

Problem 4.2. Is the Polynomial time hierarchy equal to the Linear time hierarchy?

The connection between the two problems is: if we can *prove in* $I\Delta_0$ that the polynomial time hierarchy equals to the linear time hierarchy, then every Π_1 sentence provable in $I\Delta_0 + \Omega_1$ is provable in $I\Delta_0$. Note that this does not exclude that the answer to the second problem is YES while the answer to the first one is NO. We conjecture that the answer to both problems is NO and hope that the first problem is somewhat easier. A negative answer to Problem 1 would give us at least some supporting evidence that the answer to Problem 2 is NO, namely that the equality of the two complexity classes is not provable in $I\Delta_0$.[1]

4.3

The natural approach is to try to show that the consistency of $I\Delta_0$ is provable in $I\Delta_0+\Omega_1$, while by Gödel's theorem it is not provable in $I\Delta_0$. This approach is, however, quite hopeless: even the consistency of Q is not provable in $I\Delta_0+\Omega_1$. This follows from:

Theorem 4.1. (Wilkie [39]) *There is an interpretation of* $I\Delta_0 + \Omega_1$ *in* Q.

It follows that $I\Delta_0+\Omega_1 \vdash Con_{I\Delta_0+\Omega_1} \equiv Con_Q$, whence $I\Delta_0+\Omega_1 \nvdash Con_Q$.

Some weaker versions of consistency statements have been proposed, but the general feeling is that there is some deeper reason why this cannot work. What we suspect is that such results cannot be proved by *diagonalization*. This term refers to a method which was used for the fist time by Cantor to prove that the cardinality of the power set is larger than the cardinality of the given set. A similar idea was used in recursion theory and later in complexity theory to separate some complexity classes. It seems that the use of *self-referential sentences* to separate theories in logic is just another facet of this method. As all attempts to separate pairs of complexity classes such as P and NP have failed and because of the similarity of self-reference to diagonalization, there is little hope that we can solve such problems as Problem 1 using a version of Gödel's theorem.

4.4

In order to get connections to more interesting problems in complexity theory than Problem 2, we have to introduce fragments of $I\Delta_0 + \Omega_1$. The idea

[1] We cannot speak about classes in $I\Delta_0$ directly, instead we talk about definability by formulas of certain complexity. Another option is to consider second order bounded arithmetic.

is to restrict the induction schema further to subsets of bounded formulas. From the point of view of logic, a natural hierarchy is obtained by restricting the number of alternations of bounded quantifiers. Unfortunately, if we use the usual language of arithmetic, we do not get classes which correspond to natural Turing machine definable classes. Therefore, the language must be extended further by some simple functions; in particular $\lfloor \log_2(x+1) \rfloor$ and a function whose growth rate is similar to $x^{\lfloor \log_2(x+1) \rfloor}$ are added as primitives. Furthermore, when defining the hierarchy of the formulas, the bounded quantifiers in which the outer function is log, are not counted. The particular details are not important, we only need that the classes of formulas Σ_i^b define exactly the sets in the complexity classes Σ_i^p and similarly for the Π classes.

We also have to modify the induction axioms. Instead of (2.1) we take

$$(\varphi(0) \wedge \forall x(\varphi(x) \rightarrow (\varphi(2x) \wedge \varphi(S(2x))))) \rightarrow \forall x \varphi(x). \qquad (4.1)$$

This is induction on the length of the binary representation of numbers. The intuitive explanation of why we take this weaker form of induction is that to verify an instance of $\varphi(n)$ we need to unwind the premise only $\log n$ times and $\log n$ is the size of the input when we compute with n.

The theory axiomatized by some basic open axioms and the schema (4.1) is denoted by S_2; it is a conservative extension of $I\Delta_0 + \Omega_1$. The fragments where (4.1) is restricted to Σ_i^b are denoted by S_2^i. So we can think of S_2^i as a theory for Σ_i^p. The theory S_2^1 deserves a particular attention as it is a theory for NP. The definition of these theories is due to Buss [3]; the idea of (4.1) goes back to Cook [6] where he introduced an equational theory PV.

Now you may ask, what is a theory for P? Well, one can give some suggestions, e.g. PV, but I have to warn you that the correspondence between the theories and the complexity classes is very loose and should not be taken literally. For instance, S_2^1 proves all instances of (4.1) for Π_1^b formulas, so the theories for NP and for $coNP$ coincide. We still think that this is not evidence for $NP = coNP$. In a moment we will see that there is also a good reason to associate S_2^1 with P rather than with NP.

5. Relations to computational complexity

There are not only "morphological" similarities between theories and complexity classes, but also several *provable* relations. We shall mention at least two basic results.

5.1

The first result concerns the question, how difficult is it to witness the existential quantifier in $\forall\exists$ sentences provable in a given theory. The paradigm is the classical result on the fragment of Peano arithmetic $I\Sigma_1$ (induction restricted to Σ_1 arithmetical formulas) due to Mints and Takeuti. This is the result that $I\Sigma_1 \vdash \forall x \exists y \varphi(x,y)$, for φ a Σ_1 formula, implies that there exists a primitive recursive function f such that $\varphi(n, f(n))$ is true for all n.

The most important of such "witnessing" theorems is the following one:

Theorem 5.1. (Buss [3]) *Suppose* $S^1_2 \vdash \forall x \exists y \varphi(x,y)$, *for* φ *a* Σ^b_1 *formula. Then there exists a polynomial time computable function* f *such that* $\varphi(n, f(n))$ *is true for all* n.

This theorem has a natural extension to theories S^i_2 and there are several witnessing theorems for other theories and other classes of formulas. Theorem 5.1 has an easy corollary which can be interpreted as an independence result.

Corollary 5.1. *If it is provable in* S^1_2 *for a class* X *that* $X \in NP \cap coNP$, *then* $X \in P$, *so* $NP \cap coNP \neq P$ *is not provable in* S^1_2.

It has also been shown that there exists a *model* of PV in which $NP \cap coNP \neq P$ [15]. We cannot deduce much for complexity theory from this result, since as soon as we make the theory only slightly stronger, the proof breaks down. Still it is a nice result and we would like to get more results like this.

5.2

The second result concerns the problem of the hierarchy of the theories S^i_2. For all we know they may be the same (the sets of their theorems may coincide).

Problem 5.1. Are there infinitely many i for which S^{i+1}_2 properly extends S^i_2?

It is well-known that this is equivalent to the statement that S_2 is not finitely axiomatized, and to the statement that $I\Delta_0 + \Omega_1$ is not finitely axiomatized; it also implies that $I\Delta_0$ is not finitely axiomatized. The corresponding problem about complexity classes is:

Problem 5.2. Are there infinitely many i for which $\Pi^p_i \neq \Pi^p_{i+1}$?

It has been conjectured that the answer to this problem is positive. This conjecture belongs to one of the strongest considered in complexity theory; in particular it implies $P \neq NP$ and $NP \neq coNP$. Unlike with Problems 1 and 2, one can prove:

Theorem 5.2. (Krajíček et al. [20]) *A positive answer to Problem 5.2 implies a positive answer to Problem 5.1.*

This theorem is proven using a special kind of witnessing theorems for theories S_2^i. Note that contrary to what one may expect, we need $\Pi_{i+2}^p \neq \Sigma_{i+2}^p$ for proving $S_2^i \not\equiv S_2^{i+1}$. There are several extension and variation of this result. In particular, we know that $S_2^i = S_2^{i+1} \Rightarrow S_2^i = S_2$ [4]; the corresponding fact for the polynomial time hierarchy is trivial.

To prove unconditional separation of S_2^{i+1} from S_2^i, we have to find concrete sentences which are provable in S_2^{i+1} but not in S_2^i. We do have some candidates for such sentences but we have no idea how to prove their unprovability. There is however something which is halfway between this and using conjectures from complexity theory. We can extend the language by adding an uninterpreted predicate, say α and consider variants of the theories in this language. I.e. we do not add any basic axioms about α, but we extend the induction axioms to all formulas of appropriate complexity containing α. We denote such an extension of a theory T by $T(\alpha)$. In several cases we can solve separation problems for such extensions; in particular we know that $S_2^i(\alpha) \not\equiv S_2^{i+1}(\alpha)$. For those who know basics of complexity theory it is not that surprising, since there is a similar concept for complexity classes. This is the so called *relativization*, which means that we augment Turing machines with an *oracle*, which is essentially free access to some possibly complex set. Then one can prove a lot of separations. The symmetry between theories and complexity classes is, however, again not complete: using relativizations we can make classes both unequal *and equal* (e.g. $P^A \neq NP^A$ for some A and $P^B = NP^B$ for another B), while we can only separate theories by adding the uninterpreted predicate (unless we also add additional axioms).

6. Propositional calculus

We have considered theories and complexity classes. Now we shall talk about a third kind of system related to the previous two: *proof systems for propositional calculus*. Thus we sink to the bottom of the universe of formal systems.

Most of the research into propositional calculus deals with semantical modifications (extensions of the classical propositional calculus by modalities, weakenings, such as intuitionistic logic etc). What we are interested in is something different. We use only the classical propositional calculus and we study possible ways in which the concept of the proof can be formalized. Again this is related to complexity theory, if not just a branch of it.

As the set of tautologies is fixed once for ever, we cannot classify the proof systems by what they prove. Instead we shall distinguish them by what they prove *using short proofs*. By "short" we mean, of course, polynomial in the size of the formula. Furthermore we can (quasi)order the systems by defining $P \leq Q$ for two proof systems, if for any tautology τ its shortest proof in Q is at most polynomially longer than its shortest proof in P. In all concrete cases that we have so far encountered, if $P \leq Q$, then there exists a polynomial algorithm which any proof of τ in P transforms in a proof (at most polynomially longer) of τ in Q. Then we say that Q *polynomially simulates* P.

The most common proof systems for propositional calculus are the systems based on finitely many axiom schemas and finitely many rules of the type of *Modus Ponens*; quite often *Modus Ponens* is the only rule of the system. The technical term for such systems is *Frege system*. It has been proved [7] that all Frege systems have essentially the same power, namely they polynomially simulate each other. (This is very easy to prove if the two systems use the same connectives.)

To make the system stronger we add a rule that allows us to abbreviate (long) formulas by a single variable. Formally this means that we can introduce for every formula φ the equivalence $\varphi \equiv p$, where p is a propositional variable not used in the previous part of the proof nor in the proved formula. Frege systems with this rule added are called *Extended Frege systems.* Note that in this way we may reduce the total size of a proof, but we cannot save on the number of steps.

To get substantially stronger systems we have to abandon the idea that the proof consists of propositional formulas.[2] For instance, the next natural system after Extended Frege systems is the *Quantified propositional calculus* [8]. Such a system is obtained from a Frege system by adding rules for quantifiers. A proof is a sequence of quantified propositional formulas derived according to the rules. We may use this system to derive quantified propositional formulas. If we want to use it as a proof system for propositional calculus, we think of quantified propositional formulas as auxiliary means to eventually derive a quantifier free propositional formula.

In general we require only that one can effectively check the proofs of the system in question. More precisely, there must exist a polynomial time algorithm to decide for a given sequence if it is a proof in the system. Thus the proofs need not even be structured into steps as in usual proofs.

To give an example of a very strong proof system for propositional calculus, define d to be a proof of τ, if d is a proof in ZF of the statement $Taut(\tau)$ expressing that τ is a tautology. To see why this system is strong just realize that ZF proves that Frege, Extended Frege, Quantified propositional calculus and lot of others are sound systems. Thus given e.g. an Extended Frege proof of τ we only need to check in ZF that it is an Extended Frege proof of τ and then immediately we get that τ is a tautology (provably in ZF). Hence this proof system polynomially simulates all the above systems.

We can use the same construction for any theory in which the concept $Taut$ can be reasonably formalized. So we define for a such a theory T the propositional proof system P_T to be the system where a proof of τ is a proof of $Taut(\tau)$ in T. The P_T is not only an interesting construction, but it could

[2] Of course it is always possible to use infinitely many axiom schemas, but then we have to talk about how these schemas are defined.

be useful for showing unprovability of certain Π_1 sentences, namely universal closures of Π_1^b formulas. We denote this class of sentences by $\forall\Pi_1^b$.

Let φ be a Π_1^b sentence, i.e., φ has no free variables. Then we can express φ using a propositional formula by replacing *log* bounded quantifiers by several disjunctions or conjunctions and coding the variables bounded by universal bounded quantifiers by propositional variables. Instead of going into details of this transformation, let us only note that a propositional formula is true iff it is satisfied for all possible values of the propositional variables. Thus we implicitly interpret it as if there were universal quantifiers (which we can actually add in the quantified propositional calculus). The range of this quantification is exponential in the size of the formula in the same way as it is in Π_1^b sentences.

Let $\varphi(x)$ be a Π_1^b formula. Then we get a sequence of propositional formulas τ_n from the closed instances $\varphi(n)$, $n = 0, 1, 2, \ldots$. Since we encode n in binary, the length of τ_n is polynomial in $\log n$. If $\forall x\varphi(x)$ is provable in T, then we get important information on the lengths of proofs of these propositional formulas in proof system P_T:

Theorem 6.1. *Let T be a sufficiently strong arithmetical theory and $\varphi(x)$ be a Π_1^b formula. Suppose that $\forall x\varphi(x)$ is provable in T. Then the propositional translations of sentences $\varphi(n)$ have proofs in P_T whose lengths are polynomial in the lengths of these propositions.*

Proof-sketch. Let τ_n be the translation of $\varphi(n)$. If T is sufficiently strong, then it proves $\forall x(\varphi(x) \rightarrow Taut(\tau_x))$. Hence if T proves $\forall x\varphi(x)$, then all instances τ_n have polynomial size proofs. □

To show that a given Π_1 sentence is not provable, we have to prove a superpolynomial lower bound on the lengths of proofs for the corresponding tautologies. But even if we only show that there are some tautologies which do not have polynomial size proofs in the proof system P_T, we get a very interesting independence result:

Theorem 6.2. *Suppose that the proof system P_T of a sufficiently strong theory T is not polynomially bounded, i.e., there is no polynomial upper bound on the length of the shortest P_T proofs of the propositional tautologies. Then T does not prove $NP = coNP$.*

Proof-sketch. Suppose that T does prove $NP = coNP$. Then in T the $coNP$ predicate $Taut(x)$ is equivalent to an NP predicate $\alpha(x)$. To prove some τ in P_T we need to show $Taut(\tau)$; the proof of this sentence can be longer than the shortest proof of $\alpha(\tau)$ only by a constant factor. Since α is NP we only have to take a polynomial size witness for the truth of $\alpha(\tau)$ and check it. This gives a polynomial size proof of τ. □

A possible approach for proving $NP \neq coNP$ is to prove gradually for stronger and stronger propositional proof systems that they are not polynomially bounded hoping that eventually we develop a technique allowing us to prove that no propositional proof system is polynomially bounded (which is equivalent to $NP \neq coNP$ because of the general definition of the concept of a propositional proof system). The theorem above shows that this is essentially the same as showing unprovability of $NP \neq coNP$ for stronger and stronger theories.

A successful application of this reduction clearly depends very much on how much hold we can get on the proof system P_T. This system seems to be very strong even for weak theories. There is another way of associating a propositional proof system to a theory which produces weaker systems, but it is not as universal as the construction of P_T. The system is called *the associated proof system* of T and it is defined, roughly speaking, by requiring the simulation of Theorem 6.1 to hold and that T proves its soundness (for a precise definition see [34]). The latter condition is not satisfied by P_T. Associated propositional proof systems have some nice properties, in particular Theorem 6.2 holds for them too.

For strong theories it seems hopeless to give a comprehensible combinatorial description even of the associated proof system. Fortunately, at least for some systems of bounded arithmetic we get natural associated proof systems. The most interesting particular case is the theory S^1_2 whose associated propositional proof system is, up to polynomial simulation, an Extended Frege system [6, 3]. For S^i_2 in general we can take fragments of the quantified propositional calculus obtained by an appropriate restriction on the quantifier complexity of formulas [18].

A large part of the activity is concentrated on proving lower bounds on the lengths of propositional proofs and we can report steady progress [1, 12, 13, 21, 31, 32]. Unfortunately the proof systems for which one can prove that they are not polynomially bounded are still much weaker than Extended Frege; even proving a superpolynomial lower bound for Frege systems would be a breakthrough. Thus we do not expect that concrete independent Π_1 sentences will be found for strong theories in the near future. Still we cannot exclude that somebody finds a completely new powerful method for proving independence. In particular, Gödel's theorem does not quite fit into the picture drawn above, where provability is thought of as polynomial length proofs in the associated propositional proof system. We know that consistency statements are not provable and can be formalized as $\forall\Pi^b_1$, but we do not have superpolynomial lower bounds on the lengths of proofs of its propositional translations in the associated proof system.[3]

[3] For P_T of a sufficiently strong theory T we have polynomial size *upper* bounds on the lengths of proofs of its propositional translations. Let a sufficiently strong theory T be given. The translations of Con_T are some tautologies τ_n expressing that "no $x \leq n$ is a proof of contradiction in T". Let σ_m be tautologies expressing that "there is no proof of contradiction in T of length $\leq m$"; the lengths of of σ_m are polynomial in m. If m is the length of n, then $\sigma_m \rightarrow \tau_n$ has a proof polynomial in m, i.e., in the length of τ_n. In [33] we proved that first order sentences $Con_T(m)$ expressing that there is no proof of contradiction in

7. Model theory of weak arithmetical theories

The natural numbers are a basic algebraic structure, thus we can also use the machinery of algebra and model theory. Then it is more convenient to axiomatize the integers instead of just the positive ones. The basic theory is the theory of discretely ordered commutative rings. The next stronger system which has been studied is obtained by adding induction for open formulas; it is denoted by *IOpen*. For this theory it is possible to prove independence by constructing explicitly models. Thus it has been proved that it is consistent with *IOpen* that $x^3 + y^3 = z^3$ has a nontrivial solution and that $\sqrt{2}$ is rational [36]. It is possible to get in such a way independence for slightly stronger theories (e.g. postulating that the ring is integrally closed in its fraction field, which, in particular, implies that $\sqrt{2}$ is irrational [23]), but there seems to be a serious obstacle to do it for theories which contain $I\Delta_0$ and S^1_2. It is well-known that these theories do not have nonstandard recursive models, but it is even worse: any nonstandard model contains an initial segment which is a nonstandard model of PA. Thus, in spite of using weak theories we get the whole complexity of nonstandard models of PA [25]. Even if this research does not lead directly to independence results for stronger theories, there are interesting problems in this area from the point of view of both logic and number theory [9, 26, 38].

Instead of constructing models directly one can try to modify a given nonstandard model. We mentioned in Section 3. that we can extend a model of PA only by adding elements which are larger than all the old elements. That argument is not valid for models of bounded arithmetic, there it is possible to add small elements (if the model is "short"). Several partial results have been obtained in this way [1, 19, 35]. Another possibility is to choose a submodel. In this way one can prove Theorem 5.1 [40] (for an exposition see [11][Chap. V, Sec. 4]). For the most recent applications of model theory in bounded arithmetic see [16, 37]

8. Conclusions

The reader may be a little disappointed now, because we promised to talk on foundations of mathematics, but instead most of the time we talked about computational complexity. That is not a mistake. Any reasonable definition of a formal system presupposes the concept of computability. We have to use some formalism, as pure reliance on intuition cannot be considered a foundation. When working with weak theories, natural connections with computational complexity are almost ubiquitous. Our feeling is then that we

---

T of length $\leq m$ have proofs in T of length polynomial in m. However, if T is sufficiently strong, then it proves $Con_T(m) \equiv Taut(\sigma_m)$ using polynomial size proofs. Thus also $Taut(\tau_n)$ have polynomial size proofs in T, which means that τ_n have polynomial size proofs in P_T.

cannot solve problems of foundations of mathematics without solving or at least understanding more deeply problems in complexity theory. But maybe also in order to solve fundamental problems in complexity theory we need to understand more about the foundations of mathematics.

Acknowledgement. I would like to thank to Sam Buss, Armin Haken, Jan Krajíček and Jiří Sgall for suggesting several corrections and improvements. I am also grateful to Claus Lange and Yurij Matiyasevich for informing me that Riemann's Hypothesis is a Π_1 sentence.

References

1. M. Ajtai. The complexity of the pigeonhole principle, in: *Proceedings of the 29th Annual Symposium on Foundations of Computer Science*, IEEE Computer Society, Piscataway, New Jersey, 1988, pp. 346–355.
2. P. Beame, R. Impagliazzo, J. Krajíček, T. Pitassi and P. Pudlák. Lower bounds on Hilbert's Nullstellensatz and propositional proofs, Proc. London Math. Soc., to appear.
3. S. R. Buss. *Bounded Arithmetic*, Bibliopolis 1986, Napoli. Revision of 1985 Princeton University Ph.D. thesis.
4. S. R. Buss. Relating the bounded arithmetic and polynomial time hierarchies, *Annals of Pure and Applied Logic* 75 (1995), pp. 67–77.
5. S. R. Buss. First order proof theory of arithmetic, in *Handbook of Proof Theory*, S. R. Buss ed., North Holland, to appear.
6. S. A. Cook. Feasibly constructive proofs and the propositional calculus, in: *Proceedings of the Seventh Annual ACM Symposium on the Theory of Computing*, Association for Computing Machinery 1975, New York, pp. 83–97.
7. S. A. Cook and R. A. Reckhow. The relative efficiency of propositional proof systems, *Journal of Symbolic Logic* 44, (1979) pp. 36–50.
8. M. Dowd. *Propositional Representation of Arithmetic Proofs*, PhD thesis, University of Toronto 1979.
9. Van den Dries. Which curves over **Z** have points with coordinates in a discretely ordered ring? *Transactions AMS* 264 (1990) pp. 33-56.
10. F. Ferreira. Binary models generated by their tally part, Archive for Math. Logic, to appear.
11. P. Hájek and P. Pudlák. *Metamathematics of First-order Arithmetic*, Springer-Verlag 1993, Berlin.
12. A. Haken. The intractability of resolution, *Theoretical Computer Science*, 39, (1985) pp. 297–308.
13. J. Krajíček. Lower bounds to the size of constant-depth propositional proofs, *Journal of Symbolic Logic*, 59, (1994) pp. 73–86.
14. J. Krajíček. On Frege and extended Frege proof systems, in: *Feasible Mathematics II*, J. Krajíček and J. Remmel, eds., Birkhäuser, Boston, 1994, pp. 284–319.
15. J. Krajíček. *Bounded Arithmetic, Propositional Logic and Complexity Theory*, Cambridge University Press 1995.
16. J. Krajíček. Extensions of models of PV, manuscript, 1996.
17. J. Krajíček and P. Pudlák. Propositional proof systems, the consistency of first-order theories and the complexity of computations, *Journal of Symbolic Logic*, 54, (1989) pp. 1063–1079.

18. J. Krajíček and P. Pudlák. Quantified propositional calculi and fragments of bounded arithmetic, *Zeitschrift für Mathematische Logik und Grundlagen der Mathematik*, 36, pp. 29–46.
19. J. Krajíček and P. Pudlák. Propositional provability in models of weak arithmetic, in: *Computer Science Logic'89*, E. Boerger et al. eds., Springer-Verlag LNCS 440, (1990), pp. 193-210.
20. J. Krajíček, P. Pudlák and G. Takeuti. Bounded arithmetic and the polynomial hierarchy, *Annals of Pure and Applied Logic* 52, (1991) pp. 143-153.
21. J. Krajíček, P. Pudlák, and A. Woods. An exponential lower bound to the size of bounded depth Frege proofs of the pigeonhole principle, *Random structures and Algorithms* 7/1 (1995), pp. 15–39.
22. G. Kreisel. Mathematical significance of consistency proofs. *Journal of Symbolic Logic*, 23 (1958), pp. 155-182.
23. D. Marker and A. Macintyre. Primes and their residue rings in models of open induction, *Annals of Pure and Applied Logic* 43, (1989) pp. 57-77.
24. Yu. Matiyasevich. The Riemann hypothesis from a logician's point of view. *Proc. First Conference of the Canadian Number Theory Association*, Ed. R.Mollin. Walter de Gruyter, 1990, pp. 387-400.
25. K. McAloon. On the complexity of models of arithmetic, *Journal of Symbolic Logic 47*, (1982), pp. 403-415.
26. M. Otero. Quadratic forms in normal open induction, *Journal of Symbolic Logic*, to appear.
27. R. Parikh. Existence and feasibility in arithmetic, *Journal of Symbolic Logic*, 36, (1971) pp. 494–508.
28. J. B. Paris and L. Harrington. A mathematical incompleteness in Peano arithmetic, in: *Handbook of Mathematical Logic* North-Holland 1977, pp. 1133-1142.
29. J. B. Paris and A. J. Wilkie. Counting problems in bounded arithmetic, in: *Methods in Mathematical Logic, Proceedings of the 6-th Latin American Symposium, Caracas, Venezuella*, C. A. Di Prisco, ed., Lecture Notes in Mathematics #1130, Springer-Verlag, Berlin, 1985 pp. 317–340.
30. J. B. Paris and A. J. Wilkie. Δ_0 sets and induction, in: *Proc. of the Jadwisin Logic Conf., Poland*, Leeds Univ. Press, pp. 237-248.
31. T. Pitassi, P. Beame, and R. Impagliazzo. Exponential lower bounds for the pigeonhole principle, *Computational Complexity*, 3, (1993) pp. 97–140.
32. P. Pudlák. Lower bounds for resolution and cutting planes proofs and monotone computations, *Journal of Symbolic Logic*. to appear.
33. P. Pudlák. Improved Bounds to the Lengths of Proofs of Finitistic Consistency Statements, in *Logic and Combinatorics*, S. J. Simpson editor, Contemporary Mathematic 65, American Mathematical Society 1987, pp. 309-331.
34. P. Pudlák. The lengths of proofs, in *Handbook of Proof Theory*, S. R. Buss ed., North Holland, to appear.
35. S. Riis. Making infinite structures finite in models of second order bounded arithmetic, in: *Arithmetic, Proof Theory and Computational Complexity*, P. Clote and J. Krajíček eds., Oxford Univ. Press 1993, pp. 289-319.
36. J.C. Shepherdson. A non-standard model of a free variable fragment of number theory, *Bull. Acad. Pol. Sci.*, 12, (1964) pp. 79-86.
37. G. Takeuti and M. Yasumoto. Forcing on bounded arithmetic, this procceedings.
38. A. A. Wilkie. Some results and problems on weak systems of arithmetic, in A. Macintyre et al eds., *Logic Colloquium'77*, North-Holland 1978, pp. 285-296.
39. A. A. Wilkie. On sentences interpretable in systems of arithmetic, in: *Logic Colloquium'84*, North-Holland 1986, pp. 329-342.

40. A. A. Wilkie. A model-theoretic proof of Buss's characterization of the polynomial time computable function, manuscript, 1985.
41. A. A. Wilkie and J. B. Paris. On the schema of induction for bounded arithmetical formulas, *Annals of Pure and Applied Logic* 35, (1987) pp. 261-302.

K-graph Machines: generalizing Turing's machines and arguments *

Wilfried Sieg and John Byrnes

Department of Philosophy,
Carnegie Mellon University, Pittsburg

Summary. The notion of *mechanical process* has played a crucial role in mathematical logic since the early thirties; it has become central in computer science, artificial intelligence, and cognitive psychology. But the discussion of Church's Thesis, which identifies the informal concept with a mathematically precise one, has hardly progressed beyond the pioneering work of Church, Gödel, Post, and Turing. Turing addressed directly the question: *What are the possible mechanical processes a human computor can carry out in calculating values of a number-theoretic function?* He claimed that all such processes can be simulated by machines, in modern terms, by deterministic Turing machines. Turing's considerations for this claim involved, first, a formulation of boundedness and locality conditions (for linear symbolic configurations and mechanical operations); second, a proof that computational processes (satisfying these conditions) can be carried out by Turing machines; third, the central thesis that all mechanical processes carried out by human computors must satisfy the conditions. In Turing's presentation these three aspects are intertwined and important steps in the proof are only hinted at. We introduce *K-graph machines* and use them to give a detailed mathematical explication of the first two aspects of Turing's considerations for general configurations, i.e. K-graphs. This generalization of machines and theorems provides, in our view, a significant strengthening of Turing's argument for his central thesis.

Introduction

Turing's analysis of effective calculability is a paradigm of a foundational study that (i) led from an informally understood concept to a mathematically precise notion, (ii) offered a detailed investigation of the new mathematical notion, and (iii) settled an important open question, namely the *Entscheidungsproblem.* The special character of Turing's analysis was recognized immediately by Church in his review of Turing's 1936 paper. The review was published in the first issue of the 1937 volume of the Journal of Symbolic Logic, and Church contrasted in it Turing's mathematical notion for effective calculability (via idealized machines) with his own (via λ-definability) and Gödel's general recursiveness and asserted: "Of these, the first has the advantage of making the identification with effectiveness in the ordinary (not explicitly defined) sense evident immediately...."

Gödel had noticed in his (1936) an "absoluteness" of the concept of computability, but found only Turing's analysis convincing; he claimed that Turing's work provides "a precise and unquestionably adequate definition of the

* This paper is in its final form and no similar paper has been or is being submitted elsewhere.

general concept of formal system" (1964, p. 369). As a formal system is simply defined to be a mechanical procedure for producing theorems, the adequacy of the definition rests on Turing's analysis of mechanical procedures. And with respect to the latter Gödel remarked (pp. 369-70): "Turing's work gives an analysis of the concept of 'mechanical procedure' (alias 'algorithm' or 'computation procedure' or 'finite combinatorial procedure'). This concept is *shown* [our emphasis] to be equivalent with that of a 'Turing machine'." Nowhere in Gödel's writings is there an indication of the nature of Turing's conceptual analysis or of a proof for the claim that the analyzed concept is equivalent with that of a Turing machine.

Gödel's schematic description of Turing's way of proceeding is correct: in section 9 of (Turing 1936) there is an analysis of effective calculability, and the analysis is intertwined with a sketch of an argument showing that mechanical procedures on linear configurations can be performed by very restricted machines, i.e., by deterministic Turing machines over a two-letter alphabet. Turing intended to give an analysis of mechanical processes on planar configurations; but such processes are not described, let alone proved to be reducible to computations on linear objects. This gap in Turing's considerations is the starting-point of our work. We formulate broad boundedness and locality conditions that emerge from Turing's conceptual analysis, give a precise mathematical description of planar and even more general computations, and present a detailed reductive argument. For the descriptive part we introduce *K-graph machines*; they are a far-reaching generalization of Post production systems and thus, via Post's description of Turing machines, also of Turing machines.

1. Turing's Analysis[1]

In 1936, the very year in which Turing's paper appeared, Post published a computation model strikingly similar to Turing's. Our brief discussion of Post's model is not to emphasize this well-known similarity, but rather to bring out the strikingly dissimilar methodological attitudes underlying Post's and Turing's work. Post has a worker operate in a symbol space consisting of "a two way infinite sequence of spaces or boxes ...".[2] The boxes admit two conditions: they can be unmarked or marked by a single sign, say a vertical stroke. The worker operates in just one box at a time and can perform a number of *primitive acts*: make a vertical stroke [*V*], erase a vertical stroke

[1] This section is based on (Sieg 1994) which was completed in June 1991; for details of the reconstruction of Turing's analysis and also for the broader systematic and historical context of our investigations we refer the reader to that paper.

[2] Post remarks that the infinite sequence of boxes is ordinally similar to the series of integers and can be replaced by a potentially infinite one, expanding the finite sequence as necessary.

[E], move to the box immediately to the right [M_r] or to the left [M_l] (of the box he is in), and determine whether the box he is in is marked or not [D]. In carrying out a "combinatory process" the worker begins in a special box and then follows directions from a finite, numbered sequence of instructions. The i-th direction, i between 1 and n, is in one of the following forms: (i) carry out act V, E, M_r, or M_l and then follow direction j_i, (ii) carry out act D and then, depending on whether the answer is positive or negative, follow direction j_i' or j_i''. (Post has a special stop instruction, but that can be replaced by the convention to halt, when the number of the next direction is greater than n.)

Are there intrinsic reasons for choosing this formulation as an explication of effective calculability, except for its simplicity and Post's expectation that it will turn out to be equivalent to recursiveness? An answer to this question is not clear from Post's paper, at the end of which he wrote:

> The writer expects the present formulation to turn out to be equivalent to recursiveness in the sense of the Gödel-Church development. Its purpose, however, is not only to present a system of a certain logical potency but also, in its restricted field, of psychological fidelity. In the latter sense wider and wider formulations are contemplated. On the other hand, our aim will be to show that all such are logically reducible to formulation 1. We offer this conclusion at the present moment as a *working hypothesis*. And to our mind such is Church's identification of effective calculability with recursiveness.

Investigating wider and wider formulations and *reducing* them to *Formulation 1* would change for Post this "hypothesis not so much to a definition or to an axiom but to a *natural law*". It is methodologically remarkable that Turing proceeded in *exactly* the opposite way when trying to justify that all computable numbers are machine computable or, in our way of speaking, that all effectively calculable functions are Turing computable: He did not extend a narrow notion reducibly and, in this way, obtain quasi-empirical support, but rather analyzed the intended broad concept and reduced it to a narrow one, once and for all. The intended concept was mechanical calculability by a human being, and in the reductive argument Turing exploited crucially limitations of the computing agent.

Turing's *On computable numbers* opens with a description of what is ostensibly its subject, namely, "real numbers whose expressions as a decimal are calculable by finite means". Turing is quick to point out that the problem of explicating "calculable by finite means" is the same when considering, e.g., computable functions of an integral variable. Thus it suffices to address the question: "What does it mean for a real number to be calculable by finite means?" But Turing develops first the theory of his machines.[3] A Turing

[3] Note that the presentation of Turing machinex we give is not Turing's, but rather the one that evolved from Post's formulation in (1947).

machine consists of a finite, but potentially infinite tape; the tape is divided into squares, and each square may carry a symbol from a finite alphabet, say, the two-letter alphabet consisting of 0 and 1. The machine is able to scan one square at a time and perform, depending on the content of the observed square and its own internal state, one of four operations: print 0, print 1, or shift attention to one of the two immediately adjacent squares. The operation of the machine is given by a finite list of commands in the form of quadruples $q_i s_k c_l q_m$ that express: if the machine is in internal state q_i and finds symbol s_k on the square it is scanning, then it is to carry out operation c_l and change its state to q_m. The deterministic character of the machine operation is guaranteed by the requirement that a program must not contain two different quadruples with the same first two components.

In section 9 Turing argues that the operations of his machines "include all those which are used in the computation of a number". But he does not try to establish the claim directly; he rather attempts to answer what he views as "the real question at issue": "What are the possible processes which can be carried out [by a computor[4]] in computing a number?" Turing imagines a computor writing symbols on paper that is divided into squares "like a child's arithmetic book". As the two-dimensional character of this computing space is taken—*without any argument*—not to be essential, Turing considers the one-dimensional tape divided into squares as the basic computing space and formulates one important restriction. The restriction is motivated by limits of the human sensory apparatus to distinguish *at one glance* between symbolic configurations of sufficient complexity and states that only finitely many distinct symbols can be written on a square. Turing suggests as a reason that "If we were to allow an infinity of symbols, then there would be symbols differing to an arbitrarily small extent", and we would not be able to distinguish at one glance between them. A second and clearly related way of arguing this point uses a finite number of symbols and strings of such symbols. E.g., Arabic numerals like 9979 or 9989 are seen by us at one glance to be different; however, it is not possible for us to determine immediately that 9889995496789998769 is different from 98899954967899998769. This second avenue suggests that a computor can operate directly only on a finite number of (linear) configurations.

Now we turn to the question: What determines the steps of the computor, and what kind of elementary operations can he carry out? The behavior is *uniquely* determined at any moment by two factors: (i) the symbolic configuration he observes and (ii) his internal state. This uniqueness requirement may be called the **determinacy condition (D)**; it guarantees that computations are deterministic. Internal states, or as Turing also says "states of mind", are introduced to have the computor's behavior depend possibly on

[4] Following Gandy, we distinguish between a *computor* (a human carrying out a mechanical computation) and a *computer* (a mechanical device employed for computational purposes); cf. (Gandy 1988), p. 81, in particular fn. 24.

earlier observations and, thus, to reflect his experience. Since Turing wanted to isolate operations of the computor that are "so elementary that it is not easy to imagine them further divided", it is crucial that symbolic configurations that help fix the conditions for a computor's actions are immediately recognizable. We are thus led to postulate that a computor has to satisfy two **boundedness conditions**:

(B.1) *there is a fixed bound for the number of symbolic configurations a computor can immediately recognize;*
(B.2)[5] *there is a fixed bound for the number of internal states that need be taken into account.*

For a given computor there are consequently only boundedly many different combinations of symbolic configurations and internal states. Since his behavior is, according to **(D)**, uniquely determined by such combinations and associated operations, the computor can carry out at most finitely many different operations. These operations are restricted by the following **locality conditions**:

(L.1) *only elements of observed symbolic configurations can be changed;*
(L.2) *the distribution of observed squares can be changed, but each of the new observed squares must be within a bounded distance of an immediately previously observed square.*

Turing emphasized that "the new observed squares must be immediately recognisable by the [computor]"; that means the observed configurations arising from changes according to **(L.2)** must be among the finitely many ones of **(B.1)**. Clearly, the same must hold for the symbolic configurations resulting from changes according to **(L.1)**. Since some steps may involve a change of internal state, Turing concluded that the most general single operation is a change *either* of symbolic configuration and, possibly, internal state *or* of observed square and, possibly, internal state. With this restrictive analysis of the steps a computor can take, the proposition that his computations can be carried out by a Turing machine is established rather easily.[6] Thus we have:

Theorem 1.1 (Turing's Theorem for calculable functions).
Any number theoretic function F that can be calculated by a computor satisfying the determinacy condition **(D)** *and the conditions* **(B)** *and* **(L)** *can be computed by a Turing machine.*

[5] Gödel objected in (1972) to this condition for a notion of human calculability that might properly extend mechanical calculability; for a computor it seems quite unobjectionable.

[6] Turing constructed machines that mimic the work of computors on linear configurations directly and observed: "The machines just described do not differ very essentially from computing machines as defined in § 2, and corresponding to any machine of this type a computing machine can be constructed to compute the same sequence, that is to say the sequence computed by the computer [in our terminology: comput*or*]." Cf. section 2 below for this reductive claim.

As the Turing computable functions are recursive, F is recursive. This argument for F's recursiveness does not appeal to any form of Church's Thesis; rather, such an appeal is replaced by the assumption that the calculation of F is done by a computor satisfying the conditions **(D)**, **(B)**, and **(L)**. If that assumption is to be discharged a substantive thesis is needed. We call this thesis—that a mechanical computor must satisfy the conditions **(D)** and **(B)**, and that the elementary operations he can carry out must be restricted as conditions **(L)** require—**Turing's Central Thesis**.

In the historical and systematic context in which Turing found himself, he asked exactly the right question: What are the processes a computor can carry out in calculating a number? The general problematic *required* an analysis of the idealized capabilities of a computor, and exactly this feature makes the analysis epistemologically significant. The separation of conceptual analysis (leading to the axiomatic conditions) and rigorous proof (establishing Turing's Theorem) is essential for clarifying on what the correctness of his central thesis rests; namely, on recognizing that the axiomatic conditions are true for computors who proceed mechanically. We have to remember that clearly when engaging in methodological discussions concerning artificial intelligence and cognitive science. Even Gödel got it wrong, when he claimed that Turing's argument in the 1936 paper was intended to show that "mental processes cannot go beyond mechanical procedures".

2. Post Productions & Puzzles

Gödel's misunderstanding of the intended scope of the analysis may be due to Turing's provocative, but only figurative attribution of "states of mind" to machines; it is surprising nevertheless, as Turing argues at length for the eliminability of states of mind in section 9 (III) of his paper. He describes there a modified computor and avoids the introduction of "state of mind", considering instead "a more physical and definite counterpart of it". The computor is now allowed to work in a desultory manner, possibly doing only one step of the computation at a sitting: "It is always possible for the [computor] to break off from his work, to go away and forget all about it, and later to come back and go on with it." But on breaking off the computor must leave a "note of instruction" that informs him on how to proceed when returning to his job; such notes are the "counterparts" of states of mind. Turing incorporates notes into "state formulas" (in the language of first order logic) that describe states of a machine mimicking the computor and formulates appropriate rules that transform a given state into the next one.

Post used in (1947) a most elegant way of describing Turing machines purely symbolically via his production systems (on the way to solving, negatively, the word-problem for semi-groups).[7] The configurations of a Turing machine are given by *instantaneous descriptions* of the form $\alpha q_l s_k \beta$, where α and β are possibly empty strings of symbols in the machine's alphabet; more precisely, an *id* contains exactly one state symbol, and to its right there must be at least one symbol. Such *id*'s express that the current tape content is $\alpha s_k \beta$, the machine is in state q_l, and it scans (a square with symbol) s_k. Quadruples $q_i s_k c_l q_m$ of the program are represented by rules; for example, if the operation c_l is *print* 0, the corresponding rule is:

$$\alpha q_i s_k \beta \Rightarrow \alpha q_m 0 \beta.$$

That can be done, obviously, for all the different operations; one just has to append 0 or s_0 to α (β) in case c_l is the operation *move to the left (right)* and α (β) is the empty string—reflecting the expansion of the only potentially infinite tape by a blank square. This formulation can be generalized so that machines operate directly on finite strings of symbols; operations can be indicated as follows:

$$\alpha \gamma q_l \delta \beta \Rightarrow \alpha \gamma^* q_m \delta^* \beta.$$

If in internal state q_l a *string machine* recognizes the string $\gamma\delta$ (i.e., takes in the sequence at one glance), it replaces that string by $\gamma^*\delta^*$ and changes its internal state to q_m. Calling ordinary Turing machines *letter machines*, Turing's claim reported in note 6 can be formulated as a **Reduction Lemma:** Any computation of a string machine can be carried out by a letter machine.

The rule systems describing string machines are semi-Thue systems and, as the latter, not deterministic, if their programs are just sequences of production rules. The usual non-determinism certainly can be excluded by requiring that, if the antecedents of two rules coincide, so must the consequents. But that requirement does not remove every possibility of two rules being applicable simultaneously: consider a machine whose program includes in addition to the above rule also the rule

$$\alpha \gamma^{\#} q_l \delta^{\#} \beta \Rightarrow \alpha \gamma^{\perp} q_n \delta^{\perp} \beta,$$

where $\delta^{\#}$ is an initial segment of δ, and $\gamma^{\#}$ is an end segment of γ; then both rules would be applicable to $\gamma q_l \delta$. This kind of non-determinism can be excluded in a variety of ways, for example, by ordering the rules and always using the first applicable rule; this approach was taken by Markov in his 1954 *Theory of Algorithms.*

However, as we emphasized already, Turing had intended to analyze genuine planar computations, not just string machines or letter machines operating in the plane.[8] To formulate and prove a Reduction Lemma for planar

[7] Post's way of looking at Turing machines underlies also the presentation in (Davis 1958); for a more detailed discussion the reader is referred to that classical text.

[8] Such machines are also discussed in Kleene's *Introduction to Metamathematics*, pp. 376-381, in an informed and insightful defense of Turing's Thesis. However,

computations, one has to specify the finite symbolic configurations that can be operated on and the mechanical operations that can be performed. Turing recognized the significance of Post's presentation for achieving mathematical results, but also for the conceptual analysis of calculability: as to the former, Turing extended in his (1950) Post's and Markov's result concerning the unsolvability of the word-problem for semi-groups to semi-groups with cancellation; as to the latter, we look at Turing's semi-popular and most informative presentation of *Solvable and Unsolvable Problems* (1953).

Turing starts out with a description of puzzles: square piece puzzles, puzzles involving the separation of rigid bodies or the transformation of knots; i.e., puzzles in two and three dimensions. "Linear" puzzles are described as Post systems and called *substitution puzzles.* They are viewed by Turing as a "normal" or "standard" form of describing puzzles; indeed, a form of the Church-Turing thesis is formulated as follows:

> Given any puzzle we can find a corresponding *substitution puzzle* which is equivalent to it in the sense that given a solution of the one we can easily find a solution of the other. If the original puzzle is concerned with rows of pieces of a finite number of different kinds, then the substitutions may be applied as an alternative set of rules to the pieces of the original puzzle. A transformation can be carried out by the rules of the original puzzle if and only if it can be carried out by the substitutions... (1953, p.15)

Turing admits, with some understatement, that this formulation is "somewhat lacking in definiteness" and claims that it will remain so; he characterizes its status as lying between a theorem and a definition: "In so far as we know *a priori* what is a puzzle and what is not, the statement is a theorem. In so far as we do not know what puzzles are, the statement is a definition which tells us something about what they are." Of course, Turing continues, one could define puzzle by a phrase beginning with 'a set of definite rules', or one could reduce its definition to that of 'computable function' or 'systematic procedure'. A definition of any of these notions would provide one for puzzles.

Even before we had seen Turing's marvelous 1953 paper, our attempts of describing mechanical procedures on general symbolic configurations had made use of the puzzle-metaphor. The informal idea had three distinct components: a computor was to operate on finite *connected* configurations; such configurations were to contain a unique *distinguished element* (corresponding to the scanned square); the operations were to *substitute neighborhoods* (of a bounded number of different forms) of the distinguished element by appropriate other neighborhoods resulting in a new configuration, and such substitutions were to be given by *generalized production rules.* Naturally,

---

in Kleene's way of extending configurations and operations, much stronger normalizing conditions are in place; e.g., when considering machines corresponding to our string machines the strings must be of the same length.

the question was how to transform this into appropriate mathematical concepts; referring to Turing's statement above, we were unwittingly trying to remove (as far as possible) the *lack of definiteness* in the description of general puzzles. But in contrast to Turing, we wanted to analyze deterministic procedures and follow more closely his own analysis given in 1936. For this purpose we introduced *K-graph machines*. These machines were inspired, in part, by Kolmogorov and Uspensky's 1958 analysis of algorithms. K-graph machines operate, not surprisingly, on *K-graphs*. These are finite connected graphs whose vertices are labeled by symbols and contain a uniquely labeled *central* vertex. They satisfy also the *principle of unique location*, i.e., every path of labels (starting with the label of the central vertex) determines a unique vertex. K-graph machines substitute distinguished K-subgraphs by other K-graphs; their programs are finite lists of generalized production rules specifying such substitutions. As these substitutions are local, we say that the machines satisfy the *principle of local action*. The subtle difficulties surrounding the principle of local action, even for the case of string machines, are discussed in the next section.[9]

Turing machines, when presented by production systems as above, are easily seen to be K-graph machines. Conversely, the theorem in section 4 shows that computations of K-graph machines can be carried out by Turing machines. Given this mathematical analysis, Turing's central thesis is turned into the thesis that K-graph machines, clearly satisfying the boundedness and locality conditions, subsume directly the work of computors. Our main theorem thus reduces mechanical processes carried out by computors to Turing machine computations. — We want to emphasize very forcefully that our generalization of Turing's analysis is a direct extension of the latter, both technically and conceptually. This is in striking contrast to other such generalizations, e.g., those of Friedman and Shepherdson, see (Shepherdson 1988); Gandy's penetrating analysis of *machine computability* is discussed briefly in the Concluding Remarks.

3. K-Graph Machines

To state and prove the main theorem we have to review some general concepts from graph theory and introduce some notions especially appropriate for our goals. As labeled graphs are going to be considered, we let $\mathcal{U}$ be a (potentially infinite) universe of vertices, $\mathcal{L}$ a finite set of labels, or *alphabet*, and *lb* a

[9] Thus, our K-graph machines are deterministic graph rewriting systems; there is a considerable literature in computer science that discusses such systems, see for example the survey article (Courcelle 1990). The category theoretic way of presenting rewrite systems is for our purposes, however, not suitable: the substitution operations have to be graphically concrete and direct, not indirectly obtainable through pushout diagrams. Tim Herron (1995) used the category theoretic framework to characterize K-graph machines.

labeling function from $\mathcal{U}$ to $\mathcal{L}$. *$\mathcal{L}$-labeled finite graphs* G are defined with reference to $\mathcal{U}$, $\mathcal{L}$, and lb; thus, they can be given as ordered pairs $\langle V, E\rangle$, where V is a finite subset of $\mathcal{U}$ and $E \subseteq \{\{u,v\} \mid u,v \in V\}$; a pair $\{u,v\} \in E$ is called an *edge* and may be denoted by uv. u and v are said to be *adjacent*. For a vertex $v \in V$ and an edge $uv \in E$, we write also $v \in G$ and $uv \in G$. The sets of vertices and edges of G are also denoted by V_G and E_G. Given $G = \langle V, E\rangle$ and $G' = \langle V', E'\rangle$, $G \cup G' = \langle V \cup V', E \cup E'\rangle$. We write $G' \subseteq G$ and say that G' is a *subgraph* of G if $V' \subseteq V$ and $E' \subseteq E \restriction V'$. $G \setminus G' = \langle V \setminus V', E \restriction (V \setminus V')\rangle$. Here the symbol '$\restriction$' indicates the restriction of a relation to a subset of its domain.

A *path in* G from u_1 to u_n is a sequence $u_1u_2 \ldots u_n$ of distinct vertices of G such that for every pair of consecutive vertices u_i and u_{i+1} the edge u_iu_{i+1} is in G. A vertex v belongs to the path if v is an element of the sequence; an edge uv belongs to the path in case u and v are consecutive vertices in the sequence. The length of a path is defined as the number of edges belonging to the path; $\text{len}(u,v)$ is the length of a shortest path from u to v, if any path from u to v exists. A *component* of a graph G is a maximal subgraph G' of G such that for any two vertices u and v in G', there is a path in G' from u to v; if $G = G'$ then G is called *connected*.

The remaining definitions are tailored to our purposes and allow us to solve succinctly the central issue of representing symbolic configurations in a most general way—as labeled graphs. The *structure* of such configurations is fixed by the underlying, unlabeled vertices. As the specific nature of the vertices is irrelevant, we call a label-preserving bijection π from $\mathcal{U}$ to $\mathcal{U}$ a *permutation* and use such bijections to specify isomorphisms. Clearly, two labeled graphs G and G' are *isomorphic* just in case there is a label and edge preserving bijection between G and G'; we write $G \simeq G'$. Given a graph G, a permutation π picks out a unique graph G^π isomorphic to G, defined by $V_{G^\pi} = \{\pi(u) \mid u \in V_G\}$ and $E_{G^\pi} = \{\{\pi(u), \pi(v)\} \mid \{u,v\} \in E_G\}$. If in addition π is the identity over V_k for some graph K, we write $G \simeq_K G^\pi$. Finally, to fix the analogue of the scanned square, or rather the handle for the puzzle pieces, a distinguished label $* \in \mathcal{L}$ is considered. *$\mathcal{L}_*$-labeled graphs* are those $\mathcal{L}$-labeled graphs that contain exactly one vertex v with $lb(v) = *$; this vertex is then called the graph's *central vertex* and is referred to simply as $*$ when the context is clear.

For an $\mathcal{L}_*$-labeled graph G, we let G^* be the (unique) component of G containing $*$. A sequence α of labels associated with a path from the central vertex $*$ to some vertex v is called a *label-sequence* for v; the set of such sequences is denoted by $\text{Lbs}(v)$. If labeled graphs have the property that, for any vertex v, a label sequence from $\text{Lbs}(v)$ labels a path to v and not to any other vertex, then the labeling provides a coordinate system. Notice, however, that each vertex may have a number of different "coordinates". This leads to the following definition:

Definition 3.1. *A finite connected* $\mathcal{L}_*$*-labeled graph* K *is a* Kolmogorov-graph, *or* K-graph, *over* $\mathcal{L}$ *if* $(\forall\alpha)(\forall u, v \in K)[\alpha \in Lbs(u) \cap Lbs(v) \Rightarrow u = v]$.

We refer to the above property of graphs as the *principle of unique location*; it guarantees that isomorphisms between K-graphs are uniquely determined. This principle and its relation to condition (α) in Kolmogorov and Uspensky's work is discussed in remark 1 below. — K-graphs constitute the class of finite symbolic configurations on which our machines operate, and we describe now what elementary operations are allowed on such configurations. The operations take the form of generalized production rules and are directly motivated as puzzle-piece substitutions.

Definition 3.2. *A* graph-rewrite rule, *or simply* rule, R *is an ordered pair* $\langle A, C\rangle$, *where (*R*'s* antecedent*)* A *and (*R*'s* consequent*)* C *are K-graphs. For a given* R *we let* A_R *be* A *and* C_R *be* C*. A sequence* $\mathcal{R}$ *of rules such that for every* $Q, R \in \mathcal{R}$ $[A_Q \simeq A_R \Rightarrow A_Q \cup C_Q \simeq A_R \cup C_R]$ *is called a* program.

The application of a rule R to a K-graph K substitutes C_R for A_R in K: that requires, certainly, that A_R is (isomorphic to) a subgraph of K and that C_R can be "inserted into" $K \setminus A_R$. The crucial work is done by the vertices which occur in both A_R and C_R. As an easy example, consider rule R:

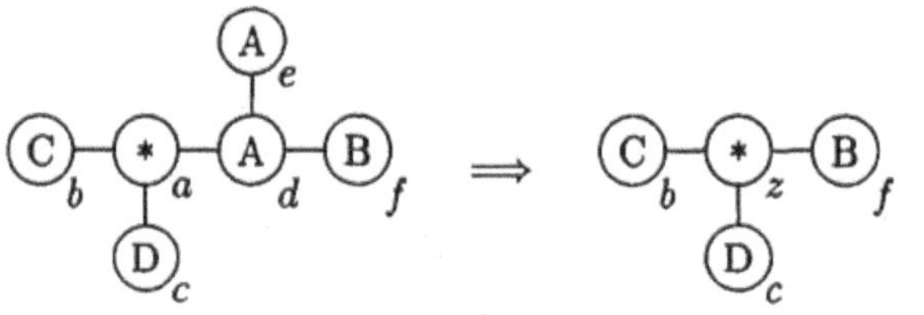

and K-graph K:

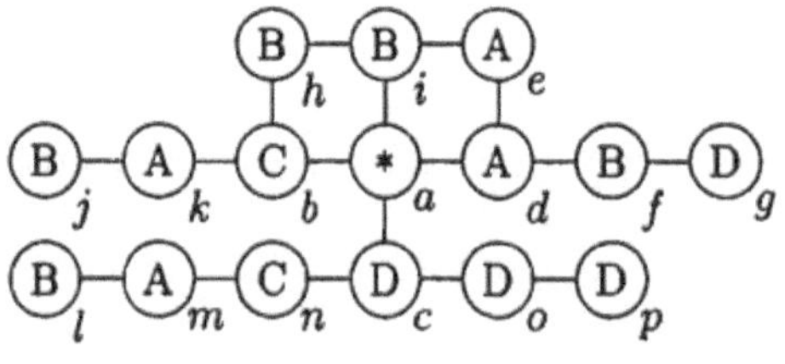

To apply R to K, we first remove A_R from K, except for those vertices which appear also in C_R; this is given by $K \setminus (A_R \setminus C_R)$:

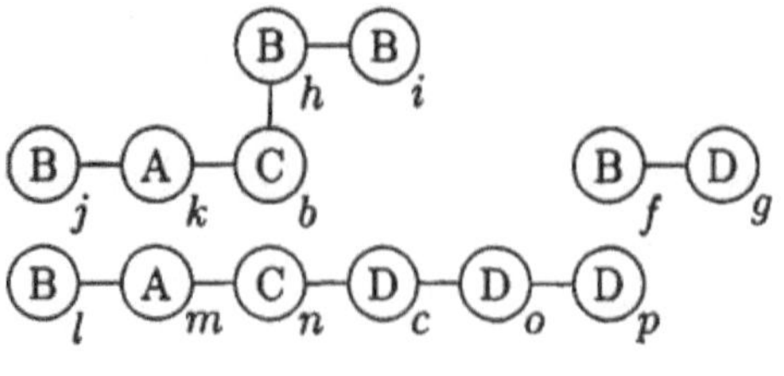

Inserting C_R into $K \setminus (A_R \setminus C_R)$ leads to the graph $(K \setminus (A_R \setminus C_R)) \cup C_R$, denoted by $R[K]$:

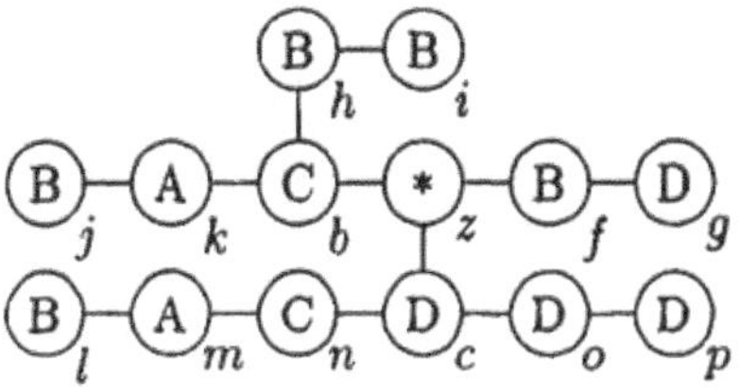

The above case is special in two respects. First, $A_R \subseteq K$. In general, we would like to apply R, if A_R is isomorphic to a subgraph of K, and, clearly, for every $K' \simeq K$, $R[K']$ should be isomorphic to $R[K]$. This can be most easily accomplished by using a permutation π such that $A_R^\pi \subseteq K$. Because of the key role played by the identity between certain vertices in A_R and in C_R, we of course also have to apply π to C_R and make a similar replacement.

The second special property of our example is this: none of the "new" vertices of C_R (i.e., vertices which did not occur in A_R), occur in $K \setminus (A_R \setminus C_R)$. In general, however, $K \setminus A_R^\pi$ may contain vertices which appear also in C_R^π; these vertices in C_R^π must be replaced as well. This second replacement, say, to C', has to satisfy two properties: (i) none of the vertices in $V_{A_R^\pi} \cap V_{C_R^\pi}$ may be replaced; (ii) all vertices which occur in C' but not in A_R^π should not occur anywhere in $K \setminus A_R^\pi$. (i) may be stated as $C' \simeq_{A_R^\pi} C_R^\pi$; (ii) is given by $V_{C'} \cap V_K \subseteq V_{A_R^\pi}$.

Definition 3.3. *Given a K-graph K and a rule R, R is said to be* applicable *to K if for some permutation π, $A_R^\pi \subseteq K$. In this case, $R[K] = (K \setminus (A_R^\pi \setminus C')) \cup C'$ for some $C' \simeq_{A_R^\pi} C_R^\pi$ such that $V_{C'} \cap V_K \subseteq V_{A_R^\pi}$.*

If we restrict ourselves to programs, we rule out the kind of non-determinism that Turing considered for his machines, but it may still be that a number of different rules can be applied to a given state K. As suggested for the case of string machines in section 2, we avoid this difficulty by ordering the rules linearly and always using the first (in that ordering) applicable rule. Finally, having defined the configurations on which our machines operate and the steps they can take, we define the machines themselves.

Definition 3.4. *Let $\mathcal{L}$ be a finite alphabet with a distinguished element $*$ and let $\mathcal{M} = \langle \mathcal{S}, \mathcal{F} \rangle$, where $\mathcal{S}$ is the set of all K-graphs over $\mathcal{L}$ and $\mathcal{F}$ is a partial function from $\mathcal{S}$ to $\mathcal{S}$. $\mathcal{M}$ is a* K-graph Machine *over $\mathcal{L}$ if and only if there is a program $\mathcal{R} = \langle R_0, \ldots, R_n \rangle$ such that:*

> *For every $S \in \mathcal{S}$, if there is an $R \in \mathcal{R}$ that is applicable to S, then $\mathcal{F}(S) = (R_i[S])^*$, where $i = \min\{j \mid R_j \in \mathcal{R}$ is applicable to $S\}$; otherwise $\mathcal{F}$ is undefined for S.*

The elements of S are the *states* of $\mathcal{M}$, and $\mathcal{F}$ is the machine's *transition function*. We say that $\mathcal{F}$ satisfies the *principle of local action.* — We can give an obviously equivalent definition of K-graph machines that brings out the special principles more directly. Let $\mathcal{M} = \langle \mathcal{S}, \mathcal{F} \rangle$, where $\mathcal{S}$ is a set of $\mathcal{L}_*$-labeled graphs and $\mathcal{F}$ is a partial function from $\mathcal{S}$ to $\mathcal{S}$; $\mathcal{M}$ is a K-graph Machine over $\mathcal{L}$ if and only if (i) $\mathcal{S}$ is the largest set of $\mathcal{L}_*$-labeled graphs that satisfy the principle of unique location, and (ii) $\mathcal{F}$ satisfies the principle of local action.

Remarks

1. The principle of unique location.. In the brief discussion preceding the definition of K-graph machines, we mentioned two sources of non-determinism present already for letter and string machines (and the standard way of circumventing them). However, there is one additional source, when graphs, even labeled ones with a central vertex, are considered as the configurations on which production rules operate: if the antecedent of a rule can be embedded into a given graph, it usually can be done in a variety of ways; the result of the rule application will in general depend on the chosen embedding. It is precisely this kind of indeterminacy that is excluded by the principle of unique location, as it guarantees that there is a unique embedding, if a K-graph can be embedded into another K-graph at all. The principle is also related to a second conceptual issue, namely, immediate recognizability. For any K-graph machine $\mathcal{M}$ and any K-graph whatsoever, we can decide in constant time, whether any rule of $\mathcal{M}$'s program is applicable. Mathematically, the principle is exploited for the reduction in section 4. It guarantees that, for any $\mathcal{L}_*$-labeled graph, paths starting at the central vertex are uniquely characterized by the sequence of symbols labeling their vertices. Thus, using the lexicographical ordering on strings of labels from $\mathcal{L}$, we can choose for each vertex a unique address which picks out that vertex in terms only of labels.

Kolmogorov and Uspensky used their condition (α) for similar purposes. That condition is formulated in our setting as follows: For every graph $S \in \mathcal{S}$, if u and v are vertices of S both adjacent to some vertex w of S, then $lb(u) \neq lb(v)$. We call a connected $\mathcal{L}_*$-labeled graph satisfying condition (α) a *Kolmogorov complex over* $\mathcal{L}$ or, briefly, a K-complex. Condition (α) implies our principle of unique location, but is not implied by it. The first claim is easily established; for the second claim one sees directly that $*$—A—A—A—A is an example of a K-graph that is not a K-complex.

2. Preservation under rule application.. We require that $\mathcal{F}$ is a partial function from $\mathcal{S}$ to $\mathcal{S}$. As matters stand, programs and K-graph machines cannot be "identified"; the reason is this: not every rule, when applied to a K-graph, yields a K-graph. As a trivial example, consider $R = *$—A—B $\Rightarrow$ $*$—C—B and $K =$ B—C—$*$—A—B—A. Then $R[K] =$ B—C—$*$—C—B—A which is certainly not a K-graph, even though K is a K-graph and R is a rule.

It is straightforward to modify any given machine program $\mathcal{R}$ to a program $\mathcal{R}'$ such that $\mathcal{R}$ and $\mathcal{R}'$ yield the same result when $\mathcal{R}$ transforms a K-graph into a K-graph, but $\mathcal{R}'$ makes the machine diverge on K-graph inputs that $\mathcal{R}$ transforms into graphs not satisfying the principle of unique location. Given this modification, $\mathcal{R}'$ defines a unique partial function from $\mathcal{S}$ to $\mathcal{S}$. — Kolmogorov and Uspensky required a particular structure on rules preserving condition (α). They use a partial function $\phi : V_A \longrightarrow V_C$ to determine those vertices that, in our definition, are in $V_A \cap V_C$. As one has to be careful only about the symbols adjacent to vertices in the image of ϕ, they imposed in effect the following condition (β) for a given rule $\langle A, C, \phi \rangle$:

$$(\forall y \in C)[(y = \phi(x) \ \& \ vy \in C) \Rightarrow [lb(v) = * \text{ or } (\exists w)(wx \in A \ \& \ lb(w) = lb(v))]].$$

Consequently, if a program $\mathcal{R}$ satisfies (β), then $\mathcal{M} = \langle \mathcal{S}, \mathcal{F} \rangle$ is a K-graph machine, where $\mathcal{S}$ is the set of all K-graphs on $\mathcal{L}$ and $\mathcal{F}$ is the unique function on $\mathcal{S}$ defined by $\mathcal{R}$.

3. Turing's conditions.. K-graph machines clearly satisfy the determinacy condition, but also the boundedness and locality conditions—when those are suitably interpreted: the number of "immediately recognizable" symbolic configurations is given by the number of distinct antecedents and consequents of the machine's program; operations are quite properly viewed as modifying observed configurations, and observed labeled vertices lie always within a fixed "radius" around the central vertex. (The radius can be read off from the program, e.g., it can be taken to be the maximal length of paths in any K-graph of the program.) We make some additional remarks about the principle of local action, as it might be thought that—even in the case of string machines—locality is violated! The reason being, that in an "implementation" of those machines, e.g., on a standard Turing machine, the total tape content is affected when using a rule that replaces a string by either a longer or a shorter one. However, this seems to be pertinent only if the tape has a rigid extrinsic coordinate system as given, for example, by the set of integers **Z**. When a different presentation of Turing machines is chosen, as suggested for example in (Gandy 1980), or when the underlying structure is flexible to insertions, as in our set-up, the concern disappears.[10] It is precisely the use of an *intrinsic coordinate system*, guaranteed through the principle of unique location, that makes for the locality of the replacement operations.

4. Subsumption and Simulation

K-graph machines capture the general starting-point of Turing's analysis in a most natural way. Consider, for example, encoding the squares in Turing's "child's arithmetic book" as follows:

[10] These two ways of dealing with the issue are two sides of the same coin.

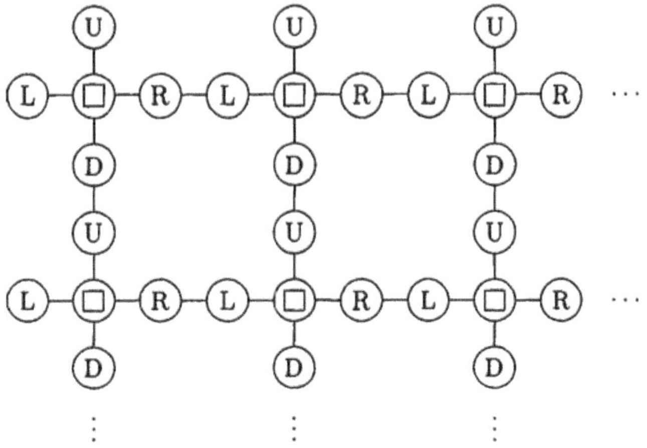

One can attach numerals to the squares (like writing them in the book), and directly encode, for example, the elementary school algorithm for column addition via K-graph operations. A rule from such a machine is shown below. (The rule collapses two digits and enters a special "carry" configuration so that succesive rule applications will move the central vertex to the top of the next column and place a '1' there before returning to continue collapsing the current column.)

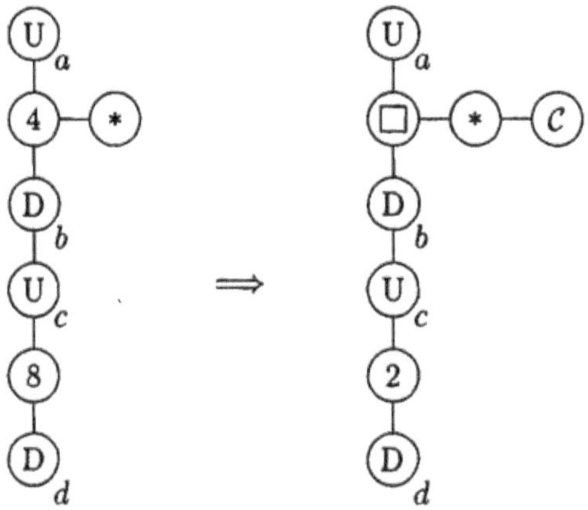

It is also immediately evident that Turing machines are K-graph machines: consider the following formulation of the rule $\alpha q_i s_k \beta \Rightarrow \alpha q_m 0 \beta$ from section 2:

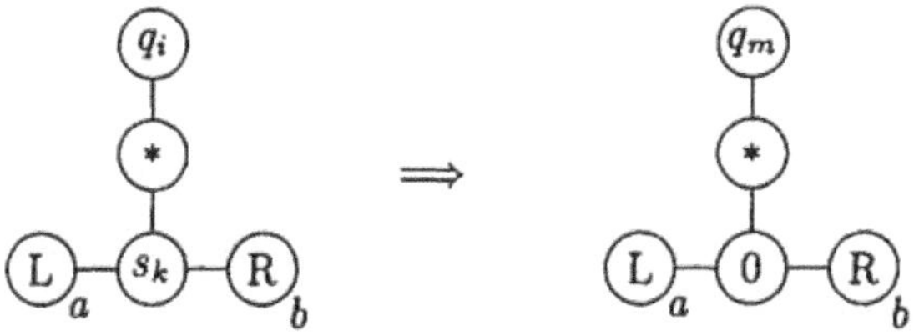

Such considerations can be given in a similarly direct way for string machines and their generalizations to higher dimensions, thus in particular for the generalized Turing machines described in (Kleene 1952). A host of other models of computations, including the *Kolmogorov machines* as defined in (Uspensky and Semenov) or the *register machines* as introduced by (Shepherdson and Sturgis), can be shown to be subsumed under K-graph machines. Joining these observations with the main result of this section, we have an *absolutely uniform way of reducing computations of a particular model to Turing machine computations*: We have only to verify that the computation model is subsumed under K-graph machines. So let us turn to the substantive task of this section, namely, to simulate K-graph machine computations by a Turing machine, i.e., for an arbitrary K-graph machine $\mathcal{M} = \langle \mathcal{S}, \mathcal{F} \rangle$ over the alphabet $\mathcal{L}$ we construct a Turing machine[11] **M** over the alphabet $\{0,1\}$ that simulates $\mathcal{M}$. The simulation requires (i) that we give linear representations of K-graphs, and (ii) that we show for every $S \in \mathcal{S}$, if $\sigma = S, S_1, \ldots, S_n$ is a computation of $\mathcal{M}$, then $\tau = T_0, T_1, \ldots, T_m$ is a computation of **M**. Here T_0 represents S, and T_m represents a K-graph isomorphic to S_n; furthermore, there exists a subsequence $T_{i_1}, T_{i_2}, \ldots, T_{i_{n-1}}$ of τ such that for $1 \leq j < n-1$, T_{i_j} represents a K-graph isomorphic to S_j. The conditions for infinite computations are similar.

Before addressing (ii) in the proof of theorem 4.1, we discuss the linear representation of K-graphs. We assume that $\mathcal{L}$ is linearly ordered and let $\prec$ be the lexicographical ordering on finite sequences of symbols from $\mathcal{L}$ induced by that linear ordering. A second ordering on finite sequences α, β of symbols is defined by $\alpha < \beta$ iff α is shorter than β or α and β are of the same length and $\alpha \prec \beta$. The *address* $\mathrm{Ad}(v)$ of a vertex $v \in V$ is the $<$-minimal element of $\mathrm{Lbs}(v)$; by connectedness such an address exists, and by the principle of unique location Ad is injective. We assume that each vertex of $\mathcal{U}$ is a natural number, and define for an arbitrary edge $uv \in E$ the *location description* $\mathrm{LD}(uv)$ by:

$$\mathrm{LD}(uv) = \begin{cases} \langle u, lb(u), v, lb(v) \rangle & \text{if} \quad \mathrm{Ad}(u) < \mathrm{Ad}(v) \\ \langle v, lb(v), u, lb(u) \rangle & \text{if} \quad \mathrm{Ad}(v) < \mathrm{Ad}(u) \end{cases}$$

When we refer to an edge uv we assume from now on that $\mathrm{Ad}(u) < \mathrm{Ad}(v)$. We define now an ordering on the location descriptions $\mathrm{LD}(E) = \{\mathrm{LD}(uv) \mid$

[11] In contrast to our earlier discussion, we are going to use a Turing machine whose tape is extendable only to the right.

$uv \in E\}$ for a given K-graph K as follows: $\mathrm{LD}(u_1v_1) \sqsubset \mathrm{LD}(u_2v_2)$ iff $\mathrm{Ad}(u_1) < \mathrm{Ad}(u_2)$ or $u_1 = u_2$ and $\mathrm{Ad}(v_1) < \mathrm{Ad}(v_2)$. $\mathrm{GR}(K)$, the *canonical graph representation* of K, is the $\sqsubset$-ordered sequence of location descriptions for K.

To obtain a *tape representation* of this sequence in the alphabet $\{0, 1\}$, we first assign natural numbers to the symbols in $\mathcal{L}$: 0 is used for the least element (in the original linear ordering of $\mathcal{L}$), 1 for the next, etc. Every vertex $v \in V$ is already a natural number, and we assume without loss of generality that a state S with n vertices consists of $\{1, \ldots, n\}$. Finally, natural numbers are represented in a modified binary form obtained from the standard one by replacing every 1 by 11 and every 0 by 10; elements of a sequence of encoded natural numbers are separated by exactly two 0's; elements of a sequence of such sequences are separated by exactly three 0's. After these preparatory considerations, we can establish:

Theorem 4.1. *Any K-graph machine $\mathcal{M}$ can be simulated by some Turing machine* **M**.

Proof. The program of the Turing machine **M** that is to simulate $\mathcal{M}$ transforms any state S into $\mathcal{F}(S)$, returns to its initial state, and, if possible, further transforms the resulting state $\mathcal{F}(S)$; the machine halts when none of the rules defining $\mathcal{F}$ can be applied. (It should be obvious that this yields the kind of simulation indicated above.) In the following more detailed description, we always use S to refer to the graph currently coded on the tape, even though there are stages when some edges are removed and others are added.

The first step in constructing **M**'s program is to modify the encoding of $\mathcal{M}$'s program $\mathcal{R}$ with rules $R_0, \ldots, R_{r-1}$.[12] For a given rule, the antecedent A and the consequent C are encoded by $\mathrm{GR}(A)$ and $\mathrm{GR}(C)$. As we are only concerned with the isomorphism class represented by each rule, we are free to encode each rule R_i $(0 \leq i < r)$ by an isomorphic rule. Let N be the maximum number of vertices occurring in any rule and replace each rule R_i by an isomorphic R'_i such that for every vertex v of R'_i, $N \cdot i < v \leq N \cdot (i+1)$. $\mathcal{R}$ will indicate now the modified program $R'_0, \ldots, R'_{r-1}$. The tape is to contain as input S, all of whose vertices v are replaced however by $v + M$ (where $M = N \cdot r$) so that all (vertex-) numbers are greater than M. Obviously, $C_R \cap (S \setminus A_R) = \emptyset$ for any R. This has also the following advantage (as

[12] An alternative to this simulation is the following: We can also represent a rule on the tape simply by writing GR(A) followed by a separator (say, '0000') and then by GR(C). To represent the program we represent each rule in order, separating them by, say, '00000'. Then we can describe a Turing machine **U** that simulates any K-graph machine conceptually in exactly the same way in which a universal Turing machine simulates any other Turing machine **M**. In the latter case the input to the machine is a Gödel number of **M** and an input to **M**. Here, **U** will take as input the coded program for the K-graph machine followed by the initial state. Then **U** will carry out systematically the coded program.

will be clear from our further discussion): by looking at a vertex one can determine immediately whether it occurred in the original state or whether it was written by a rule; in the latter case one can decide which rule wrote it.

Now **M** examines each rule in the given order via a subroutine CHECK-R that determines whether R can be applied to S. If R is applicable, then CHECK-R tries to modify S to an isomorphic S', such that A_R is a subgraph of S'. If R is not applicable, CHECK-R may nevertheless have changed some vertices on the tape, but the modified graph is isomorphic to S. Note that this way of proceeding is essentially "dual" to that in the definition of K-graph machines: we replace vertices occuring in the original state rather than those in the rule being applied.

CHECK-R starts with the head leftmost on the tape and proceeds by moving to the right, searching for appropriate edges one at time. Let min be the least and max the greatest vertex of R. In the ith step, we are looking for an edge in S which matches the ith edge uv of A_R. u is the second vertex of some edge (already matched to an edge in S) and occurs consequently in S. We search for w in S such that the edge uw is in S and such that $lb(w) = lb(v)$. If such a vertex is found, we distinguish two cases: (i) if $min \leq w \leq max$ and $w \neq v$, then w was written by R and is the image of some vertex in A_R other than v, and the search for the correct vertex has to be continued; (ii) if $w < min$ or $w > max$ or $w = v$, then w is the correct vertex, and **M** substitutes v for w everywhere on the tape and proceeds to search for the next matching edge in S.

In sum, if **M** reaches the end of the tape before finding such a vertex, **M** fails for this R; if CHECK-R fails for every rule R, **M** halts; if CHECK-R finds a matching edge for every edge in A_R, R is applicable. (Note that the principle of unique location allows us to avoid backtracking in case the algorithm fails for a particular vertex.)

If a rule R is applicable, **M** is going to modify its tape appropriately. For each rule R there is a subroutine APPLY-R which applies R to the current state: all edges which contain any vertex from $V_{A_R} \setminus V_{C_R}$ are erased and all those from C_R are inserted—leftmost onto the tape (in the order of their appearance in the canonical encoding GR(C_R)). The tape contains now a representation of $R[S]$; recall that the next state $\mathcal{F}(S)$ is defined as the connected component of $R[S]$ containing the central vertex. However, the remaining computation will not be interfered with by unconnected edges left on the tape. Finally, the head is returned to the left end of the tape, and the state is set to **M**'s initial state. □

We want to determine now, in a rough way, the number of steps **M** needs to transform a K-graph S with n vertices into $\mathcal{F}(S)$. Assume that the language for $\mathcal{M}$ contains l symbols, and that $\mathcal{M}$'s program has r rules of size at most N (i.e., at most N vertices occur in $A \cup C$). For a given state S with n vertices, the maximal degree of each vertex is $l + 1$; otherwise, the principle of unique location would be violated. Thus at most $n(l+1)/2$ edges have to be represented. The largest (vertex-) number to be represented is n, which has length $2\log n$ in our modified binary notation. The largest LD has length of order $\log(2n+2l)$. Thus the representation of S is of length $O(n\log n)$.

The renumbering step must traverse the entire tape. Since we wish to increase each vertex by at least M, we take M' to be the least power of two greater than or equal to M and add M' to each vertex. This operation requires shifting all of the cells right of the vertex being updated up to $\log M'$ cells to the right. This requires rewriting up to $O(n\log n)$ cells for each vertex. Since this operation is done to all occurrences (of which there may be up to $l+1$ many) of each of the n vertices, it is an $O(n^2\log n)$ operation. (The rewriting itself can be done in a single pass over the number and requires $\log n$ steps.)

The further rewriting operations required for finding the applicable rule all involve replacing numbers greater than M' by numbers smaller than M'; thus, no shifting is involved in these operations, since we allow extra 0's to occur between integers. Attempting to match a given edge in a rule to one on the tape might require looking at the entire tape and is an $O(n\log n)$ operation; that may have to be done for every edge in every antecedent.

If we succeed in finding an applicable rule, we apply it; i.e., we transform the tape by erasing all edges from A_R and inserting all edges from C_R at the beginning of the tape. This may require shifting all $O(n\log n)$ symbols by at most the length of the largest $\mathrm{GR}(A_R)$. Hence only $O(n\log n)$ many steps are required for rule selection and application. But once this has been accomplished, the entire transformation is complete, so the complexity of the simulation is $O(n^2\log n)$.

Now let us consider simulating a full computation of M. If we let $k = \max\{|C_R| - |A_R| \mid R\in\mathcal{R}\}$, then for any $S\in\mathcal{S}$, $|\mathcal{F}(S)|\le|S|+k$; here $|K|$ is the cardinality of the set of vertices of K. Let h be a natural number such that $hn^2\log n$ is the complexity of the "step-simulation" for **M** of M we just discussed. Assume, in a first example, that $\mathcal{M}$ runs in constant time, say, in m steps. Then the length of the computation of **M** for input of size n is bounded by

$$s = hn^2\log n + h(n+k)^2\log(n+k) + \cdots + h(n+km)^2\log(n+km).$$

Clearly,

$$n^2\log n \le s \le mh(n+km)^2\log(n+km) = O(n^2\log n)$$

so $s = O(n^2\log n)$.

If $\mathcal{M}$ runs in higher order time, however, the step-complexity of **M** is not preserved. Assume, for example, that $\mathcal{M}$ runs in mn^c-many steps, for some m and c. Then the complexity of **M** for input of size n is bounded by

$$hn^2\log n + h(n+k)^2\log(n+k) + \cdots + h(n+kmn^c)^2\log(n+kmn^c) = O(n^{3c}\log n)$$

This analysis illustrates the quite obvious point that the complexity of computing a given function depends on the machine used to carry out the computation. An interesting example is multiplication: it can be computed in linear time by RAM's and SMM's (Schönhage), but it is also known that Turing machines cannot multiply in linear time (Cook and Aanderaa; Paterson e.a.); we do not know, whether K-graph machines can. This question and related ones raise many interesting issues about complexity, particularly whether one model allows a more fundamental analysis of the complexity of algorithms than another.

5. Concluding Remarks

For Turing the ultimate justification for his restrictive conditions lies in the *necessary limitation* of human memory, and that can be directly linked to physical limitations also for machines; cf. (Mundici and Sieg), section 3. Church in his review of Turing's paper seems to have mistaken Turing's analysis as an analysis of machine computations. Church's apparent misunderstanding is common: see, e.g., (Mendelson 1990). So it is worthwhile to point out that machine computability was analyzed only much later by Gandy (1980). Gandy followed Turing's three-step-procedure of analysis, axiomatic formulation of general principles, and proof of a reduction theorem, but for "discrete deterministic mechanical devices", not computors.

Gandy showed that everything computable by a device satisfying his principles, a *Gandy machine*, can already be computed by a Turing machine. To see clearly the difference between Turing's analysis and Gandy's, note that Gandy machines incorporate parallelism: they compute directly Conway's game of life and operate, in parallel, on bounded parts of symbolic configurations of possibly unbounded size. The boundedness conditions for Gandy machines and the *principle of local causation* are motivated by physical considerations. We have been concerned, in contrast, with an explication and generalization of Turing's arguments for his thesis, that all mechanical processes can be simulated by (Turing) machines. We are coming back to this starting-point of our considerations through three remarks.

First, Turing analyzed *mechanical* processes of a human computor. The reduction of string machines or of K-graph machines to letter machines over a two-element alphabet does not show that *mental* processes cannot go beyond mechanical ones; it only shows that Turing machines can serve as a "normal form" for machines, because of the simplicity of their description.[13] The

[13] For this reason Turing machines are most suited for theoretical investigations. This state of affairs is analogous to that involving logical calculi: natural deduction calculi reflect quite directly the structure of ordinary arguments, but have a somewhat involved metamathematical description; in contrast, axiomatic logical systems are not suited as frameworks for direct formalizations, but—due to their simple description—are most suitable for metamathematical investigations.

question, whether (different kinds of) machines are adequate mathematical models for mental processes, is left completely open. That is an empirical issue!

Second, the formulation of the boundedness and locality conditions for mechanical processes and the design of general machine models allow us to give uniform reductions. A natural generalization of K-graph machines, not giving up these broad conditions, captures parallel computations of "discrete deterministic mechanical devices", not computors. A future paper of ours gives such a generalization, based on the presentation of machines by Gandy (1980) discussed above.

Third, support for Turing's thesis is best given in two distinct steps: (i) mechanical processes satisfying boundedness and locality conditions can be recognized "directly", without coding or other effective transformations, as computations of a general model; (ii) computations of the general model can be simulated by Turing machines. The plausibility of Turing's thesis rests exclusively on the plausibility of the *modified central thesis* (i); after all, (ii) is a mathematical fact. Our modification of Turing's *central thesis* states that mechanical processes are easily seen to be computations of K-graph machines; in our view, this is a most plausible claim.

References

1. A. Church. Review of (Turing 1936). *J. Symbolic Logic 42-3*, 1 (1937).
2. S. Cook and S. Aanderaa. On the minimum computation time of functions. *Trans. AMS 142* (1969), 291–314.
3. B. Courcelle. Graph rewriting: An algebraic and logic approach. In *Handbook of Theoretical Computer Science*, J. van Leeuwen, Ed. Elsevier, Amsterdam, 1990, pp. 195–242.
4. M. Davis. *Computability and Undecidability*. McGraw-Hill, 1958.
5. R. Gandy. Church's thesis and principles for mechanisms. In *The Kleene Symposium* (Amsterdam, 1980), Barwise, Keisler, and Kunen, Eds., North-Holland, pp. 123–48.
6. R. Gandy. The confluence of ideas in 1936. In *The Universal Turing Machine - A Half Century Review*, R. Herken, Ed. Oxford Univ. Press, 1988, pp. 55–111.
7. K. Gödel. On undecidable propositions of formal mathematical systems (1934). In *Gödel's Collected Works I*, S. Feferman et al., Eds. Oxford Univ. Press, New York, 1986, pp. 346–71.
8. K. Gödel. Über die Länge von Beweisen (1936). In *Gödel's Collected Works I*. Oxford Univ. Press, New York, 1986, pp. 396–399.
9. K. Gödel. Some remarks on the undecidability results (1972). In *Gödel's Collected Works II*. Oxford Univ. Press, New York, 1990, pp. 305–6.
10. T. Herron. An alternative definition of pushout diagrams and their use in characterizing K-graph machines. Carnegie Mellon University, May 1995.
11. S. Kleene. *Introduction to Metamathematics*. P. Noordhoff N.V., 1952.
12. A. Kolmogorov and V. Uspensky. On the definition of an algorithm. *AMS Translations 21*, 2 (1963), 217–245.

13. A. Markov. *Theory of Algorithms.* Academy of Sciences of the USSR, Moscow, 1954.
14. E. Mendelson. Second thoughts about Church's thesis and mathematical proofs. *J. Phil. 87*, 5 (1990), 225–33.
15. D. Mundici and W. Sieg. Paper machines. *Philosophia Mathematica 3* (1995), 5–30.
16. M. Paterson, M. Fischer, and A. Meyer. An improved overlap argument for on-line multiplication. In *Complexity of Computation* (Providence, RI, 1974), R. Karp, Ed., AMS, pp. 97–111.
17. E. Post. Finite combinatory processes—formulation 1. *J. Symbolic Logic 1* (1936), 103–5.
18. E. Post. Recursive unsolvability of a problem of Thue. *J. Symbolic Logic 12* (1947), 1–11.
19. A. Schönhage. Storage modification machines. *SIAM J. on Computing 9* (1980), 490–508.
20. J. Shepherdson. Mechanisms for computing over arbitrary structures. In *The Universal Turing Machine—A Half Century Review*, R. Herken, Ed. Oxford Univ. Press, 1988, pp. 581–601.
21. J. Shepherdson and H. Sturgis. Computability of recursive functions. *J. ACM 10* (1963), 217–55.
22. W. Sieg. Mechanical procedures and mathematical experience. In *Mathematics and Mind*, A. George, Ed. Oxford Univ. Press, 1994, pp. 71–117.
23. A. Turing. On computable numbers, with an application to the Entscheidungsproblem. *Proc. London Math. Soc., series 2 42* (1936-7), 230–265.
24. A. Turing. The word problem in semi-groups with cancellation. *Ann. of Math. 52* (1950), 491–505.
25. A. Turing. Solvable and unsolvable problems. *Science News 31* (1953), 7–23.
26. V. Uspensky. Kolmogorov and mathematical logic. *J. Symbolic Logic 57* (1992), 385–412.
27. V. Uspensky and A. Semenov. What are the gains of the theory of algorithms: Basic developments connected with the concept of algorithm and with its application in mathematics. In *Algorithms in Modern Mathematics and Computer Science* (Berlin, 1981), A. Ershov and D. Knuth, Eds., Springer-Verlag, pp. 100–235.

Forcing on Bounded Arithmetic

Gaisi Takeuti and Masahiro Yasumoto *

[1] Gasi Takeuti
Department of Mathematics,
University of Illinois,
Urbana, Illinois 61801, U.S.A.
email: takeutimath.uiuc.edu

[2] Masahiro Yasumoto
Graduate School of Polymathematics
Nagoya University
Chikusa-ku,
Nagoya, 464-01,
JAPAN
email: D42985A nucc.cc.nagoya-u.ac.jp

Forcing method on Bounded Arithmetic was first introduced by J. B. Paris and A. Wilkie in [10]. Then M. Ajtai started in [1], [2] and [3] elaborate use of the method to get excellent results on the pigeon hole principle and the module p counting principles. Ajtai's work were followed by many works by Beame et als, Krajíček and Riis in [4], [5], [8], [9], [11].

In this paper, we develop a Boolean valued version of forcing on Bounded Arithmetic using big Boolean algebra, and discuss its relation with $NP = co - NP$ problem and $P = NP$ problem.

As is well known, Gödel raised the problem closely related to $P = NP$ problem in his letter to von Neumann in 1956. We believe that Gödel would greatly contribute to it if the complexity theory would have started at the time.

We also would like to mention about Gödel's close felling to Boolean valued models. Forcing and Boolean valued model theory are equivalent. But Gödel was much more impressed by Boolean valued models than forcing in the following reason. Gödel did have a systematic reinterpretation of the logical operations with a view to a formal independence proof, but it was too messy for his taste. He realized that the Boolean valued models are a straightforward model-theoretic variant of his earlier reinterpretation.

When one of the authors started Boolean valued analysis by using Boolean algebras of projections in Hilbert space, he received a strong encouragement from Professor Gödel. We feel that our work is in the line of Gödel's vision.

1. The generic models

Let N be a countable nonstandard model of the true arithmetic $Th(\mathbb{N})$ where $\mathbb{N}$ is the standard model of arithmetic. Let n be a nonstandard element in N and $M = \{x \in N \mid \text{there exists some } n\#\cdots\#n \text{ such that } x \leq n\#\cdots\#n\}$.

* This is the final version of the paper which will not be published elsewhere.

Obviously M is a model of Buss' theory S_2. Let $n_0 = |n|$ and $M_0 = \{|x| \mid x \in M\}$. M_0 is an initial part of M and $x \in M_0$ iff there exists a polynomial p such that $x \leq p(n_0)$.

M can be considered as a first order structure as described above but also can be considered a second order structure over M_0 as follows. Let a second order object X be a pair of (a, b) where $a \in M$ and $b \in M_0$. Then by X we express the set defined by

$$X = \{i < b \mid \mathrm{Bit}(i, a) = 1\}.$$

In this case b is denoted by $|X|$. The second order structure thus obtained is denoted by (M_0, M).

In (M_0, M), the first order variables denote the member of M_0. The second order variables $X, Y, Z, \ldots$ denote sets of members of M_0. For X, Y, Z, $|X|$, $|Y|$, $|Z|, \ldots$ denote members of M_0.

The language of (M_0, M) is described as follows.

First order variables a, b, $c, \ldots,$ x, y, $z, \ldots$.

Second order variables X, Y, $Z, \ldots$.

First order constants 0, 1,

First order function constants $+$, $\cdot$, $\lfloor \frac{1}{2} \rfloor$, $|\ |$

Second order function constants $|\ |$

First order predicate $\leq$, $=$

Second order predicate $\in$.

Terms.

1. 0, 1, the first order free variables a, b, $c, \ldots$ and $|X|$, $|Y|, \ldots$ are terms where X, $Y, \ldots$ are second order free variables.
2. If $t_1, \ldots, t_n$ are terms and f is a function constant, then $f(t_1 \cdots, t_n)$ is a term.
3. All terms are obtained by (1) and (2). In the structure (M_0, M), every term expresses a member of M_0.

Formulas.

1. If t_1 and t_2 and terms and X is a second order free variable, then $t_1 \leq t_2$, $t_1 = t_2$ and $t_1 \in X$ is a formula.
2. If φ and ψ are formula, then $\neg\varphi$, $\varphi \wedge \psi$, and $\varphi \vee \psi$ are formulas.
3. If $\varphi(a)$ is a formula and t is a term and X is a second order free variable, then
$$\forall x \varphi(x), \exists x \varphi(x), \forall x \leq t \varphi(x), \exists x \leq t \varphi(x), \forall x \in X \varphi(x)$$
and $\exists x \in X \varphi(x)$ are formulas, where x is a bound variable not occurring in $\varphi(a)$.
4. If $\varphi(X)$ is a formula and t is a term, then
$$\forall X \varphi(X), \exists X \varphi(x), \forall X \leq t \varphi(x), \exists X \leq t \varphi(x)$$
are formulas.

5. Every formula is obtained by (1)-(4).
   The meaning of $\forall X \leq t$ and $\exists X(\leq t$ are $\forall X(|X| \leq t \to \cdots)$ and $\exists X(|X| \leq t \wedge \cdots)$ respectively.

Definition 1.1. *In the second order language of (M_0, M), $\forall x \leq t$, $\exists x \leq t$, $\forall x \in X$, $\exists x \in X$ are called first order bounded quantifiers. These correspond to sharply bounded quantifiers in the first order language of M.*

Corresponding to hierachies of bounded formulas Σ_1^b, Π_i^b on M, we define hierachies of second order bounded formulas on (M_0, M) as follows.

$\Sigma_0^1(BD) = \Pi_0^1(BD)$ is the class of formulas in which every quantifier is a first order bounded quantifier.

For $i > 0$, $\Sigma_i^1(BD)$ and $\Pi_i^1(BD)$ are defined to be the smallest class of formulas satisfying the following conditions.

a) Both $\Sigma_1^1(BD)$ and $\Pi_i^1(BD)$ are subclass of $\Sigma_{i+1}^1(BD) \cap \Pi_{i+1}^1(BD)$.
b) If $\varphi \in \Sigma_{i+1}^1(BD)$, then $\exists X \leq t\varphi(X)$, $\forall x \leq t\varphi(x)$ and $\exists x \leq t\varphi(x)$ belong to $\Sigma_{i+1}^1(BD)$.
c) If $\varphi \in \Pi_{i+1}^1(BD)$, then $\forall X \leq t\varphi(X)$, $\exists X \leq t\varphi(X)$ and $\forall x \leq t\varphi(x)$ belong to $\Pi_{i+1}^1(BD)$.
d) If φ and ψ belong to $\Sigma_{i+1}^1(BD)$ then both $\varphi \wedge \psi$ and $\varphi \vee \psi$ belong to $\Sigma_{i+1}^1(BD)$. If φ and ψ belong to $\Pi_{i+1}^1(BD)$, then both $\varphi \wedge \psi$ and $\varphi \vee \psi$ belong to $\Pi_{i+1}^1(BD)$.
e) If $\varphi \in \Sigma_{i+1}^1(BD)$.
If $\varphi \in \Pi_{i+1}^1(BD)$, then $\neg\varphi \in \Sigma_{i+1}^1(BD)$ then $\neg\varphi \in \Sigma_{i+1}^1(BD)$.

A formula is said to be a bounded formula if it belongs to $\bigcup_i \Sigma_i^1(BD) = \bigcup_i \Pi_i^1(BD)$ A bounded formula in the second order language of (M_0, M) corresponds to a bounded formula in the first order language of M.

(M_0, M) satisfies the following axioms.

1. Basic axioms on the first order function constants and the first order predicate constants.
2. Axiom on $|X|$ $\forall x \forall X(x \in X \supset x < |X|)$
3. Comprehension Axioms

$$\forall a \exists X \leq a(|x| = a \wedge \forall x < a(x \in X \leftrightarrow \varphi(x)))$$

   where φ is a bounded formula.
4. The least number principleLNP

$$\forall X(X \neq \emptyset \supset \exists x \in X \forall y \in X(x \leq y)).$$

This axiom is equivalent to the Induction Axiom

$$\varphi(0) \wedge \forall x(\varphi(x) \supset \varphi(x+1)) \supset \forall x \varphi(x)$$

where φ is a bounded formula.

Now we are going to define a Boolean algebra. First we introduce Boolean variables $p_0, p_1, p_2, \cdots p_{n_0-1}$ and its negation $\bar{p}_0, \bar{p}_1, \bar{p}_2, \ldots, \bar{p}_{n_0-1}$. More precisely we define some coding of these literals. Now we generate free Boolean algebra from these literals.

In [6], S. Buss developed the theory of sequence in S_2^1. By RSUV-Isomorphisms in [13], the second order theory of sequences hold in (M_0, M) and "X is a sequence" is a $\Delta_1^1(BD)$ predicate where $\Delta_1^1(BD) = \Pi_1^1(BD) \cap \Sigma_1^1(BD)$.

The Boolean algebra B is the set of b which is a sequence $(X_0, X_1, \cdots, X_r)$ with $r \in M_0$ satisfying one of the following conditions.

1. X_i is p_j with $j < n_0$.
2. X_i is $\bar{p}_j$ with $j < n_0$.
3. X_i is $(\wedge, Y_0, Y_1, \cdots, Y_s)$ or $(\vee, Y_0, \cdots, Y_s)$ where $Y_j (j \leq s)$ is one of $(X_0, X_1, \cdots, X_{i-1}$, where the intended meaning of $(\wedge, Y_0, Y_1, Y_s)$ and $(V, Y_0, Y_1, \cdots, Y_s)$ are $Y_0 \wedge Y_1 \wedge \cdots \wedge Y_s$ and $Y_0 \vee Y_1 \vee \cdots \vee Y_s$ respectively.

It is easily seen that there exists a $\Delta_1^1(BD)$ formula φ such that

$$b \in B \quad iff \quad \varphi(b).$$

B is not definable in N since M is not definable in N. However $b \in B$ implies $b \in N$.

Let $b = (X_0, X_1, \cdots, X_s) \in B$. Then $\neg b$ is defined to be $(\bar{X}_0, \bar{X}_1, \cdots, \bar{X}_s)$ where $\bar{X}_i$ is defined by the following rules.

1. If X_i is p_j, then $\bar{X}_i$ is $\bar{p}_j$. If X_i is $\bar{p}_j$, then $\bar{X}_i$ is p_j.
2. If $X_i = (\vee, Y_0, \cdots, Y_t)$, then $\bar{X}_i = (\wedge, \bar{Y}_0, \cdots, \bar{Y}_t)$. If $X_i = (\wedge, Y_0, \cdots, Y_t)$, then $\bar{X}_i = (\wedge, \bar{Y}_0, \cdots, \bar{Y}_t)$. If $X_i = (\wedge, Y_0, \cdots, Y_t)$, then $\bar{X}_i = (\vee, \bar{Y}_0, \cdots, \bar{Y}_t)$.

Let for $i \leq t$ $b_i = (X_0^i, \cdots, X_{s_i}^i) \in B$. Then $\bigvee_{i \leq t} b_i$ is defined to be

$$(X_0^0, \cdots, X_{s_0}^0, X_s^1, \cdots, X_{s_1}^1, \cdots, X_0^t, \cdots, X_{s_t}^t, Z)$$

where $Z = (\vee, X_{s_0}^0, \cdots, X_{s_t}^t)$.

In the same way $\bigwedge_{i \leq t} b_i$ is defined to be

$$(X_0^0, \cdots, X_{s0}^0, X_0^1, \cdots, X_{s_1}^1, \cdots, X_{s_t}^t, Z^1)$$

where $Z^1 = (\wedge, X_{s_0}^0, \cdots, X_{s_t}^t)$.

Now let $A \in M$ be a subset of $\{0, \cdots, n_0 - 1\}$. Then A gives a truth value to $p_0, \cdots, p_{n_0-1}$ in the following way.

If $i \in A$, then it assigns 1 to p_i. If $i \notin A$, then it assigns 0 to p_i. So we call A an atom evaluation. Therefore for each $b \in B$, A makes an evaluation of b denoted by eval(A, b). eval(A, b) satisfies the following rules.

1. eval(A, b) is either 0 or 1.

2. $\mathrm{eval}(A, p_i) = 1$ iff $i \in A$.
3. $\mathrm{eval}(A, \neg b) = 1$ iff $\mathrm{eval}(A, b) = 0$.
4. $\mathrm{eval}(A, \wedge_i b_i) = \wedge_i \mathrm{eval}(A, b_i)$
5. $\mathrm{eval}(A, \vee_i b_i) = \vee_i \mathrm{eval}(A, b_i)$.

For $b_1, b_2 \in B$, we define $b_1 \overset{B}{=} b_2$ to be $\forall A$ atom evaluation $(\mathrm{eval}(A, b_1) = \mathrm{eval}(A, b_2))$.

Then

$b_1 \overset{B}{=} b_2$ is $\Pi_1^1(BD)$ in (M_0, M) and Π_1^b in M.

The Boolean algebra we use is $B/\overset{B}{=}$ though we use B in the place of $B/\overset{B}{=}$ for simplicity.

Now we define M^B as follows. $M^B = \{X \in M \mid \exists y \in M_0(X : y \to B)\}$. Now let $x, y, z, \cdots \in M_0$ and $X \in M^B$. We define the truth value of formulas on (M_0, M^B) by the following rules.

$$[[x + y = z]] = 1 \quad \text{iff} \quad x + y = z$$
$$[[x \cdot y = z]] = 1 \quad \text{iff} \quad x \cdot y = z$$

In the same way for every atomic formula φ

$$[[\varphi]] = 1 \quad \text{iff} \quad M_0 \models \varphi \quad \text{and}$$
$$[[\varphi]] = 0 \quad \text{iff} \quad M_0 \not\models \varphi.$$

$$[[x \in M]] = \begin{cases} X(x) & : \text{ if } X : y \to B \text{ and } x < y \\ 0 & : \text{ otherwise} \end{cases}$$

$$\begin{aligned}
[[\varphi \vee \psi]] &= [[\varphi]] \vee [[\psi]] \\
[[\varphi \wedge \psi]] &= [[\varphi]] \wedge [[\psi]] \\
[[\neg\varphi]] &= \neg[[\varphi]] \\
[[\exists x \le t \varphi(x)]] &= \bigvee_{x \le t} [[\varphi(x)]] \\
[[\forall x \le t \varphi(x)]] &= \bigwedge_{x \le t} [[\varphi(x)]] \\
[[\forall x \in X \varphi(x)]] &= [[\forall x < |X|(x \in X \supset \varphi(x)]] \\
&= \bigwedge_{X < |X|} ([[x \in X]] \supset [[\varphi(x)]]) \\
[[\exists x \in X \varphi(x)]] &= [[\exists x < |X|(x \in X \wedge \varphi(x)]] \\
&= \bigvee_{x < |X|} ([[x \in X]] \wedge [[\varphi(x)]]).
\end{aligned}$$

The following lemma is obvious from the definition.

Lemma 1.1. *Let* $\varphi \in \Sigma_0^1(BD)$. *Then*

$$[[\varphi]] \in B$$

For $\varphi \notin \Sigma_0^1(BD)$, $[[\varphi]]$ *is not defined.*

Definition 1.2. *For* $X \in M$, $\check{X}$ *is defined by the equation*

$$\begin{aligned} \check{X} &= \{< x, 1 >| x < |X| \wedge x \in X\} \\ &\cup \{< x, 0 > | x < |X| \wedge x \notin X\} \end{aligned}$$

where $< a, b >$ *expresses the ordered pair of* a *and* b.

Definition 1.3. *A subset* $I \subseteq B$ *is said to be an ideal if* $0 \in I$, $1 \notin I$, *and* I *is closed under* $\vee$ *and satisfies* $\forall b \in I \forall b' \in B(b' \leq b \supset b' \in I)$.

A subset $D \subseteq B$ *is said to be dense over* I *if the following condition is satisfied.*

$$\forall X \in B - I \exists Y \in D - I(Y \leq X).$$

D *is said to be definable if there exist a formula* $\varphi(b)$ *such that*

$$D = \{b \in M | N \models \varphi(b)\}$$

where φ *may contain members of* N *as parameters.*

Let M *be defined by the equation*

$$\mathcal{M} = \{D \subseteq B | D \quad \text{is definable}\}.$$

Since N *is countable,* $\mathcal{M}$ *is also countable and enumerated as*

$$D_0, D_1, D_2, \ldots.$$

Definition 1.4. *Let* $G \subseteq B$. G *is said to be* $\mathcal{M}$*-generic over* I *if the following condition is satisfied.*

For every $D \in \mathcal{M}$ *if* D *is dense over* I, *then*

$$G \cap (D - I) \neq \emptyset.$$

Let $I \supseteq B$ *is an ideal of* B. I *is said to be* M_0*- complete if the following condition is satisfied.*

$$\forall X \in M(\exists x \in M_0(X : x \to I) \supset \bigvee_{y<x} X(y) \in I).$$

"I *is* M_0*-complete" belongs to* $\Pi_1^1(BD)$.

Let $K \subseteq 2^{n_0}$ *and be definable in* N. *Then for* $b \in B$ *we define* $m_K(b)$ *by the equation*

$$m_K(b) = |\{a \in K | eval(a, b) = 1\}|$$

where $|\{a \in K|\cdots\}|$ *is the cardinality of* $\{a \in K|\cdots\}$ *calculated in* N.

Let $|K| \notin M_0$ *hold. then* I_k *is defined by the equation*

$$I_K = \{b \in B | m_K(b) \in M_0\}.$$

When $k = 2^{n_0}$, I_k *is denoted by* I_0.

Theorem 1.1. I_k *is* M_0 *complete.*

Proof. This is immediately implied by the following obvious property.

$$m_K(\bigvee_i b_i) \leq \Sigma_i m_K(b_i).$$

Example 1.1. Let $n_0 = a \cdot b$ and define $< x, y >$ such that for every $z < n_0$ there exists unique $x < a$ and $y < b$ such that $z =< x, y >$ and $\forall x < a \forall y < b(< x, y >< n_0)$. Define K by

$$K = \{f \in N | f : a \to b\}$$

where the meaning of the function is defined by $< x, y >$. Then $|K| \notin M_0$.

Definition 1.5. *Let* $F \subseteq B$. F *is said to be a filter over* I *if* $F \subseteq B - I$, $1 \in F$,

$$\forall b_1, b_2 \in F(b_1 \wedge b_2 \in F), \quad and \quad \forall b \in F \forall b' \in B(b \leq b' \to b' \in F).$$

F *is said to be a maximal filter over* I *if* F *is maximal among filters over* I.

Theorem 1.2. *Let* I *be a* M_0*-complete ideal of* B. *Then there exists and* $\mathcal{M}$*-generic maximal filter over* I,

Proof. Let D_0, D_1, D_2, $\cdots$ be an enumeration of all dense sets of $\mathcal{M}$ over I and b_0, b_1, b_2, $\cdots$ be an enumeration of all members of B. We define b_0^1, b_1^1, b_2^1, $\cdots$, as follows.

1. If $b_0 \notin I$, then define $b'_0 = b_0$. Otherwise define $b'_0 = 1$.
2. Let $b'_0 \geq b'_1 \geq \cdots \geq b'_{2i} \notin I$ have been defined. Since D_i is dense over I, there exists $b'_{2i+1} \leq b'_{2i}$ such that $b'_{2i+1} \in D_i - I$.
3. Let $b'_0 \geq b'_1 \geq \cdots \geq b'_{\geq i+1} \notin I$ have been defined. If $b'_{2i+1} \wedge b_i \notin I$, then define $b'_{2i+2} = b_{2i+1} \wedge b_i$. Otherwise define $b'_{2i+2} = b'_{2i+1}$.

After all b'_i are defined, define G by the equation

$$G = \{b \in B | \exists i(b'_i \leq b)\}.$$

Then G is obviously an $\mathcal{M}$-generic maximal filter over I.

Theorem 1.3. *Let I be an M_0-complete ideal of B, $X \in M$ satisfy $X : y \to B \wedge \bigvee_{x<y} X(x) = 1$, and G be an $\mathcal{M}$-generic maximal filter over I. Then the following holds.*

$$\exists x < y(X(x) \in G).$$

Proof. Let $D = \{b \in B | \exists x < y(b \leq X(x))\}$. We claim that D is dense over I. For this, let $b' \in B - I$. Then

$$\bigvee_{x<y} X(x) \wedge b' = b' \notin I.$$

Since I is M_0-complete, $\exists x < y(X(x) \wedge b' \in D - I)$. Since $X(x) \wedge b' \leq b'$, we have proved our claim. Since G is $\mathcal{M}$-generic, $\exists b' \in G \cap D$. Therefore $\exists x < y(b' \leq X(x))$ and $X(x) \in G$.

Let G be an $\mathcal{M}$-generic maximal filter over I and $X : y \to B$. Then we define $i_G(X)$ by the equation

$$i_G(X) = \{x < y | X(x) \in G\}.$$

Then we define $M[G]$ as follows.

$$M[G] = \{i_G(X) | \exists y \in M_0(X : y \to B)\}.$$

The following theorem is obvious.

Theorem 1.4. $i_G(\check{X}) = X$ *for every* $X \in M$.

Corollary 1.1. $M \subseteq M[G]$

Theorem 1.5. *Let $x_1, \cdots, \in M_0$, $X_1, \cdots \in M^B$, $\varphi \in \Sigma_0^1(BD)$, I M_0-complete, and G be an $\mathcal{M}$-generic maximal over I. Then we have the following equivalence.*

$$(M_0, M[G]) \models \varphi(x_1, \cdots, i_G(X_1), \cdots)$$

iff $[[\varphi(x_1, \cdots, X_1, \cdots)]] \in G$.

Proof. We prove this by the induction on the number of logical symbols in $\varphi(x_1, \cdots, X_1, \cdots)$. We treat only the nontrivial cases.

Remark 1.1 (Case 1). φ is of the form $x \in X$. Let $X : y \to B$. Then we have

$$\begin{aligned} [[x \in X]] \in G &\leftrightarrow X(x) \in G \\ &\leftrightarrow x \in i_G(X). \end{aligned}$$

Remark 1.2 (Case 2). φ is of the form $\exists x \le t\varphi(x)$.

$$
\begin{aligned}
[[\exists x \le t\varphi(x)]] \in G \quad & \text{iff} \quad \bigvee_{x \le t} [[\varphi(x)]] \in G \\
& \text{iff} \quad \exists x \le t([[\varphi(x)]] \in G) \\
& \text{iff} \quad \exists x \le t(M_0, M[G]) \models \varphi(x) \\
& \text{iff} \quad (M_0, M[G]) \models \exists x \le t\varphi(x).
\end{aligned}
$$

Let I be M_0-complete, G an M-generic maximal filter over I, $\tilde{X} = i_G(X)$, and $X : y \to B$. Then we define $|\tilde{X}|$ to be y.

Theorem 1.6. *Let* $\tilde{X} \in M[G]$. *Then the following LNP holds on* $(M_0, M[G])$.

$$\exists x \le t(x \in \tilde{X}) \to \exists x \le t(x \in \tilde{X} \wedge \forall y < x \neg y \in \tilde{X}).$$

Proof. Let $\tilde{X} = i_G(X)$ where $X : y \to B$ Then $Y : y \to B$ is defined by the equation

$$Y(x) = X(x) - \bigvee_{z < x} X(z).$$

Then we have $\bigvee_{x<t} X(x)$. Since the following two equations hold

$$
\begin{aligned}
&[[\exists x \le t(x \in X)]] = \bigvee_{x \le t} X(x) \\
&[[\exists x \le t(x \in X \wedge \forall y < x \neg y \in X)]] = \bigvee_{x \le t} (X(x) - \bigvee_{y < x} (X(y)),
\end{aligned}
$$

the following holds.

$$[[\exists x \le t(x \in X)]] \in G \to [[\exists x \le t(x \in X \wedge \forall y < x \neg y \in X)]] \in G.$$

Theorem 1.7. $\Sigma_0^1(BD)$*-Comprehension Axioms hold in* $(M_0, M[G])$.

Proof. Let $\varphi(x) \in \Sigma_0^1(BD)$. If suffices to show that for every $a \in M_0$ the following holds.

$$\{x \le a | (M_0, M[G]) \models \varphi(x)\} \in M[G].$$

Define $Y \in M$ by the conditions $Y : a \to B$ and

$$x \le a \to Y(x) = [[\varphi(x)]]$$

Then the proof follows from the following equivalencies

$$
\begin{aligned}
x \in i_G(Y) \quad & \text{iff} \quad x \le a \wedge [[\varphi(x)]] \in G \\
& \text{iff} \quad x \le a \wedge (M_0, M[G]) \models \varphi(x).
\end{aligned}
$$

So far we have considered the second order version (M_0, M) of the first order structure M. In the same way, we will consider the first order version $M[G]$ of the second order structure $(M_0, M[G])$.

For M^B, we can add every polynomial time computable function since every polynomial time computable function can be expressed by a polynomial size circuit and the Boolean algebra B is closed by any polynomial size circuit. ¿From this follows that we can introduce all polynomial time computable functions in the structure of $M[G]$. Therefore from now on we always assume that all polynomial time computable functions are defined on the first order structure $M[G]$.

Theorem 1.8. *Let $\varphi(x)$ be sharply bounded. Then if $M \models \forall x\varphi(x)$, then $M[G] \models \forall x\varphi(x)$.*

Proof. Let a be an atom evaluation. (Previously an atom evaluation was denoted by A since we consider it in the second order structure (M_0, M^B). We are now considering it in the first order structure M^B. Therefore it is now denoted by a.) Let x be expressed by $X : y \to B$ and $X^a : y \to \{0,1\}$ be defined by

$$x < y \to X^a(x) = eval(a, X(x)).$$

We also denote the first order expression of X^a by x^a. Then eval$(a, [[\varphi(x)]]) = 1$ iff $M \models \varphi(x^a)$. Therefore $M \models \forall x\varphi(x)$ implies $\forall a($eval$(a, [[\varphi(x)]]) = 1)$. Therefore $[[\varphi(x)]] \overset{B}{=} 1$. Therefore for every $x \in M[G]$, $M[G] \models \varphi(x)$ and we have $M[G] \models \forall x\varphi(x)$.

Every polynomial time computable function f can be defined by successive function equations from basic functions. This defining equation is called the defining axiom of f.

Theorem 1.9. *Every polynomial time computable function f satisfies the defining axiom of f in $M[G]$.*

Proof. The defining axiom of f can be expressed by a form $\forall \mathbf{x}\varphi(\mathbf{x})$, where $\varphi(\mathbf{x})$ is sharply bounded. Therefore the theorem is immediately implied by Theorem 1.8.

Corollary 1.2. *Let f be a polynomial time computable function and $a \in M_0$. Let $f(a) = b$ in M. Then $f(a) = b$ in $M[G]$.*

Proof. This is immediate from Theorem 1.8.

Definition 1.6. *A sequent $\Gamma \to \Delta$ is said to be Σ_1^b if every formula in Γ or Δ is Σ_1^b.*

Theorem 1.10. *Let $\Gamma \to \Delta$ be Σ_1^b and provable in S_2^1. Then $M[G] \models \Gamma \to \Delta$.*

Proof. This is immediately implied by Buss' Witness Theorem in [6].

It is very difficult to prove that $M[G]$ is a model of Bounded Arithmetic stronger than Σ_1^b-part of S_2^1. One reason is that $[[\varphi]]$ has no reasonable definition when φ is not sharply bounded. In this situation the development of forcing in set theory suggests us that $M[G]$ is probably not a model of S_2.

Let $K \subseteq 2^{n_0}$ satisfy $|K| \notin M_0$. In order to investigate I_k, first we prove the following lemma.

Lemma 1.2. *Let G be an m-generic maximal filter over I_k, $A = \{i < n_0 | p_i \in G\}$, $C \in M$, and D defined by the following equation*

$$D = \{b \in B | \exists i < n_0(i \in C \wedge b \leq \bar{p}_i) \vee \exists i < n_0(i \notin C \wedge b \leq p_i)\}.$$

Then D is dense over I_k

Proof. Let $b \in B - I_k$. Then $m_K(b) \notin M_0$. Define b_i for $i < n_0$ as follows.

$$b_i = \begin{cases} b \wedge \bar{p}_i & : \\ mboxif \;\; i \in C & : \\ b \wedge p_i & : \text{ otherwise} \end{cases}$$

Then we have

$$\begin{aligned} m_K(\bigvee_{i<n_0} b_i) &= m_K(b \wedge (\bigvee_{i \in C} \bar{p}_i \vee \bigvee_{i \notin C} p_i)) \\ &= m_K(b \wedge \neg(\bigvee_{i \in C} p_i \wedge \bigwedge_{i \notin C} \bar{p}_i)) \geq m_K(b) - 1 \notin M_0. \end{aligned}$$

Therefore $\bigvee_{i<n_0} b_i \notin I_k$. Hence follows $\exists i < n_0(b_i \notin I_k)$. Since $b_i \leq b \wedge b_i \in D - I_k$, the proof is completed.

Theorem 1.11. *Let G be an $\mathcal{M}$-generic maximal filter over I_k. Then $M[G] \notin M$.*

Proof. Let A and D be defined in Lemma 1.2. By Lemma 1.2 D is dense over I_k. Therefore we have

$$G \cap D \neq \emptyset.$$

Let $b \in G \cap D$

Remark 1.3 (Case 1). $\exists i < n_0(i \in C \wedge b \leq \bar{p}_i)$. In this case $\bar{P}_i \in G$. Therefore $i \notin A$ and $A \neq C$.

Remark 1.4 (Case 2). $\exists i < n_0(i \notin C \wedge b \leq p_i)$

In this case $p_i \in G$. Therefore $i \in A$ and $A \neq C$.

Since $b \in D$, either Case 1) or Case 2) holds. Therefore $A \neq C$. Since $C \in M$ is arbitrary, we can conclude that $A \notin M$ and $M \neq M[G]$.

Remark 1.5. So far we have assumed the definability of K. For general (non definable) $X \subseteq 2^{n_0}$ and $a \in N$, we define $|X|$ as follows.

$$|X| \leq a \quad \text{iff} \quad \forall Y \subseteq X(Y \text{ is definable in } N \to |Y| \leq a)$$

Then we define I_K by the equation

$$I_K = \{b \in B \| \{a \in K | \text{eval}(a,b) = 1\}| < a \quad \text{for all} \quad a \in N - M_0\}.$$

We can generalize our theory for this generalized case.

Let 2^{2h+1} be in M_0. We consider the set $\{1, 2, \cdots, 2^{2h+1} - 1\}$ to be a tree with the height $2h$ i.e. we stipulate that 1 is the root and $2^{2h}, 2^{2h} + 1 - 1$ are leaves. In this tree, we call $1, \cdots, 2^{2h} - 1$ nodes and if a is a node, then $2a$ and $2a + 1$ are called its successors. We also define the height of $2^i, 2^i + 1, \cdots, 2^{i+1} - 1$ to be i.

A function

$$f : \{1, 2, \cdots, 2^{2h+1} - 1\} \to \{\vee, \wedge, 0, 1, p_0, \cdots, p_{n_0-1}, \bar{p}_0, \cdots, \bar{p}_{n_0-1}\}$$

is said to be a formula if the following conditions are satisfied.

1. If a is a node with an even height, then $f(a) = \vee$.
2. If a is a node with an odd height, then $f(a) = \wedge$.
3. If a is a leaf, then $f(a)$ is one of $0, 1, p_0, \cdots, p_{n_0-1}, \bar{p}_0, \cdots, \bar{p}_{n_0-1}$.

Obviously f can be interpreted as a Boolean formula of $p_0, \cdots, p_{n_0-1}$ in the usual sense. E.g. let f be defined on $\{1, 2, \cdots, 6, 7\}$ and $f(4) = p_3, f(5) = \bar{p}_4, f(6) = \bar{p}_5$ and $f(7) = p_6$. then f represents $(p_3 \wedge \bar{p}_4) \vee (\bar{p}_5 \wedge p_6)$.

For a theory of the thus formalized formulas see the discussion on complete normal $(\vee, \wedge)-$ formulas in [15].

Let B_0 be the set of all formulas. We make B_0 a Boolean algebra by defining the operations $\neg, \vee$ and $\wedge$ on B_0 as in [15].

Then we embed B_0 into B in the natural way and consider B_0 to be subalgebra of B.

Now we assume $NC^1 \neq P$, where NC^1 and P are non uniform NC^1 and P respectively. Then we have $B_0 \neq B$. Let $x \in B - B_0$. Define I_x by the equation

$$I_x = \{y \in B \mid \exists z \in B_o(z < x \wedge y \leq z)\}.$$

Then I_x is an M_0-complete ideal of B. We define $\tilde{I}_x = I_x \vee I_{\neg x} = \{z_1 \vee z_2 \mid z_1 \in I_x \text{ and } z_2 \in I_{\neg x}\}$. Then $\tilde{I}_x$ is again M_0-complete and $1 \notin \tilde{I}_x$.

Lemma 1.3. *Let* $C \in M$ *and* $b_0 = \bigwedge_{i \in C} p_i \wedge \bigwedge_{i \notin C} \bar{p}_i$.

Then $b_0 \in \tilde{I}_x$.

Proof. Since b_0 is a minimal nonzero element of B, $b_0 \wedge x = b_0$ or $b_0 \wedge x = 0$.

Remark 1.6 (Case 1). $b_0 \wedge x = b_0$. In this case we have $b_0 \leq x$ and $b_0 \in I_x$.

Remark 1.7 (Case 2). $b_0 \wedge x = 0$. In this case we have $b_0 \leq \neg x$ and $b_0 \in I_{\neg x}$.

Lemma 1.4. *Let $C \in M$ and $b_0 = \bigwedge_{i \in C} p_i \wedge \bigwedge_{i \notin C} \bar{p}_i$.*

If $b \in B - \tilde{I}_n$, then $b \wedge \neg b_0 \in B - \tilde{I}_x$.

Proof. Since $b \leq (b \wedge \neg b_0) \vee b_0$ we have

$$\begin{aligned} (b \wedge \neg b_0) \in \tilde{I}_x &\rightarrow (b \wedge \neg b_0) \vee b_0 \in \tilde{I}_x \\ &\rightarrow b \in \tilde{I}_x \end{aligned}$$

Lemma 1.5. *Let $D = \{b \in B \mid \exists i < n_0(i \in C \wedge b \leq \bar{p}_i) \vee \exists i < n_0(i \notin C \wedge b \leq p_i)\}$. Then D is dense over $\tilde{I}_x$*

Proof. Define b_i by the following equation

$$b_i = \begin{cases} b \wedge \bar{p}_i & : \text{ if } i \in C \\ b \wedge p_i & : \text{ otherwise} \end{cases}.$$

Then we have

$$\bigvee_{i < n_0} b_i = b \wedge \neg b_0 \notin \tilde{I}_x.$$

Therefore we have $\exists i < n_0(b_i \notin \tilde{I}_x)$.

Now let G be an $\mathcal{M}$-generic maximal filter over $\tilde{I}_x$. Then we have $G \notin M$ in the same way as in Theorem 1.11.

2. $M[G]$ and $NP = co - NP$.

In this section we consider M and $M[G]$ as first order structures. Let ψ be a set of formulas with parameters from M.

Let I be an M_0-complete ideal of B and G be an $\mathcal{M}$- generic maximal filter over I.

Definition 2.1. *$M[G]$ is said to be a Ψ-extension of M if for every formula $\varphi(\mathbf{a})$ in Ψ the following property holds.*

$$\forall \mathbf{a} \in M(M \models \varphi(\mathbf{a}) \rightarrow M[G] \models \varphi(\mathbf{a})).$$

When Ψ is the set of all sharply bounded formulas, we denote Ψ-extension by sb-extension. When Ψ is the set of all bounded formulas, we denote Ψ-extension by bounded-extension. The following theorem is immediate from Theorem 1.8 in 1.

Theorem 2.1. *$M[G]$ is a sb-extension of M.*

As we discussed in 1., we can hardly expect that $M[G]$ is a model of S_2 and we conjecture that $M[G]$ is not a model of S_2. In the same way, we conjecture that $M[G]$ is not a bounded-extension of M.

In the following we shall show that our conjectures imply $NP \neq co-NP$ therefore $P \neq NP$.

Theorem 2.2. *If $M[G]$ is not a model of S_2, then $NP \neq co-NP$ and therefore $P \neq NP$.*

Proof. Suppose that $NP = co{-}NP$ holds. Then there exists an NP-complete predicate $\exists x \leq t(a)A(x,a)$ with sharply bounded $A(x,a)$ and a sharply bounded $B(y,a)$ such that $\exists x \leq t(a)A(x,a) \leftrightarrow \forall y \leq s(a)B(y,a)$. Then there exists polynomial time computable functions f and g such that the following two sequents hold.

$$b \leq t(a), c \leq s(a), A(b,a) \rightarrow B(c,a)$$
$$f(a) \leq s(a) \supset B(f(a),a) \rightarrow g(a) \leq t(a) \wedge A(g(a),a).$$

It follows from Theorem 1.8 in 1. that these sequents also hold on $M[G]$. Therefore every bounded formula on $M[G]$ is equivalent to Σ_1^b formula on $M[G]$. This implies that $M[G]$ is a model of S_2, since $M[G]$ is a model of Σ_1^b-part of S_2^1.

Theorem 2.3. *If $M[G]$ is not a bounded-extension of M, then $NP \neq co-NP$.*

Proof. Suppose $NP = co - NP$. Then every bounded formula is equivalent to Π_1^b formula. From the proof of Theorem 2.2 it follows that $NP = co - NP$ also holds on $M[G]$. From Theorem 1.8 in 1. it follows that $M[G]$ is an Π_1^b-extension of M. Therefore $M[G]$ is a bounded-extension which is a contradiction.

Definition 2.2. *A predicate $A(x)$ is said to be sparse, if there exists a term $t(a)$ satisfying the following condition.*

$$\mid \{x \mid A(x) \wedge x < a\} \mid \leq |t(a)|$$

where $\mid \{x \mid \varphi(x)\} \mid$ is the number of all x satisfying $\varphi(x)$. In this definition we are considering some structure e.g. M or $M[G]$ and notions defined on them.

Let $A(x)$ be a formula of S_2. We say that "$A(x)$ is sparse" is provable in S_2, if there exists a term in S_2 and the following formula is provable in S_2.

$$\exists w \leq BdSq(a, t(a))(Seq(w) \wedge \beta(1, w) = \mu x < aA(x)$$
$$\wedge Len(w) = |t(a)|$$
$$\wedge \forall i <| t(a) | (0 < i \supset \beta(i+1, w) = \mu x < a(\beta(i, w) < x \wedge A(x))$$
$$\wedge \forall x < a(A(x) \supset \exists i <| t(a) | (0 < i \wedge x = \beta(i, w)))$$

where BdSq, Seq, $\beta(i, w)$, Len are notations in [6] and the intended meaning of Seq(w), $\beta(i, w)$, Len(w) and BdSq$(a, t(a))$ are "w is a number expressing a sequence", "i-th member of the sequence w", "the length of the sequence w", and an upperbound of all sequences whose members $\leq a$ and whose length $\leq |t(a)|$.

The meaning of the above formula is that one can emunerate all x satisfying $x < a \wedge A(x)$ according to its order. We denote the formula by

$$\exists w \leq \mathrm{BdSq}(a, t(a)) B(w, a).$$

If A is a bounded formula, then B is also a bounded formula.

Theorem 2.4. *Let a bounded formula $A(a)$ be sparse and "$A(a)$ is sparse" be provable in S_2. If $a \in M[G] - M$ and $M[G] \models A(a)$, the $NP \neq co - NP$.*

Proof. Take $b \in M$ such that $a < b$. If $NP = co - NP$ then $M[G] \models S_2$. Therefore we have

$$M[G] \models \exists w \leq \mathrm{BdSq}(b, t(b))(B(w, b) \wedge \exists k < |t(b)|(a = \beta(k, w)).$$

Therefore there exists $k < |t(b)|$ satisfying

$$M[G] \models \exists w \leq \mathrm{BdSq}(b, t(b))(B(w, b) \wedge a = \beta(k, w))$$

Since M is a model of S_2, there exists $c \in M$ satisfying

$$M \models c < b \wedge \exists w \leq \mathrm{BdSq}(b, t(b))(B(w, b) \wedge c = \beta(k, w)).$$

Therefore there exists $w \in M$ satisfying

$$M \models w \leq \mathrm{BdSq}(b, t(b)) \wedge B(w, b) \wedge c = \beta(k, w),$$

If $NP = Co - NP$, then $M[G]$ is a bounded extension of M. Therefore the following holds

$$M[G] \models w \leq \mathrm{BdSq}(b, t(b)) \wedge B(w, b) \wedge c = \beta(k, w).$$

This implies that $c = a$ holds on $M[G]$ which is a contradiction.

3. Proper class forcing.

Now we shall consider a bigger Boolean algebra. The Boolean algebra $\tilde{B}$ is the set of b which is a sequence $(X_0, X_1, \cdots, X_v)$ with $r \in M_0$ satisfying one of the following conditions.

1. X_i is p_j with $j \in M_0$.
2. X_i is $\bar{p}_i$ with $j \in M_0$.
3. X_i is $(\wedge, Y_0, \ldots, Y_s)$ or $(V, Y_0, \ldots, Y_s)$.

where $Y_j(j \leq s)$ is one of $X_0, X_1, \ldots, X_{i-1}$. The difference between B and $\tilde{B}$ is that p_j or $\bar{p}_i$ are restricted to $j < n_0$ in B but there are no such restriction in $\tilde{B}$. Even for $\tilde{B}$. $b \in \tilde{B}$ is $\Delta_1^1(BD)$ and $b \in \tilde{B}$ implies $b \in N$.

We can define $\neg b$, $\bigvee_{i \leq t} b_i$ and $\bigwedge_{i \leq t} b_i$ as before for members b, b_i in $\tilde{B}$.

For every $b \in \tilde{B}$, there exists $\delta \in M_0$ such that if p_i or $\bar{p}_i$ occurs in b, then $i < \delta$. Such δ is called a bound for b. Let δ be a bound for b and $A \in M$ be a subset of $\{0, \ldots, \delta - 1\}$. Then A gives a truth value to $p_0, \ldots, p_{\delta-1}$ as before and is called an atom evaluation of b.

As before we define $b_1 \overset{\tilde{B}}{=} b_2$ for $b_1, b_2 \in \tilde{B}$ by $\forall A$ atom evaluation $(\text{eval}(A, b_1) = \text{eval}(A, b_2))$. We can take only A which is a subset of $\{0, 1, \ldots, \delta - 1\}$ and δ is a bound for both b_1 and b_2. Therefore $b_1 \overset{\tilde{B}}{=} b_2$ is $\Pi_1^1(BD)$ in (M_o, M).

We define $[[\varphi]]$ for $\Sigma_1^1(BD)$ formula φ, $\check{X}$ for $X \in M$, an ideal I of $\tilde{B}$, a dense definable set over I, M_0-completeness of an ideal I, and M in the same way as before.

Now we are going to define M_0 complete ideals $\tilde{I}_0$ and $\tilde{I}_k$ of $\tilde{B}$.

For $\delta \in M_0$, B_δ be the subset of $\tilde{B}$ which consists of the element b whose bound is δ. Then $\tilde{B} = \bigcup_{\delta \in M_0} B_\delta$. Now for $b \in B_\delta$, we define $\tilde{m}(b)$ by

$$\tilde{m}(b) = \frac{|\{a < 2^\delta | \text{eval}(a, b) = 1|\}|}{2^\delta}$$

Then the value $\tilde{m}(b)$ does not depend on δ if δ is bound for b.

We define $\tilde{I}_0$ by

$$\tilde{I}_0 = \{b \in \tilde{B} \mid \forall \alpha \in M_0(\alpha \tilde{m}(b) < 1)\} \quad \text{and} \quad I_\delta = \tilde{I}_0 \cap B_\delta.$$

We are going to show that $\tilde{I}_0 = \bigcup_{\delta \in M_0} I_\delta$ is M_0-complete.

Let $X : y \to \tilde{I}_0$. Then $\forall x < y(X(x) \in \tilde{I}_0)$. Then for every $x < y$, define $\alpha(x)$ to be the minimum α such that $\alpha \tilde{m}(X(x)) \geq 1$. Then $\alpha(y) \notin M_0$. Define $\alpha_0 = \min\{\alpha(x) \mid x < y\} - 1$. Then $\alpha_0 \notin M_0$ and $\forall x < y(\alpha_0 m(X(x)) < 1)$.

For any $\alpha \in M_0$ we have

$$\alpha \tilde{m}(\bigvee_{x<y} X(x)) \leq \alpha \sum_{x<y} \tilde{m}(X(x)) < \alpha y \frac{1}{\alpha_0} < 1.$$

Therefore $\bigvee_{x<y} X(x) \in \tilde{I}_0$.

Now we are going to generalize $\tilde{I}_0$ to $\tilde{I}_k$. Let K be definable in N.

Let $\mu \in N - M$ be fixed. We define

$$K_\mu = \{a \in K \mid a < 2^\mu\}.$$

For $b \in \tilde{B}$, we define $\tilde{m}_K(b)$ by the equation

$$\tilde{m}_K(b) = \frac{\mid \{a \in K_\mu \mid \text{eval}(a,b) = 1\} \mid}{|K_\mu|}$$

and $\tilde{I}_K$ by the equation

$$\tilde{I}_K = \{b \in \tilde{B} | \exists \alpha \notin M_0 (\tilde{m}_K(b) \leq \frac{1}{\alpha})\}$$

For $K = N$, $\tilde{I}_k$ coincides with $\tilde{I}_0$. Now we are going to show that $\tilde{I}_k$ is M_0-complete.

Let $X : y \to \tilde{B}$ and $\forall x < y(X(x) \in \tilde{I}_k)$.

Consider the following value for $x < y$

$$\mid \{a \in K_\mu \mid \text{eval}(a, X(x)) = 1\} \mid.$$

Let m be the maximum of them. Then there exists $\alpha_0 \notin M_0$ such that

$$\frac{m}{|K\mu|} \leq \frac{1}{\alpha_0}$$

Therefore there exists $\alpha \notin M_0$ such that

$$\alpha y \leq \alpha_0.$$

then we have $\tilde{m}_k(\bigvee_{x<y} X(x)) \leq \frac{1}{\alpha}$.

Remark 3.1. As before the definability of K is not necessary. For general $K \subseteq N$, we define

$$\begin{aligned}\tilde{I}_K = \{b \in \tilde{B} \mid \forall V, W \subseteq 2^\mu \exists \alpha \notin M_o (V \subseteq K_\mu \subseteq W \\ \wedge (V, W \quad \text{are definable in} N) \supset \\ \frac{\mid \{a \in V \mid \text{eval}(a,b) = 1\} \mid}{|w|} \leq \frac{1}{\alpha})\}\end{aligned}$$

Everything goes in the same way as in the definable case.

We define $M^{\tilde{B}}, [[\varphi]], \check{X}, \tilde{\mathcal{M}}, G$ etc. in the same way as before. Then the theorems in 1. and 2. can be proved in the same way by just changing B to $\tilde{B}$ and $\mathcal{M}$ to $\tilde{\mathcal{M}}$.

Let $\delta_0, \delta_1 \in M_0$ and $\delta_0 < \delta_1$. We define

$$\Omega(\delta_0, \delta_1) = \{b \in \tilde{B} \mid b = q_{\delta_0} \wedge q_{\delta_0+1} \wedge \ldots \wedge q_{\delta_1 - 1} \text{ where } q_i \text{ is } p_i \text{or } \bar{p}_i\}$$

The following lemma is obvious.

Lemma 3.1. *If $2^{\delta_1 - \delta_0} \in M_0$, then*

$$\forall b \in \Omega(\delta_0, \delta_1)(b \in \tilde{I}_0).$$

Theorem 3.1. *Let G be an $\tilde{\mathcal{M}}$ generic maximal filter over $\tilde{I}_k$. Then we have*

$$\forall b \in \Omega(\delta_0, \delta_1)(b \in \tilde{I}_k) \to A = \{i \mid \delta_0 \le i < \delta_1 \wedge p_i \in G\} \notin M$$

Proof. Let $C \in M$ an $C \subseteq [\delta_0, \delta_1)$.
We define

$$D = \{Y \in \tilde{B} \mid \exists i \in C(Y \le \neg p_i) \vee \exists i \notin C(Y \le p_i)\}.$$

We claim that D is dense over $\tilde{I}_k$. Let $b \in \tilde{B} - \tilde{I}_k$. Then we have

$$\exists \alpha \in M_0(\tilde{m}_K(b) > \frac{1}{\alpha}).$$

Define b_i by the equation

$$b_i = \begin{cases} b \wedge \neg p_i & : \text{ if } i \in C \\ b \wedge p_i & : \text{ otherwise} \end{cases}.$$

Define $C' = [\delta_0, \delta_1) - C$.

$$\begin{aligned} \tilde{m}_K(\bigvee_{\delta_0 \le i < \delta_1} b_i) &= \tilde{m}_K(b \wedge (\bigvee_{i \in C} \bar{p}_i \vee \bigvee_{i \in C'} p_i)) \\ &= \tilde{m}_K(b \wedge \neg(\bigwedge_{i \in C} p_i \wedge \bigwedge_{i \in C'} \bar{p}_i)) \\ &\ge \tilde{m}_K(b) - \tilde{m}_K(\bigwedge_{i \in C} p_i \wedge \bigwedge_{i \in C'} \bar{p}_i) \end{aligned}$$

Since $\bigwedge_{i \in C} p_i \wedge \bigwedge_{i \in C'} \bar{p}_i \in \tilde{I}_k$, we have

$$\exists \alpha \notin M_0(\tilde{m}_K(\bigwedge_{i \in C} Pi \wedge \bigwedge_{i \in C'} \bar{p}_i) < 1/\alpha).$$

Therefore we have $\bigvee_{\delta_0 \le i < \delta_1} b_i \notin \tilde{I}_K$. Since $\tilde{I}_k$ is M_0-complete, we have $\exists i \in [\delta_0, \delta_1)(b_i \notin \tilde{I}_k)$. Since $b_1 \le b$ and $b_1 \in D$, D is dense over $\tilde{I}_K$.

Since G is $\tilde{m}$-generic, $G \cap D \neq \emptyset$. Let $a \in G \cap D$.

Remark 3.2 (Case 1). $\exists i \in C(a \leq \bar{p}_i)$. Then $\bar{p}_i \in G$. Therefore $i \notin A$ and $A \neq C$.

Remark 3.3 (Case 2). $\exists i \in C'(a \leq p_i)$. In this case $p_i \in G$ and $i \in A$. Therefore $A \neq C$. Therefore $A \neq C$ for any $C \in M$. Therefore $A \notin M$.

References

1. M. Ajtai, *The complexity of the pigeon hole principle,* Proc. IEEE 29th Annual Symp. Foundation of Computer Science, 1988, 346-355,
2. M. Ajtai, *Parity and the pigeon hole principle,* Feasible Mathematics, editors: S.R. Buss and P.J. Scott, Birkhauser, 1-24 1990
3. M. Ajtai, *The independence of the modulo p counting principles,* Proc. of the 26th Annual ACM Symp. on Theory of Computing, 402-417, ACM Press, 1994,
4. P. Beame, R. Impagliazo, J. Krajíček, T. Pitassi, and P. Pudlák, Lower bounds on Hilbert's Nullstellensatz and, propositional proofs, to appear.,
5. P. Beame, R. Impagaliazzo, J. Krajíček, T. Pitassi, and P. Pudlák and A. Woods, *Exponential lower bounds for the pigeon hole principle,* Proc. of the 24th Annual ACM Symp. on Theory of Computing, 200-221, ACM Press, 1992,
6. S. Buss, Bounded Arithmetic, Bibliopolis, Napoli, 1986,
7. K. Gödel, *A letter to von Neumann,* Arithmetic, Proof Theory, and Computational Complexity, editors: P. Clote and J. Krajíček, Oxford University Press, 1993,
8. J. Krajíček, *On Frege and Extended Frege Proof Systems,* Feasible Mathematics II., editors: P. Clote and J.B. Remmel, Birkhaüser, 1995, 284-319,
9. J. Krajíček, *Bounded, Propositional Logic, and Complexity Theory,* Cambridge University Press, 1995,
10. J.B. Paris and A. Wilkie, *Counting problems in bounded arithmetic,* Methods in Mathematical Logic, LNM 1130, 317-340, Springer Verlag, 1985,
11. S. Riis, *Making Infinite Structures Finite in Models of Second Order Bounded Arithmetic,* Arithmetic, Proof Theory, and Computational Complexity, editors: P. Clote and J. Krajíček, Oxford University Press, 289-319,
12. G. Takeuti, Two Applications of Logic to Mathematics, Princeton University Press, 1978,
13. G. Takeuti, *RSUV Isomorphisms,* Arithmetic, Proof Theory, and Computational Complexity, editors: P. Clote and J. Krajíček, Oxford University Press, 364-386,
14. G. Takeuti, *RSUV Isomorphisms for* TAC^i, TNC^i *and* TLS, Arch. Math. Logic, 427-453, 1995,
15. G. Takeuti, *Frege proof System and* TNC^0, to appear in J. Symbolic Logic.

Uniform Interpolation and Layered Bisimulation *

Albert Visser

Department of Philosophy, University of Utrecht, Heidelberglaan 8, 3584CS Utrecht

Summary. In this paper we give perspicuous proofs of Uniform Interpolation for the theories IPC, K, GL and S4Grz, using bounded bisimulations. We show that the uniform interpolants can be interpreted as propositionally quantified formulas, where the propositional quantifiers get a semantics with bisimulation extension or bisimulation reset as the appropriate accessibility relation. Thus, reversing the conceptual order, the uniform interpolation results can be viewed as quantifier elimination for bisimulation extension quantifiers.

1. Introduction

Bisimulation and bounded bisimulation can be used to 'visualize' proofs. The aim of this paper is to present proofs for uniform interpolation results as clearly and perspicuously as possible using bounded bisimulation. Ordinary interpolation for a given theory T says that if $\mathsf{T} \vdash A \to B$, then there is a formula $I(A, B)$ in the language containing only the shared propositional variables, say $\mathbf{q}$, such that $\mathsf{T} \vdash A \to I$ and $\mathsf{T} \vdash I \to B$. Uniform interpolation is a strengthening of ordinary interpolation in which the data in terms of which the interpolant is to be specified are weaker: the interpolant can be found from either A and $\mathbf{q}$ or from $\mathbf{q}$ and B. Thus, if uniform interpolation holds, there is, for every A and $\mathbf{q}$, a 'post-interpolant' $I(A, \mathbf{q})$ such that $\mathsf{T} \vdash A \to I(A, \mathbf{q})$ and, for all B such that $\mathsf{T} \vdash A \to B$ and such that the shared propositional variables of A and B are among $\mathbf{q}$, we have $\mathsf{T} \vdash I(A, \mathbf{q}) \to B$. Similarly for the 'pre-interpolant'. As we will see, uniform interpolation can be viewed as quantifier elimination for certain propositional quantifiers in T: the quantifiers that correspond to the trans-model accessibility relation *bisimulation extension* (or: *bisimulation reset*).

In this paper we prove Uniform Interpolation for IPC (Intuitionistic Propositional Calculus), for K, for GL (Löb's Logic) and for S4Grz.[1] Uni-

* The present paper is in its final form and no similar paper has been or is being submitted elsewhere.

[1] Uniform Interpolation for IPC was first proved by Pitts using proof theoretical methods. It was proved by the present method by Ghilardi and Zawadowski and, independently but later, by the author. Uniform Interpolation for K is due to Ghilardi. Uniform Interpolation for GL was first proved by Shavrukov. It was proved by the present method by the author. To give the due credit it should be pointed out that the method here is similar to the one used by Ghilardi and Zawadowski and, independently, the author, to prove the result for IPC. The result for S4Grz is, as far as I know, new in this paper.

form interpolation for S4Grz is rather surpising, since it fails for the closely related theory S4, as was shown by Ghilardi and Zawadowski in their paper [4].

2. Models

We start with introducing the notion of Kripke model and specifying some notations. A (Kripke) model is a structure $\mathbb{K} = \langle K, \prec, \models, \mathcal{P} \rangle$. Here:

- K is a non-empty set of nodes
- $\prec$ is a binary relation on K
- $\mathcal{P}$ is a (possibly empty) set of propositional variables[2]
- $\models$ is a relation between K and $\mathcal{P}$

We can, alternatively, view a model $\mathbb{K}$ as a function that assigns to a fixed set of pairwise disjoint labels $\{\underline{K}, \underline{\preceq}, \underline{\models}, \underline{\mathcal{P}}\}$ the appropriate objects. In this style we will write e.g. $\mathcal{P}_{\mathbb{K}}$ for $\mathbb{K}(\underline{\mathcal{P}})$. We will say that $\mathbb{K}$ is a $\mathcal{P}$-model if $\mathcal{P}_{\mathbb{K}} = \mathcal{P}$. Similarly for $K, \models$-model, etcetera. Similar conventions will be employed for other kinds of models. Define: $\mathsf{PV}_{\mathbb{K}}(k) := \{p \in \mathcal{P}_{\mathbb{K}} \mid k \models_{\mathbb{K}} p\}$. Note that $\models$ and PV are interdefinable. $\mathbf{p}, \mathbf{q}, \mathbf{r}$ will range over finite sets of propositional variables. A model $\mathbb{K}$ is finite if both $K_{\mathbb{K}}$ and $\mathcal{P}_{\mathbb{K}}$ are finite. We will call the class of models Mod.

It is often pleasant to think in terms of *a node in a model*. It is worthwile to make this notion explicit. A *pointed* model is a structure $\mathbb{K} = \langle \mathbb{K}_0, k \rangle$, where $\mathbb{K}_0$ is a model, and k is a node of $\mathbb{K}_0$. A pointed model $\langle \mathbb{K}, b \rangle$ is called *rooted* if for all $k \in K$: $b \prec^* k$[3] b is called the *root*. We can confuse a class of models with its disjoint union, taking as new nodes the pointed models corresponding to the models of the class. We define, e.g., $\langle \mathbb{K}, k \rangle \prec \langle \mathbb{K}', k' \rangle :\Leftrightarrow \mathbb{K} = \mathbb{K}'$ and $k \prec_{\mathbb{K}} k'$. Thus, we can confuse a pointed model $\langle \mathbb{K}, k \rangle$ with a 'free floating' node k. Note that the disjoint union of all models is not strictly speaking a model in our sense. The set of popositional variables that is declared to be present need not be constant in different 'nodes'. It is essential for our purposes for this to be so, since we want to study transitions between nodes in different models that do not leave the set of variables present constant. The totality of pointed models will be called Pmod and the totality of rooted models Rmod.

Suppose $\mathbb{K}$ is a —possibly pointed— $\mathcal{P}$-model. Then $\mathbb{K}[\mathcal{Q}]$ is the $\mathcal{P} \cap \mathcal{Q}$-model obtained by restricting $\models_{\mathbb{K}}$ to $\mathcal{P} \cap \mathcal{Q}$. For any $k \in K$, $\mathbb{K}[k]$ is the rooted model $\langle K', k, \prec', \models', \mathcal{P} \rangle$, where $K' := {\uparrow} k := \{k' \in K \mid k \prec^* k'\}$ and where $\prec'$ and $\models'$ are the restrictions of $\prec$ respectively $\models$ to K'. (We will often

[2] We take the set of propositional variables as 'internal' to the models (and the languages), because we want to think about model extensions, which involve changing the set of variables of the model.

[3] $\prec^*$ is the transitive reflexive closure of $\prec$.

simply write $\prec$ and $\models$ for $\prec'$ and $\models'$.) In case we are using the convention of confusing a node k with its pointed model, $\langle \mathbb{K}, k\rangle$, we will, e.g., write $k[Q]$ for $\langle \mathbb{K}[Q], k\rangle$.

We will consider several properties of models. $\mathbb{K}$ will be said to be *transitive* if $\prec_{\mathbf{K}}$ is transitive, etcetera. $\mathbb{K}$ is *persistent* if $\mathsf{PV}_{\mathbf{K}}$ is monotonic w.r.t. $\prec_{\mathbf{K}}$ and $\subseteq$.

It will be convenient to extend the natural numbers ω with an extra element ∞. Let ω^∞ be $\omega \cup \{\infty\}$. We let $\alpha, \beta, \ldots$ range over ω^∞. ω^∞ is equipped with the obvous ordering $\leq$. We extend addition by: $\infty + \alpha = \alpha + \infty = \infty$. We extend cut-off substraction in our structure by: $\infty - n = \infty$. We will avoid the question of what $\infty - \infty$ is.

Transitive models are going to play a special role in this paper so we will need some some special notions concerned with transitive models. Consider any *transitive* model $\mathbb{K}$. Define:

- $k \prec^+ k' :\Leftrightarrow k \prec k'$ and not $k' \prec k$
- $k \approx k' :\Leftrightarrow k = k'$ or ($k \prec k'$ and $k \prec k'$). So $\approx$ means *being in the same cluster.*
- $d_{\mathbf{K}}(k) := \sup(\{(d_{\mathbf{K}}(k')+1) \in \omega^\infty \mid k' \succ^+ k\})$
- If $\mathbb{K}$ is pointed with designated node k, we put: $d(\mathbb{K}) := d_{\mathbf{K}}(k)$

Note that if $k \prec^+ k'$, then $d_{\mathbf{K}}(k') < d_{\mathbf{K}}(k)$. k is a top node if it is a top node w.r.t. $\prec^+$. Note that k is a top node precisely if $d_{\mathbf{K}}(k) = 0$.

3. Layered Bisimulation

In this section we introduce bisimulation and bounded bisimulation. To avoid formulating most definitions and theorems twice —once for bounded and once for ordinary bisimulation— we make use of a portmanteau notion: *layered bisimulation.*[4]

Consider $\mathcal{P}$-models $\mathbb{K}$ and $\mathbb{M}$. We write $K := K_{\mathbf{K}}$ and $M := K_{\mathbf{M}}$. A *layered bisimulation* or *ℓ-bisimulation* $\mathcal{Z}$ between $\mathbb{K}$ and $\mathbf{M}$ is a ternary relation between K, ω^∞ and M, satisfying the conditions specified below. We will consider $\mathcal{Z}$ also as an ω^∞-indexed set of binary relations between K and M writing $k\mathcal{Z}_\alpha m$ for $\langle k, \alpha, m\rangle \in \mathcal{Z}$. We often write $k\mathcal{Z}m$ for $k\mathcal{Z}_\infty m$. We give the conditions:

1. $k\mathcal{Z}_\alpha m \Rightarrow \mathsf{PV}_{\mathbf{K}}(k) = \mathsf{PV}_{\mathbf{M}}(m)$
2. $k' \succ_{\mathbf{K}} k\mathcal{Z}_{\alpha+1} m \Rightarrow$ there is an m' with $k'\mathcal{Z}_\alpha m' \succ_{\mathbf{M}} m$;
   i.o.w. $\succ_{\mathbf{K}} \circ \mathcal{Z}_{\alpha+1} \subseteq \mathcal{Z}_\alpha \circ \succ_{\mathbf{M}}$.

[4] Bisimulation is used both in computer science and modal logic. See e.g. the papers in [11] for an impression. In model theory bisimulation and bounded bisimimulation appears in the guise of Ehrenfeucht games and back-and-forth equivalence. See [7].

3. $k\mathcal{Z}_{\alpha+1}m \prec_{\mathbf{M}} m' \Rightarrow$ there is a k' with $k \prec_{\mathbf{K}} k'\mathcal{Z}_\alpha m'$;
   i.o.w. $\mathcal{Z}_{\alpha+1} \circ \prec_{\mathbf{M}} \subseteq \prec_{\mathbf{K}} \circ \mathcal{Z}_\alpha$

Note that we allow ℓ-bisimulations to be undefined on some nodes. They may even be empty. Note also that ℓ-bisimulations occur only between models for the same set of variables. We call (2) the *zig*$_{\alpha+1}$*-property* (see Fig. 3.1) and (3) the *zag*$_{\alpha+1}$*-property*. If $\alpha = \infty$ we simply speak of the zig- and the zag-property. A binary relation $\mathcal{Z}$ between $\mathbb{K}$ and $\mathbb{M}$ is a *bisimulation* between

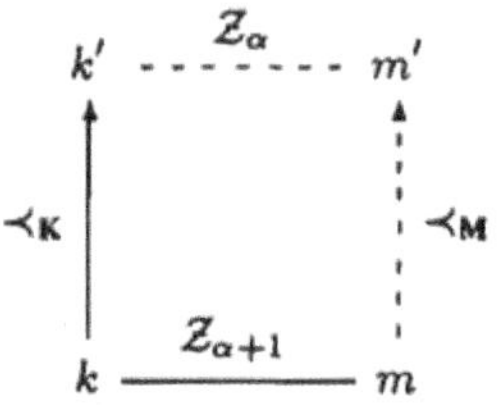

Fig. 3.1. *The zig*$_{\alpha+1}$*-property*

$\mathbb{K}$ and $\mathbb{M}$ iff $\{\langle k, \infty, m\rangle \mid k\mathcal{Z}m\}$ is an ℓ-bisimulation. We will simply confuse bisimulations $\mathcal{Z}$ with the corresponding ℓ-bisimulations. An ℓ-bisimulation $\mathcal{Z}$ is a *bounded bisimulation* if for some natural number n: $k\mathcal{Z}_\alpha m \Rightarrow \alpha \leq n$.

Let $\mathsf{ID}_{\mathbf{K}} := \{\langle k, \alpha, k\rangle \mid k \in K,\ \alpha \in \omega^\infty\}$. Suppose $\mathcal{Z}$ is an ℓ-bisimulation between $\mathbb{K}$ and $\mathbb{M}$ and that $\mathcal{U}$ is an ℓ-bisimulation between $\mathbb{M}$ and $\mathbb{N}$. We define $\mathcal{Z} \circ \mathcal{U}$ by: $(\mathcal{Z} \circ \mathcal{U})_\alpha := \mathcal{Z}_\alpha \circ \mathcal{U}_\alpha$, and $\widehat{\mathcal{Z}}$ by $(\widehat{\mathcal{Z}})_\alpha := \widehat{(\mathcal{Z}_\alpha)}$, where $\widehat{(.)}$ is the usual inverse on binary relations. $\mathcal{Z}^\alpha$ is the relation given by: $\mathcal{Z}^\alpha_\beta := \mathcal{Z}_{\alpha+\beta}$. We say that $\mathcal{Z}$ is *downward closed* if for all $\alpha \prec \beta$: $\mathcal{Z}_\beta \subseteq \mathcal{Z}_\alpha$, The *downward closure* $\mathcal{Z}{\downarrow}$ of $\mathcal{Z}$ is the smallest downwards closed relation extending $\mathcal{Z}$. In the following theorem we collect the necessary elementary facts.

Theorem 3.1. *1.* $\mathsf{ID}_{\mathbf{K}}$ *is an ℓ-bisimulation.*

2. $\mathcal{Z} \circ \mathcal{U}$ *is an ℓ-bisimulation between* $\mathbb{K}$ *and* $\mathbb{N}$.
3. $\widehat{\mathcal{Z}}$ *is an ℓ-bisimulation between* $\mathbb{M}$ *and* $\mathbb{K}$.
4. $\mathcal{Z}^\alpha$ *is an ℓ-bisimulation.*
5. *The downward closure of* $\mathcal{Z}$ *is an ℓ-bisimulation.*
6. *Suppose* Z *is a set of ℓ-bisimulations between* $\mathbb{K}$ *and* $\mathbb{M}$. *Then* $\bigcup \mathsf{Z}$ *is again an ℓ-bisimulation between* $\mathbb{K}$ *and* $\mathbb{M}$. *It follows that there is always a maximal ℓ-bisimulation,* $\simeq^{\mathbf{K},\mathbf{M}}$ *between two models. (1)-(5) imply that for any* α*:*
   - $\mathsf{ID}_{\mathbf{K}} \subseteq\ \simeq^{\mathbf{K},\mathbf{M}}$
   - $\simeq^{\mathbf{K},\mathbf{M}} \circ \simeq^{\mathbf{M},\mathbf{N}} \subseteq\ \simeq^{\mathbf{K},\mathbf{N}}$
   - $\widehat{\simeq^{\mathbf{K},\mathbf{M}}} \subseteq\ \simeq^{\mathbf{M},\mathbf{K}}$
   - $\simeq^{\mathbf{K},\mathbf{M}}$ *is downward closed.*

   Note that, by the above, each of the $\simeq^{\mathbf{K},\mathbf{M}}_\alpha$ *is an equivalence relation.*

7. *Consider $k \in K$ and $m \in M$. Let $\mathcal{Z}[k,m]$ be the restriction of $\mathcal{Z}$ to $\uparrow k \times \uparrow m$. Then $\mathcal{Z}[k,m]$ is an ℓ-bisimulation between $\mathbb{K}[k]$ and $\mathbb{M}[m]$.*
8. *Consider two transitive models $\mathbb{K}$ and $\mathbb{M}$. Consider the relation $\mathcal{W}$, given by:*

$$k\mathcal{W}_\alpha m :\Leftrightarrow \text{for some } k', m' : k \approx k'\mathcal{Z}_\alpha m' \approx m \text{ and } k \simeq_0 m.$$

We have: $\mathcal{W}$ is an ℓ-bisimulation. It follows, e.g., taking $\mathbb{M} := \mathbb{K}$ and $\mathcal{Z} := \mathsf{ID}_K$, that $\approx \cap \simeq_0$ is an ℓ-bisimulation on $\mathbb{K}$.

We will often drop the superscript of $\simeq^{\mathbf{K,M}}$ In case $\alpha = \infty$, we will drop the subscript of $\simeq_\alpha^{\mathbf{K,M}}$ (if no confusion is possible). We will say that k and m (considered as pointed models) *n-simulate* if $k \simeq_n m$ and that k and m *bisimulate* if $k \simeq m$. The following theorem tells us that the number of $\simeq_n$-equivalence classes on a model has a fixed finite bound that only depends on n.

Theorem 3.2. *Define $F(N,0) := 2^N$, $F(N,n+1) := 2^{F(N,n)+N}$. Suppose $|\mathcal{P}| = \mathcal{N}$, then the number of possible $\simeq_n$ equivalence classes is smaller or equal to $F(N,n)$.*

Proof. By a simple induction on n, noting that the $n+1$-equivalence class of a node k is fully determined by the atoms forced in k and the n-equivalence classes of the nodes 'seen' by k. □

In this paper we are particularly interested in things like extending or even changing the forcing of the propositional variables on nodes. We introduce the relevant notions. Let $k, k', m, m' \ldots$ be pointed models.

- $k \simeq_{\alpha,\mathcal{Q}} m :\Leftrightarrow \mathcal{P}_k \cap \mathcal{Q} = \mathcal{P}_m \cap \mathcal{Q}$ and $k[\mathcal{Q}] \simeq_\alpha m[\mathcal{Q}]$. So, roughly, this means that k and m α-bisimulate w.r.t. the variables in $\mathcal{Q}$.
- $k \simeq_{\alpha,[\mathcal{Q}]} m :\Leftrightarrow k \simeq_{\alpha,\mathcal{Q}^c} m$ and $\mathcal{Q} \subseteq \mathcal{P}_m$. So, roughly, this means that k differs from m *modulo* $\simeq_\alpha$ only at $\mathcal{Q}$ and m is at least a $\mathcal{Q}$-node. We will say that m is a *$\mathcal{Q},\alpha$-bisimulation reset* of k. In case $\alpha = \infty$, we will speak of a *$\mathcal{Q}$-bisimulation reset.*
- $k \sqsubseteq_{\alpha,\mathcal{Q}} m :\Leftrightarrow k \simeq_{\alpha,\mathcal{P}_k} m$ and $\mathcal{Q} \cap \mathcal{P}_k = \emptyset$ and $\mathcal{Q} \cup \mathcal{P}_k = \mathcal{P}_m$. We will say that m is a *$\mathcal{Q},\alpha$-bisimulation extension* of k. In case $\alpha = \infty$, we will speak of a *$\mathcal{Q}$-bisimulation extension.*

If we are studying persistent models it is often more natural to think in terms of certain orderings related to layered bisimulation, than in terms of layered bisimulation itself. We can think of these orderings as a kind of extension of the ordering in the model. For the rest of this section we think about persistent pointed $\mathcal{P}$-models. We let $k, k', m, m' \ldots$ range over such models.

- $k \preceq_0 m :\Leftrightarrow \mathsf{PV}(k) \subseteq \mathsf{PV}(m)$
- $k \preceq_{\alpha+1} m :\Leftrightarrow \mathsf{PV}(k) \subseteq \mathsf{PV}(m)$, $\forall m' \succ m\, \exists k' \succ k'\; k' \simeq_\alpha m'$

In case $\alpha = \infty$, we will drop the subscript.

Theorem 3.3. *1.* $\preceq_\alpha$ *is a partial preordering on pointed, persistent* $\mathcal{P}$-*models.*

2. $k \prec k' \Rightarrow k \preceq_\alpha k'$.

3. $\alpha \leq \beta \Rightarrow \preceq_\beta \subseteq \preceq_\alpha$.

4. $k \simeq_\alpha m \Leftrightarrow k \preceq_\alpha m$ *and* $m \preceq_\alpha k$.

5. $k \preceq_\infty m \Leftrightarrow$ *for some* $k' \succeq k$ $k' \simeq m$.

Proof. We prove (4). For $\alpha = 0$ this is easy. Suppose $\alpha > 0$. "$\Rightarrow$" Easy. "$\Leftarrow$" Suppose $k \preceq_\alpha m$ and $m \preceq_\alpha k$. We show that $\mathcal{U} :=\simeq \cup \{\langle k, \alpha, m\rangle\}$ is an ℓ-bisimulation, and, hence, that $k \simeq_\alpha m$. Clearly $\mathsf{PV}(k) = \mathsf{PV}(m)$. The zig-property for $\mathcal{U}$ follows from the fact that $m \preceq_\alpha k$. The zag-property for $\mathcal{U}$ follows from the fact that $k \preceq_\alpha m$. □

4. Basic Facts for IPC

In this section, $\vdash$ will stand for derivability in IPC. Consider any set of propositional variables, $\mathcal{P}$. We define $\mathcal{L}^i(\mathcal{P})$ as the smallest set such that:

- $\mathcal{P} \subseteq \mathcal{L}^i(\mathcal{P})$, $\bot, \top \in \mathcal{L}^i(\mathcal{P})$
- if $A, B \in \mathcal{L}^i(\mathcal{P})$, then $(A \wedge B), (A \vee B), (A \to B) \in \mathcal{L}^i(\mathcal{P})$.

$\mathsf{PV}(A)$ is the set of propositional variables occurring in A. $\mathsf{Sub}(A)$ is the set of subformulas of A. A model is an IPC-*model* if it is transitive, reflexive, antisymmetric and persistent. In this section all models will be IPC models. Consider a $\mathcal{P}$-model $\mathbb{K}$ we take $\models_i$ to be the smallest relation between K and $\mathcal{L}^i(\mathcal{P})$ such that:

- $k \models_i p :\Leftrightarrow k \models p$, $k \models_i \top$
- $k \models_i A \wedge B :\Leftrightarrow k \models_i A$ and $k \models_i B$
- $k \models_i A \vee B :\Leftrightarrow k \models_i A$ or $k \models_i B$
- $k \models_i A \to B :\Leftrightarrow \forall k' \succ k\, (k' \models_i A \Rightarrow k' \models_i B)$

We will omitt the subscript i, as long as it is sufficiently clear from the context that the persistent case is intended. Note that, by transitivity, the persistence for $\mathcal{P}$ extends to the persistence for $\mathcal{L}^i(\mathcal{P})$. Define further:

- $k \models \Gamma :\Leftrightarrow$ for all $A \in \Gamma : k \models A$
- $\mathbb{K} \models A :\Leftrightarrow$ for all $k \in K$ $k \models A$

A set X is $\mathcal{P}$-*adequate* if $X \subseteq \mathcal{L}^i(\mathcal{P})$ and X is closed under subformulas. A set Γ is X-saturated (for IPC) if for any subset Y of X: $\Gamma \vdash \bigvee Y \Rightarrow Y \cap \Gamma \neq \emptyset$. Note that it follows that Γ is consistent (the case that Y is $\emptyset$) and that Γ is closed under X-consequences (the case that Y is a singleton).

We describe the Henkin construction for IPC. To lighten our notational burdens we will assume in this section that we work with some fixed $\mathcal{P}$. Consider a $\mathcal{P}$-adequate set X. The Henkin model for X is the model $\mathbb{H} := \mathbb{H}_X$, where:

- $K_{\mathbb{H}} := \{\Delta \mid \Delta \text{ is } X\text{-saturated}\}$
- $\Gamma \prec \Delta :\Leftrightarrow \Gamma \subseteq \Delta$
- $\mathcal{P}_{\mathbb{H}} := \mathcal{P} \cap X$
- $\Gamma \models p :\Leftrightarrow p \in \Gamma$

It is easily verified that $\mathbb{H}$ is an IPC-model.

Theorem 4.1. *for all* $A \in X$: $\Gamma \models_{\mathbb{H}} A \Leftrightarrow A \in \Gamma$.

If X is finite, then $\mathbb{H}_X$ is finite. We say that $\mathbb{M}$ is a *rooted Henkin model* if it is of the form $\mathbb{H}_X[\Delta]$ for some X-saturated Δ. We have:

Theorem 4.2 (Kripke Completeness). *For* $\Gamma \subseteq \mathcal{L}^i(\mathcal{P})$ *and* $A \in \mathcal{L}^i(\mathcal{P})$:

$$\Gamma \vdash_{\mathcal{P}} A \Leftrightarrow \text{ for all } \mathcal{P}\text{-models } \mathbb{K} : \Gamma \models_{\mathbb{K}} A.$$

In case Γ *is finite, we can improve this to:*

$$\Gamma \vdash_{\mathcal{P}} A \Leftrightarrow \text{ for all finite } \mathcal{P}\text{-models } \mathbb{K} : \Gamma \models_{\mathbb{K}} A.$$

For IPC we have a distinctive result involving downward extensions of models. We first introduce the necessary machinery. Let K be a set of IPC-models. $\mathbb{M} := \mathsf{M}(\mathsf{K})$ is the IPC-model with :

- $M := \{\langle k, \mathbb{K}\rangle \mid k \in K_{\mathbb{K}} \text{ and } \mathbb{K} \in \mathsf{K}\}$
- $\langle k, \mathbb{K}\rangle \prec \langle m, \mathbb{M}\rangle :\Leftrightarrow \mathbb{K} = \mathbb{M} \text{ and } k \prec_{\mathbb{K}} m$
- $\mathcal{P}_{\mathbb{M}} := \bigvee\{\mathcal{P}_{\mathbb{K}} \mid \mathbb{K} \in \mathsf{K}\}$.
- $\langle k, \mathbb{K}\rangle \models p :\Leftrightarrow k \models_{\mathbb{K}} p$

In practice we will forget the second components of the new nodes, pretending the domains to be disjoint already. Let $\mathbb{K}$ be a IPC $\mathcal{P}$-model. $\mathsf{B}(\mathbb{K})$ is the (rooted) IPC $\mathcal{P}$-model obtained by adding a new bottom $\mathfrak{b}$ to $\mathbb{K}$ and by taking: $\mathfrak{b} \models p :\Leftrightarrow \mathbb{K} \models p$. Finally we define $\mathsf{Glue}(\mathsf{K}) := \mathsf{B}(\mathsf{M}(\mathsf{K}))$.

Theorem 4.3 (Push Down Lemma). *Let* X *be adequate. Suppose* Δ *is* X*-saturated and* $\mathbb{K}$ *is an* IPC*-model with* $\mathbb{K} \models \Delta$. *Then* $\mathsf{Glue}(\mathbb{H}_X[\Delta], \mathbb{K}) \models \Delta$.

Proof. We show by induction on $A \in X$ that $\mathfrak{b} \models A \Leftrightarrow A \in \Delta$. The cases of atoms, conjunction and disjunction are trivial. If $(B \to C) \in X$ and $\mathfrak{b} \models (B \to C)$, then $\Delta \models (B \to C)$ and, hence, $(B \to C) \in \Delta$. Conversely suppose $(B \to C) \in \Delta$. If $\mathfrak{b} \not\models B$, we are easily done. If $\mathfrak{b} \models B$, then, by the Induction Hypothesis: $B \in \Delta$, hence $C \in \Delta$ and, by the induction hypothesis: $\mathfrak{b} \models C$. □

Instead of using the Push Down Lemma we could have employed the Kleene slash. We say that Δ is $\mathcal{P}$*-prime* if it is consistent and for every $(C \vee D) \in \mathcal{L}^i(\mathcal{P})$: $\Delta \vdash (C \vee D) \Rightarrow \Delta \vdash C$ or $\Delta \vdash D$. A formula A is $\mathcal{P}$*-prime* if $\{A\}$ is $\mathcal{P}$-prime. As usual, we will suppress the $\mathcal{P}$.

Theorem 4.4. *Suppose* X *is adequate and* Δ *is* X*-saturated. then* Δ *is prime.*

Proof. Δ is consistent by definition. Suppose $\Delta \vdash C \vee D$ and $\Delta \nvdash C$ and $\Delta \nvdash D$. Suppose $\mathbb{K} \models \Delta$, $\mathbb{K} \not\models C$, $\mathbb{M} \models \Delta$ and $\mathbb{M} \not\models D$. Consider $\mathsf{Glue}(\mathbb{H}_X(\Delta), \mathbb{K}, \mathbb{M})$. By the Push Down Lemma (Theorem 4.3) we have: $\mathfrak{b} \models \Delta$. On the other hand by persistence: $\mathfrak{b} \not\models C$ and $\mathfrak{b} \not\models D$. Contradiction. □

We turn to the consideration of fragments and model descriptions for IPC. Define $\mathsf{i} : \mathcal{L}^i(\mathcal{P}) \to \omega$, by:

- $\mathsf{i}(p) := \mathsf{i}(\bot) := \mathsf{i}(\top) := 0$
- $\mathsf{i}(A \wedge B) := \mathsf{i}(A \vee B) := max(\mathsf{i}(A), \mathsf{i}(B))$
- $\mathsf{i}(A \to B) := max(\mathsf{i}(A), \mathsf{i}(B)) + 1$
- $I_n(\mathcal{P}) := \{A \in \mathcal{L}^i(\mathcal{P}) \mid \mathsf{i}(A) \leq n\}$
- $I_\infty(\mathcal{P}) := \mathcal{L}^i(\mathcal{P})$

By an easy induction on n we may prove the following theorem.

Theorem 4.5. *$I_n(\mathrm{p})$ is finite modulo* IPC*-provable equivalence.*

Define for $X \subseteq \mathcal{L}^i(\mathcal{P})$:

- $\mathsf{Th}_X(k) := \{A \in X \mid k \models A\}$
- For $\mathbb{K}$ pointed with point k: $\mathsf{Th}_X(\mathbb{K}) := \mathsf{Th}_X(k)$
- $\mathsf{Th}(k) := \mathsf{Th}_{\mathcal{L}^i(\mathcal{P})}(k)$

Theorem 4.6. *Suppose that $\mathcal{Z}$ is an ℓ-simulation between the $\mathcal{P}$-models $\mathbb{K}$ and $\mathbb{M}$. Then: $k\mathcal{Z}_\alpha m \Rightarrow \mathsf{Th}_{I_\alpha(\mathcal{P})}(k) = \mathsf{Th}_{I_\alpha(\mathcal{P})}(m)$.*

Proof. By induction on A in I_α. Suppose $k\mathcal{Z}_\alpha m$. The cases of atoms, conjunction and disjunction are trivial. Suppose, e.g., $k \not\models (B \to C)$. Then, for some $k' \succ k$, $k' \models B$ and $k' \not\models C$. There is an $m' \succ m$, such that $k'\mathcal{Z}_{\alpha-1}m'$ and hence by the induction hypothesis (applied for $\alpha - 1$, noting that if $A \in I_\alpha(\mathcal{P})$, then $B, C \in I_{\alpha-1}(\mathcal{P})$): $m' \models B$ and $m' \not\models C$. Ergo $m \not\models (B \to C)$. □

Theorem 4.7. $k \preceq_\alpha m \Rightarrow \mathsf{Th}_{I_\alpha(\mathcal{P})}(k) \subseteq \mathsf{Th}_{I_\alpha(\mathcal{P})}(m)$, *for $\mathcal{P}$-nodes k and m.*

Proof. In case $\alpha = 0$, this is trivial. Suppose $\alpha > 0$ and $k \preceq_\alpha m$. The proof is a simple induction on $A \in I_\alpha(\mathcal{P})$. The cases of atoms, $\wedge$, $\vee$ are trivial. Suppose $A = (B \to C)$ and $m \not\models (B \to C)$. Then for some $m' \succeq m$: $m' \models B$ and $m' \not\models C$. There is a $k' \succeq k$, such that $k' \simeq_{\alpha-1} m'$ and, hence, by Theorem 4.6: $k' \models B$ and $k' \not\models C$. Ergo $k \not\models (B \to C)$. □

We formulate a partial converse for Theorem 4.7. It is well known that the converse for the case of ∞, i.e. for the case where one would like to infer bisimulation from the relation of forcing the same formulas of the full language, does not go through. There is a lot of work (for the analogous case of modal logic) on better converses than the one given here. We refer the reader to [6] and [8].

Theorem 4.8. $\mathsf{Th}_{I_n(\mathrm{p})}(k) \subseteq \mathsf{Th}_{I_n(\mathrm{p})}(m) \Rightarrow k \preceq_n m$, *for* p*-nodes k and m.*

Proof. Suppose k and m are **p**-nodes, and $\mathsf{Th}_{I_n(\mathbf{p})}(k) \subseteq \mathsf{Th}_{I_n(\mathbf{p})}(m)$. We want to prove: $k \preceq_n m$. In case $n = 0$ this is trivial. Suppose $n > 0$. Define, for k' in the model corresponding to k and m' in the model corresponding to m:

$$k'\mathcal{Z}_i m' :\Leftrightarrow \mathsf{Th}_{I_i(\mathbf{p})}(k') = \mathsf{Th}_{I_i(\mathbf{p})}(m').$$

We check that $\mathcal{Z}$ is an ℓ-simulation and that for every $k' \succeq k$ there is an $m' \succeq m$ with $k'\mathcal{Z}_n m'$.

Suppose $i > 0$ and $k'\mathcal{Z}_i m'$. Clearly k' and m' force the same atoms. We verify e.g. the zig-property. Suppose $k' \preceq k''$. Let:

$$\eta_i(k'') := (\bigwedge\{B \in I_{i-1}(\mathbf{p}) \mid k'' \models B\} \to \bigvee\{C \in I_{i-1}(\mathbf{p}) \mid k'' \not\models C\}).$$

Clearly $k' \not\models \eta_i(k'')$ and $\eta_i(k'') \in I_i(\mathbf{p})$. Ergo $m' \not\models \eta_i(k'')$. But then for some $m'' \geq m'$:

$$m'' \models \bigwedge\{B \in I_{i-1}(\mathbf{p}) \mid k'' \models B\} \text{ and } m'' \not\models \bigvee\{C \in I_{i-1}(\mathbf{p}) \mid k' \not\models C\}.$$

It follows that $k''\mathcal{Z}_{i-1}m''$.

To show that for any $m' \succeq m$ there is a $k' \succeq k$ with $k'\mathcal{Z}_n m'$. Note that $m \not\models \eta_n(m')$, ergo $k \not\models \eta_n(m')$, and, thus, for some k':

$$k' \models \bigwedge\{B \in I_{n-1}(\mathbf{p}) \mid m' \models B\} \text{ and } k' \not\models \bigvee\{C \in I_{n-1}(\mathbf{p}) \mid m' \not\models C\}.$$

Hence: $k\mathcal{Z}_{n-1}m$. □

Let k be a **p**-node. Define:

- $\mathsf{Y}_{n,k} := \mathsf{Y}_{n,k}(\mathbf{p}) := \bigwedge\{C \in I_n(\mathbf{p}) \mid k \models C\}$
- $\mathsf{N}_{n,k} := \mathsf{N}_{n,k}(\mathbf{p}) := \bigvee\{D \in I_n(\mathbf{p}) \mid k \not\models D\}$

Theorem 4.9. *$k \models \mathsf{Y}_{n,k}$ and $k \not\models \mathsf{N}_{n,k}$.*

Let m be a **p**-node. We have:

Theorem 4.10. $k \preceq_n m \Leftrightarrow m \models \mathsf{Y}_{n,k} \Leftrightarrow k \not\models \mathsf{N}_{n,m}$.

Theorem 4.11. *For $n \leq n'$: $\mathsf{IPC} \vdash \mathsf{Y}_{n',k} \to \mathsf{Y}_{n,k}$ and $\mathsf{IPC} \vdash \mathsf{N}_{n,k} \to \mathsf{N}_{n',k}$.*

Theorem 4.12. $k \preceq_n m \overset{1}{\Leftrightarrow} \mathsf{IPC} \vdash \mathsf{Y}_{n,m} \to \mathsf{Y}_{n,k} \overset{2}{\Leftrightarrow} \mathsf{IPC} \vdash \mathsf{N}_{n,m} \to \mathsf{N}_{n,k}$

Proof. (1) "$\Rightarrow$" Suppose $k \preceq_n m$. Let r be any **p**-node with $r \models \mathsf{Y}_{n,m}$. It follows that $m \preceq_n r$ and, hence, $k \preceq_n r$. Ergo, $r \models \mathsf{Y}_{n,k}$. "$\Leftarrow$" Suppose $\mathsf{IPC} \vdash \mathsf{Y}_{n,m} \to \mathsf{Y}_{n,k}$. Since $m \models \mathsf{Y}_{n,m}$, it follows that $m \models \mathsf{Y}_{n,k}$, and, hence, $k \preceq_n m$.

(2) "$\Rightarrow$" Suppose $k \preceq_n m$. Let r be any **p**-node with $r \not\models \mathsf{N}_{n,k}$. It follows that $r \preceq_n k$ and, hence, $r \preceq_n m$. Ergo: $r \not\models \mathsf{N}_{n,m}$. "$\Leftarrow$" Suppose $\mathsf{IPC} \vdash \mathsf{N}_{n,m} \to \mathsf{N}_{n,k}$. Since $k \not\models \mathsf{N}_{n,k}$, it follows that $k \not\models \mathsf{N}_{n,m}$ and hence: $k \preceq_n m$. □

Theorem 4.13. *$\mathsf{Y}_{n,k}$ is a prime formula.*

Proof. It is easily seen that $\mathsf{Y}_{n,k}$ is $I_n(\mathbf{p})$-saturated. Apply Theorem 4.4. □

5. Uniform Interpolation for IPC

Uniform Interpolation was proved for GL by V. Shavrukov (see: [12]). Shavrukov used the method of characters as developed by Z. Gleit and W. Goldfarb, who proved the Fixed Point Theorem of Provability Logic and the ordinary Interpolation Theorem employing characters (see: [5]). The methods of Gleit & Goldfarb and later of Shavrukov can be viewed as model theoretical. For IPC, A. Pitts proved Uniform Interpolation by proof theoretical methods, using proof systems allowing efficient cut-elimination (see: [10]), developed, independently, by J. Hudelmaier (see: [9]) and R. Dyckhoff (see: [1]). Later S. Ghilardi and M. Zawadowski (see: [3]), and, independently but later, A. Visser, found a model theoretical proof for Pitt's result using bounded bisimulations.

In this section, we will use $\preceq$ for the weak partial orderings and $\prec$ for the associated strict orderings. We prove an amalgamation lemma.

Lemma 5.1. *Consider disjoint sets of propositional variables $\mathcal{Q}$, $\mathbf{p}$ and $\mathcal{R}$. Let $X \subseteq \mathcal{L}^i(\mathcal{Q},\mathbf{p})$ be a finite adequate set. Let $\langle \mathbb{K}, k_0\rangle \in \mathsf{Pmod}(\mathcal{Q},\mathbf{p})$, $\langle \mathbb{M}, m_0\rangle \in \mathsf{Pmod}(\mathbf{p},\mathcal{R})$. Let:*

$$\nu := |\{C \in X \mid C \text{ is a propositional variable or an implication}\}|.$$

Suppose that $k_0 \simeq_{2.\nu+1,\mathbf{p}} m_0$. Then there is a $\mathcal{Q}$-extension $\langle \mathbb{N}, n_0\rangle$ of $\langle \mathbb{M}, m_0\rangle$ such that $\mathsf{Th}_X(n_0) = \mathsf{Th}_X(k_0)$.

Proof. Let $\mathcal{Z}$ be a downwards closed witness of $k_0 \simeq_{2.\nu+1,\mathbf{p}} m_0$. Define Φ_X from $\mathbb{K}$ to the Henkin model $\mathbb{H} := \mathbb{H}_X$ as follows: $\Phi_X(k) := \Delta(k) := \{B \in X \mid k \models B\}$. Define further for k in $\mathbb{K}$: $d_X(k) := d_{\mathbb{H}}(\Delta(k))$. Note that: $d_X(k) \leq \nu$.

Consider a pair $\langle \Delta, m\rangle$ for Δ in $\mathbb{H}$ and m in $\mathbb{M}$. We say that k', k, m' is a *witnessing triple* for $\langle \Delta, m\rangle$ if:

$$\Delta = \Delta(k) = \Delta(k'),\ k' \preceq k,\ m' \preceq m,\ k'\mathcal{Z}_{2.d_X(k')+1}m',\ k\mathcal{Z}_{2.d_X(k')}m.$$

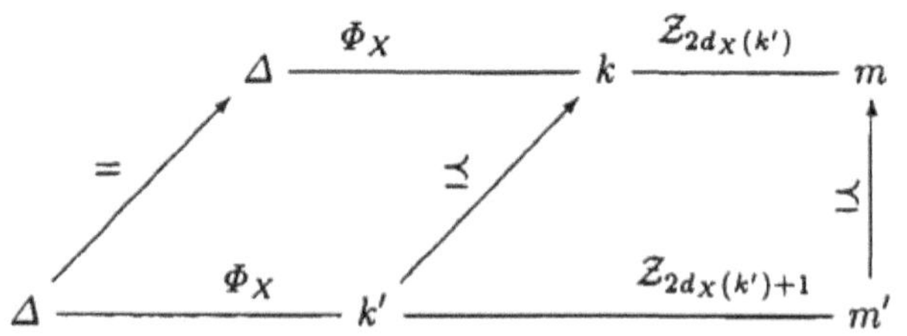

Define:

- $N := \{\langle \Delta, m\rangle \mid$ there is a witnessing triple for $\langle \Delta, m\rangle\}$
- $n_0 := \langle \Delta(k_0), m_0\rangle$
- $\langle \Delta, m\rangle \preceq_{\mathbb{N}} \langle \Gamma, n\rangle :\Leftrightarrow \Delta \preceq_{\mathbb{H}} \Gamma$ and $m \preceq_{\mathbb{M}} n$
- $\langle \Delta, m\rangle \models_{\mathbb{N}} s :\Leftrightarrow \Delta \models_{\mathbb{H}} s$ or $m \models_{\mathbb{M}} s$

Note that by assumption $k_0 \mathcal{Z}_{2\nu+1} m_0$. Moreover: $2.d_X(k_0)+1 \leq 2.\nu+1$. Hence: $k_0 \mathcal{Z}_{2d_X(k_0)+1} m_0$. So we can take k_0, k_0, m_0 as witnessing triple for n_0. Let k', k, m' be a witnessing triple for $\langle \Delta, m\rangle$. Note that for $p \in \mathbf{p} \cap X$: $\Delta \models p \Leftrightarrow k \models p \Leftrightarrow m \models p$, and hence: $\langle \Delta, m\rangle \models p \Leftrightarrow \Delta \models p \Leftrightarrow m \models p$. We claim:

Claim 1 $n_0 \simeq_{\mathbf{p},\mathcal{R}} m_0$.
Claim 2 For $B \in X$: $\langle \Delta, m\rangle \models B \Leftrightarrow B \in \Delta$.

Evidently the lemma is immediate from the claims.

We prove Claim 1. Take as bisimulation $\mathcal{B}$ with $\langle \Delta, m\rangle \mathcal{B} m$. It is evident that $\mathsf{Th}_{\mathbf{p},\mathcal{R}}(\langle \Delta, m\rangle) = \mathsf{Th}_{\mathbf{p},\mathcal{R}}(m)$. Moreover, $\mathcal{B}$ has the zig-property. We check that $\mathcal{B}$ has the zag-property. Suppose $\langle \Delta, m\rangle \mathcal{B} m \preceq n$. We are looking for a pair $\langle \Gamma, n\rangle$ in N such that $\Delta \preceq \Gamma$. Let k', k, m' be a witnessing triple for $\langle \Delta, m\rangle$. Since $k' \mathcal{Z}_{2.d_X(k')+1} m' \preceq n$, there is a h such that $k' \preceq h \mathcal{Z}_{2.d_X(k')} n$. We take $\Gamma := \Delta(h)$. We need a witnessing triple k'^*, k^*, m'^* for $\langle \Gamma, n\rangle$ We distinguish two possibilities. First, $\Delta = \Gamma$. In this case we can take: $k'^* := k'$, $k^* := h$, $m'^* := m'$.

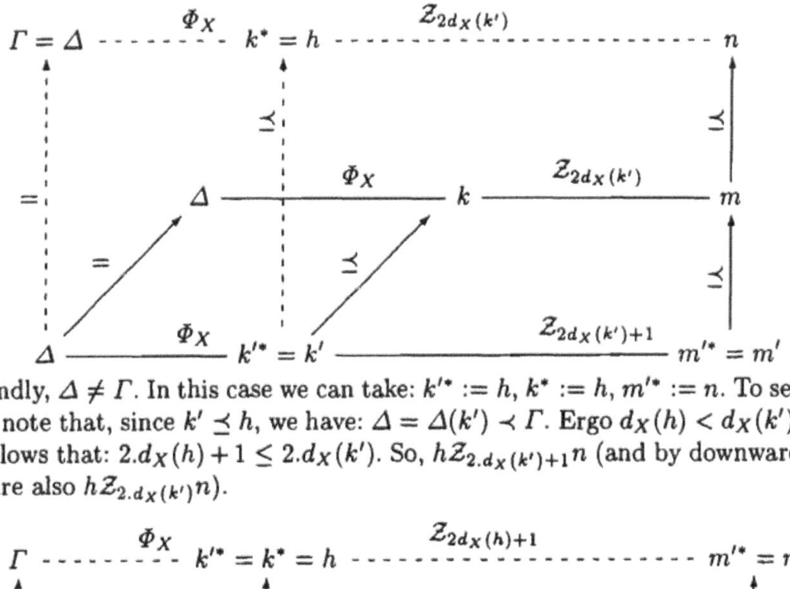

Secondly, $\Delta \neq \Gamma$. In this case we can take: $k'^* := h$, $k^* := h$, $m'^* := n$. To see this, note that, since $k' \preceq h$, we have: $\Delta = \Delta(k') \prec \Gamma$. Ergo $d_X(h) < d_X(k')$. It follows that: $2.d_X(h)+1 \leq 2.d_X(k')$. So, $h \mathcal{Z}_{2.d_X(k')+1} n$ (and by downward closure also $h \mathcal{Z}_{2.d_X(k')} n$).

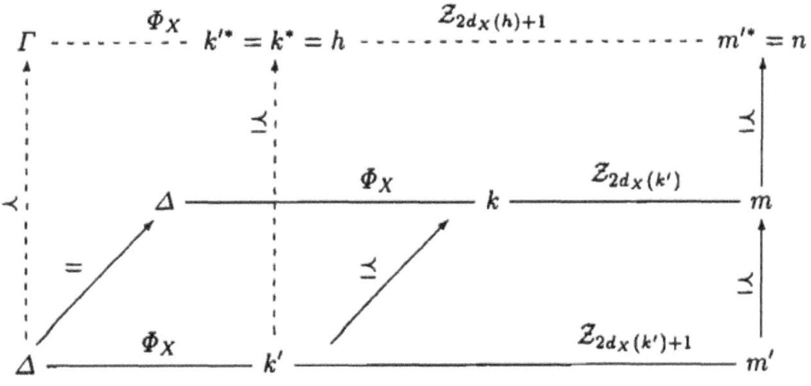

Finally, clearly, $b_{\mathbf{N}} \mathcal{B} b_{\mathbf{M}}$.

We prove Claim 2. The proof is by induction on X. The cases of atoms, conjunction and disjunction are trivial. We treat the case of implication. Suppose $(C \to D) \in X$. Consider the node $\langle \Delta, m \rangle$ with witnessing triple k', k, m'.

Suppose $(C \to D) \notin \Delta$. In case $C \in \Delta$ and $D \notin \Delta$, by the Induction Hypothesis, $\langle \Delta, m \rangle \models C$ and $\langle \Delta, m \rangle \not\models D$. So, $\langle \Delta, m \rangle \not\models (C \to D)$. Suppose $C \notin \Delta$. Clearly, $k \not\models (C \to D)$, so there is an $h \succeq k$ with $h \models C$ and $h \not\models D$. Let $\Gamma := \Delta(h)$. Since, $C \notin \Delta$, we find: $\Delta \prec \Gamma$ and, thus, $k \prec h$. Note that it follows that $2.d_X(k') \geq 2$. Since $k \mathcal{Z}_{2.d_X(k')} m$ and $k \preceq h$, there is an $n \succeq m$ with $h \mathcal{Z}_{2.d_X(k')-1} n$. Moreover: $2.d_X(h) + 1 \leq 2.d_X(k') - 1$. Ergo: $h \mathcal{Z}_{2.d_X(h)+1} n$. So h, h, n is a witnessing triple for $\langle \Gamma, n \rangle$. Clearly, $\langle \Delta, m \rangle \preceq \langle \Gamma, n \rangle$. By the Induction Hypothesis: $\langle \Gamma, n \rangle \models C$ and $\langle \Gamma, n \rangle \not\models D$. Hence, $\langle \Delta, m \rangle \not\models (C \to D)$.

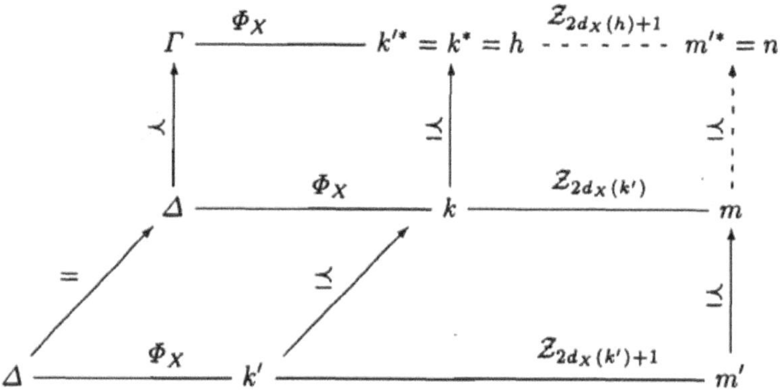

Suppose $\langle \Delta, m \rangle \not\models (C \to D)$. There is a $\langle \Gamma, n \rangle$ in $\mathbf{N}$ with $\langle \Delta, m \rangle \preceq \langle \Gamma, n \rangle$ and $\langle \Gamma, n \rangle \models C$ and $\langle \Gamma, n \rangle \not\models D$. Clearly $\Delta \preceq \Gamma$. By the Induction Hypothesis $C \in \Gamma$ and $D \notin \Gamma$. Ergo $(C \to D) \notin \Delta$. Thus we have proved Claim 2. □

Theorem 5.1 (Pitts' Uniform Interpolation Theorem). *Here is our version of Pitts' Uniform Interpolation Theorem.*

1. *Consider any formula A and any finite set of variables $\mathbf{q}$. Let*

   $$\nu := |\{C \in \mathsf{Sub}(A) \mid C \text{ is a propositional variable or an implication}\}|$$

   There is a formula $\exists \mathbf{q}.A$ such that:
   a) $\mathsf{PV}(\exists \mathbf{q}.A) \subseteq \mathsf{PV}(A) \setminus \mathbf{q}$
   b) $\mathsf{i}(\exists \mathbf{q}.A) \leq 2.\nu + 2$
   c) For all $B \in \mathcal{L}^i$ with $\mathsf{PV}(B) \cap \mathbf{q} = \emptyset$, we have:

   $$\mathsf{IPC} \vdash A \to B \Leftrightarrow \mathsf{IPC} \vdash \exists \mathbf{q}.A \to B.$$

2. *Consider any formula B and any finite set of variables $\mathbf{q}$. Let $\nu := \nu_{\mathsf{Sub}(B)}$. There is a formula $\forall \mathbf{q}.B$ such that:*

a) $\mathsf{PV}(\forall \mathbf{q}.B) \subseteq \mathsf{PV}(B) \setminus \mathbf{q}$
b) $\mathrm{i}(\forall \mathbf{q}.B) \leq 2.\nu + 1$
c) For all $A \in \mathcal{L}^i$ *with* $\mathsf{PV}(A) \cap \mathbf{q} = \emptyset$*, we have:*

$$\mathsf{IPC} \vdash A \to B \Leftrightarrow \mathsf{IPC} \vdash A \to \forall \mathbf{q}.B.$$

Proof. (1) Consider A and $\mathbf{q}$. Let $\mathbf{p} := \mathsf{PV}(A) \setminus \mathbf{q}$. Take:

$$\exists \mathbf{q}.A := \bigwedge \{C \in I_{2.\nu+2}(\mathbf{p}) \mid \mathsf{IPC} \vdash A \to C\}.$$

Clearly $\exists \mathbf{q}.A$ satisfies (a) and (b). Moreover, $\mathsf{IPC} \vdash A \to \exists \mathbf{q}.A$. Hence all we have to prove is that for all B with $\mathsf{PV}(B) \cap \mathbf{q} = \emptyset$:

$$\mathsf{IPC} \vdash A \to B \Rightarrow \mathsf{IPC} \vdash \exists \mathbf{q}.A \to B.$$

Suppose, to the contrary, that for some B: $\mathsf{PV}(B) \cap \mathbf{q} = \emptyset$ and $\mathsf{IPC} \vdash A \to B$ and $\mathsf{IPC} \nvdash \exists \mathbf{q}.A \to B$. Take $\mathbf{r} := \mathsf{PV}(B) \setminus \mathbf{p}$. Note that $\mathbf{p}, \mathbf{q}, \mathbf{r}$ are pairwise disjoint, $\mathsf{PV}(A) \subseteq \mathbf{q} \cup \mathbf{p}$ and $\mathsf{PV}(B) \subseteq \mathbf{p} \cup \mathbf{r}$.

Let m be any $\mathbf{p}, \mathbf{r}$-node with $m \models \exists \mathbf{q}.A$ and $m \not\models B$. Let $\mathsf{Y} := \mathsf{Y}_{2.\nu+1, m[\mathbf{p}]}$ and $\mathsf{N} := \mathsf{N}_{2.\nu+1, m[\mathbf{p}]}$ (see Sect. 4.). We claim that: $A, \mathsf{Y} \nvdash \mathsf{N}$. If it did, we would have: $A \vdash \mathsf{Y} \to \mathsf{N}$. And hence by definition: $\exists \mathbf{q}.A, \mathsf{Y} \vdash \mathsf{N}$. Quod non, since $m \models \exists \mathbf{q}.A, \mathsf{Y}$ and $m \not\models \mathsf{N}$. Let k be any $\mathbf{q}, \mathbf{p}$-node such that: $k \models A, \mathsf{Y}$ and $k \not\models \mathsf{N}$. We find that $k \simeq_{2.\nu+1, \mathbf{p}} m$. Apply Lemma 5.1 with $\mathsf{Sub}(A)$ in the role of X to find a $\mathbf{q}, \mathbf{p}, \mathbf{r}$-node n with: $m \simeq_{\mathbf{p}, \mathbf{r}} n$ and $\mathsf{Th}_{\mathsf{Sub}(A)}(k) = \mathsf{Th}_{\mathsf{Sub}(A)}(n)$. It follows that $n \not\models B$, but $n \models A$. A contradiction.

(2) Consider B and $\mathbf{q}$. Let $\mathbf{p} := \mathsf{PV}(B) \setminus \mathbf{q}$. Take:

$$\forall \mathbf{q}.B := \bigvee \{D \in I_{2.\nu+1}(\mathbf{p}) \mid \mathsf{IPC} \vdash D \to B\}.$$

Clearly $\forall \mathbf{q}.B$ satisfies (a) and (b). Moreover $\mathsf{IPC} \vdash \forall \mathbf{q}.B \to B$. Hence all we have to prove is that for all A with $\mathsf{PV}(A) \cap \mathbf{q} = \emptyset$:

$$\mathsf{IPC} \vdash A \to B \Rightarrow \mathsf{IPC} \vdash A \to \forall \mathbf{q}.B.$$

Suppose that, to the contrary, for some A: $\mathsf{PV}(A) \cap \mathbf{q} = \emptyset$ and $\mathsf{IPC} \vdash A \to B$ and $\mathsf{IPC} \nvdash A \to \forall \mathbf{q}.B$. Take $\mathbf{r} := \mathsf{PV}(A) \setminus \mathbf{p}$. Note that $\mathbf{p}, \mathbf{q}, \mathbf{r}$ are pairwise disjoint, $\mathsf{PV}(B) \subseteq \mathbf{q}, \mathbf{p}$ and $\mathsf{PV}(A) \subseteq \mathbf{p}, \mathbf{r}$.

Let m be any $\mathbf{p}, \mathbf{r}$-node with $m \models A$ and $m \not\models \forall \mathbf{q}.B$. Let $\mathsf{Y} := \mathsf{Y}_{2.\nu+1, m[\mathbf{p}]}$ and $\mathsf{N} := \mathsf{N}_{2.\nu+1, m[\mathbf{p}]}$. We claim that: $\mathsf{Y} \nvdash \mathsf{N} \vee B$. Note that, by Theorem 4.13, Y is prime. So if $\mathsf{Y} \vdash \mathsf{N} \vee B$, then $\mathsf{Y} \vdash \mathsf{N}$ or $\mathsf{Y} \vdash B$. Since $\mathsf{Y} \nvdash \mathsf{N}$, it follows that $\mathsf{Y} \vdash B$. But then by definition: $\mathsf{Y} \vdash \forall \mathbf{q}.B$. Quod non, since $m \models \mathsf{Y}$ and $m \not\models \forall \mathbf{q}.B$. Let k be any $\mathbf{q}, \mathbf{p}$-node such that: $k \models \mathsf{Y}$ and $k \not\models \mathsf{N} \vee B$. We find that $k \simeq_{2.\nu+1, \mathbf{p}} m$. Apply Lemma 5.1 with $\mathsf{Sub}(B)$ in the role of X to find a $\mathbf{q}, \mathbf{p}, \mathbf{r}$-node n with: $m \simeq_{\mathbf{p}, \mathbf{r}} n$ and $\mathsf{Th}_{\mathsf{Sub}(B)}(k) = \mathsf{Th}_{\mathsf{Sub}(B)}(n)$. It follows that $n \models A$, but $n \not\models B$. A contradiction. □

Note that in (1) of the above theorem we have estimate $2.\nu + 2$ and in (2) $2.\nu + 1$. With some extra work one can get the marginal improvement to $2.\nu + 1$ also for (1). We will not derive the sharper estimate here.

Theorem 5.2 (Semantics of Pitts' Quantifiers). *Consider a node m. Suppose $A \in \mathcal{L}^i$. We have:*

1. $m \models \exists q.A \Leftrightarrow \exists n\ m \simeq_{[q]} n$ *and* $n \models A$.
2. $m \models \forall q.A \Leftrightarrow$ *for all n with* $m \simeq_{[q]} n$, $n \models A$.

Proof. (1) "$\Leftarrow$" Trivial. "$\Rightarrow$" Let $\mathbf{p} := \mathsf{PV}(A) \setminus \{q\}$ and $\nu := \nu_{\mathsf{Sub}(A)}$. Suppose $m \models \exists q.A$, where m is an $\mathcal{R}$-node with $\mathbf{p} \subseteq \mathcal{R}$. Let $\mathsf{Y} := \mathsf{Y}_{2.\nu+1,m[\mathbf{p}]}$ and $\mathsf{N} := \mathsf{N}_{2.\nu+1,m[\mathbf{p}]}$. As in Theorem 5.1(2), $A, \mathsf{Y} \not\vdash \mathsf{N}$. Let k be any $q, \mathbf{p}$-node such that: $k \models A, \mathsf{Y}$ and $k \not\models \mathsf{N}$. We find that $k \simeq_{2.\nu+1,\mathbf{p}} m$. Apply Lemma 5.1 to k and $m[\mathcal{R} \setminus \{q\}]$ with $\mathsf{Sub}(A)$ in the role of X, $\{q\}$ in the role of $\mathcal{Q}$, $\mathbf{p}$ in the role of $\mathbf{p}$, $\mathcal{R} \setminus (\mathbf{p} \cup \{q\})$ in the role of $\mathcal{R}$, to find a $q, \mathbf{p}, \mathcal{R}$-node n with: $m \simeq_{[q]} n$ and $\mathsf{Th}_{\mathsf{Sub}(A)}(k) = \mathsf{Th}_{\mathsf{Sub}(A)}(n)$, and, thus, $n \models A$. The proof of (2) is similar. □

Theorem 5.1 is not formulated entirely in terms of ℓ-simulations. The reason is that such a form does not provide a very sharp estimate on uniform interpolants. But if we do not want to worry about precise complexities a watered down version can be pleasant to have. By applying Theorem 5.1 to $X := I_n(\mathbf{p}, \mathbf{q})$ we find:

Corollary 5.1. *For all disjoint $\mathbf{q}, \mathbf{p}$ and numbers s, there is an N (multi-exponential in $|\mathbf{q}, \mathbf{p}| + s$), such that: for all $k \in \mathsf{Pmod}(\mathbf{q}, \mathbf{p})$, and all $m \in \mathsf{Pmod}$ with $\mathbf{q} \cap \mathcal{P}_{\mathsf{M}} = \emptyset$ and $\mathbf{p} \subseteq \mathcal{P}_m$, we have:*

$$k \simeq_{N,\mathbf{p}} m \Rightarrow \text{ there is an } n \in \mathsf{Pmod}(\mathbf{q}, \mathcal{P}_m) \text{ with } n \simeq_{s,\mathbf{q},\mathbf{p}} k \text{ and } n \simeq_{\mathcal{P}_m} m.$$

We repeat a result from [13]. We illustrate that the increase of implicational complexity in going to a uniform interpolant is unavoidable. It is an interesting problem to find both better upper and lower bounds.

Theorem 5.3. *Every formula of $\mathcal{L}^i$ is equivalent to an I_2-formula preceded by existential quantifiers and to an I_3-formula preceded by universal quantifiers.*

Proof. Suppose $A \in \mathcal{L}^i(\mathbf{p})$. Let $\mathbf{q}$ be a set of variables disjoint from $\mathbf{p}$ that is in 1-1 correspondence with the subformulas of the form $(B \to C)$ of A. Let the correspondence be q. We define a mapping $\mathcal{T}$ as follows:

- $\mathcal{T}$ commutes with atoms, conjunction and disjunction
- $\mathcal{T}(B \to C) := \mathsf{q}(\mathfrak{B} \to \mathfrak{C})$

Define:

- $\mathsf{EQ} := \bigwedge\{\mathsf{q}(\mathfrak{B} \to \mathfrak{C}) \leftrightarrow (\mathcal{T}(\mathfrak{B}) \to \mathcal{T}(\mathfrak{C})) \mid (\mathfrak{B} \to \mathfrak{C}) \in \mathsf{Sub}(\mathfrak{A})\}$

Note that EQ is I_2. Finally we put:

- $A^{\#} := \exists \mathbf{q}(\mathsf{EQ} \wedge \mathcal{T}(A))$, $A^{\$} := \forall \mathbf{q}(\mathsf{EQ} \rightarrow \mathcal{T}(A))$

By elementary reasoning in second order propositional logic we find: $\vdash A \leftrightarrow A^{\#}$ and $\vdash A \leftrightarrow A^{\$}$. □

We end this section by verifying semantically a striking principle (present in Pitts' paper) valid for the Pitts interpretation.

Theorem 5.4. $k \models \forall p(B \vee C) \Rightarrow k \models \forall p.B$ *or* $k \models \forall p.C$.

Proof. We reason by contraposition. Suppose $k \not\models \forall p.B$ and $k \not\models \forall p.C$. It follows that there are nodes m and n, such that $k \simeq_{[p]} m \not\models B$ and $k \simeq_{[p]} n \not\models C$. Let $\mathbb{M}$ and $\mathbb{N}$ be the models of, respectively, m and n. Let $\mathbb{P} := \mathsf{Glue}(\mathbb{M}[m], \mathbb{N}[n])$. Let b be the new root. It is easily seen that $k \simeq_{[p]} b$ and $b \not\models (B \vee C)$ □

6. Uniform Interpolation for K

We first survey the connection between modal propositional formulas and bounded bisimulations. Since these facts are similar to, but simpler than the corresponding facts for IPC, we just state the results without the proofs. Let $\mathfrak{b}(A)$ be the box-depth of a formula. $B_k(\mathbf{p})$ is the set of formulas in the variables $\mathbf{p}$ with box-depth $\leq k$. $B_k(\mathbf{p})$ is finite *modulo provable equivalence.* Consider $\mathbf{p}$-nodes k and m. Then: $k \simeq_n m \Leftrightarrow \mathsf{Th}_{B_n(\mathbf{p})}(k) = \mathsf{Th}_{B_n(\mathbf{p})}(m)$. Define: $\mathsf{Y}_{n,k} := \bigwedge \mathsf{Th}_{B_n(\mathbf{p})}(k)$. Clearly, $k \simeq_n m \Leftrightarrow m \models \mathsf{Y}_{n,k} \Leftrightarrow \mathsf{K} \vdash \mathsf{Y}_{n,m} \leftrightarrow \mathsf{Y}_{n,k}$.

Before considering uniform interpolation for more complicated modal systems like S4Grz, we do the relatively easy proof for K. This theorem was first proved by Silvio Ghilardi, see [2]. Uniform interpolation for K follows from the amalgamation lemma below.

Lemma 6.1. *Consider pairwise disjoint sets of propositional variables* $\mathcal{Q}$, $\mathbf{p}$ *and* $\mathcal{R}$. *Let* $\langle \mathbb{K}, k_0 \rangle \in \mathsf{Pmod}(\mathcal{Q}, \mathbf{p})$ *and* $\langle \mathbf{M}, m_0 \rangle \in \mathsf{Pmod}(\mathbf{p}, \mathcal{R})$. *Suppose that* $k_0 \simeq_{\alpha,\mathbf{p}} m_0$. *Then there is a* $\mathcal{Q}$*-extension* $\langle \mathrm{N}, n_0 \rangle$ *of* $\langle \mathbf{M}, m_0 \rangle$ *such that* $n_0 \simeq_\alpha k_0$.

Proof. Let $\mathcal{Z}$ be a downwards closed witness of $k_0 \simeq_{\alpha,\mathbf{p}} m_0$. We add a 'virtual top' $\top$ to $\mathbb{K}$ and stipulate that $\top$ satisfies no atoms. Let's call the new model $\mathbb{K}^\top$. We extend ω^∞ with a new bottom $\perp$ to $\omega^{\infty,\perp}$. Define $\mathsf{Pd}(n+1) := n$, $\mathsf{Pd}(0) := \mathsf{Pd}(\perp) := \perp$, $\mathsf{Pd}(\infty) = \infty$. Now define the following model N:

- $N := \mathcal{Z} \cup \{\langle \top, \perp, m \rangle \mid m \in M\}$
- $\langle k, \alpha, m \rangle \prec_{\mathrm{N}} \langle k', \alpha', m' \rangle :\Leftrightarrow k \prec_{\mathbb{K}^\top} k'$ and $\alpha' = \mathsf{Pd}(\alpha)$ and $m \prec_{\mathbf{M}} m'$
- $\langle k, \alpha, m \rangle \models s :\Leftrightarrow k \models_{\mathbf{K}} s$ or $m \models_{\mathbf{M}} s$

We claim:

Claim 1 $n_0 \simeq_{\mathbf{p},\mathcal{R}} m_0$.
Claim 2 $n_0 \simeq_{\alpha,(\mathcal{Q},\mathbf{p})} k_0$.

We prove Claim 1. Take as bisimulation $\mathcal{B}$, with $\langle k,\alpha,m\rangle\mathcal{B}m' :\Leftrightarrow m = m'$. Clearly, if $n\mathcal{B}m$ then $\mathsf{Th}_{\mathbf{p},\mathcal{R}}(n) = \mathsf{Th}_{\mathbf{p},\mathcal{R}}(m)$. Moreover, $\mathcal{B}$ trivially has the zig-property. We check that $\mathcal{B}$ has the zag-property. Suppose $\langle k,\alpha,m\rangle\mathcal{B}m \prec m'$. If $\alpha \in \{0,\bot\}$, we can finish the diagram with $\langle\top,\bot,m'\rangle$. If $\alpha = \alpha' + 1$ for $\alpha' \in \omega^\infty$, we have $k\mathcal{Z}_\alpha m$ and, hence, there is a k' such that $k \prec_{\mathbf{K}} k'$ and $k'\mathcal{Z}_{\alpha'}m'$. So we can finish the diagram with $\langle k',\alpha',m'\rangle$.

We prove Claim 2. Take as layered bisimulation $\mathcal{S}$, with $\langle k,\alpha,m\rangle\mathcal{S}_\alpha k' \Leftrightarrow k = k'$ (for $\alpha \in \omega^\infty$). Clearly, if $n\mathcal{S}_\alpha k$ then $\mathsf{Th}_{\mathcal{Q},\mathbf{p}}(n) = \mathsf{Th}_{\mathcal{Q},\mathbf{p}}(k)$. We check that $\mathcal{S}$ has the zag-property. The zig-property is analogous. Suppose $\langle k,\alpha+1,m\rangle\mathcal{S}_{\alpha+1}k \prec k'$. Since $k\mathcal{Z}_{\alpha+1}m$, there exists $m' \succ m$ such that $k'\mathcal{Z}_\alpha m'$. Hence $\langle k',\alpha,m'\rangle \succ \langle k,\alpha+1,m\rangle$, and $\langle k',\alpha,m'\rangle\mathcal{S}_\alpha k'$. □

Theorem 6.1 (Uniform Interpolation). *We prove uniform interpolation for* K

1. *Consider any formula A and any finite set of variables $\mathbf{q}$. Let $\nu := \mathfrak{b}(A)$. There is a formula $\exists\mathbf{q}.A$ such that:*
   a) $\mathsf{PV}(\exists\mathbf{q}.A) \subseteq \mathsf{PV}(A) \setminus \mathbf{q}$
   b) $\mathfrak{b}(\exists\mathbf{q}.A) \leq \nu$
   c) *For all $B \in \mathcal{L}^m$ with $\mathsf{PV}(B) \cap \mathbf{q} = \emptyset$, we have:*
   $$\mathsf{K} \vdash A \to B \Leftrightarrow \mathsf{K} \vdash \exists\mathbf{q}.A \to B.$$
2. *Consider any formula B and any finite set of variables $\mathbf{q}$. Let $\nu := \mathfrak{b}(B)$. There is a formula $\forall\mathbf{q}.B$ such that:*
   a) $\mathsf{PV}(\forall\mathbf{q}.B) \subseteq \mathsf{PV}(B) \setminus \mathbf{q}$
   b) $\mathfrak{b}(\forall\mathbf{q}.B) \leq \nu$
   c) *For all $A \in \mathcal{L}^m$ with $\mathsf{PV}(A) \cap \mathbf{q} = \emptyset$, we have:*
   $$\mathsf{K} \vdash A \to B \Leftrightarrow \mathsf{K} \vdash A \to \forall\mathbf{q}.B.$$

Proof. We just prove (1). The proof of (2) is analogous. (Alternatively, we may take $(\forall\mathbf{q}.B) := (\neg\exists\mathbf{q}\neg B)$.) Consider A and $\mathbf{q}$. Let $\mathbf{p} := \mathsf{PV}(A) \setminus \mathbf{q}$. Take:

$$\exists\mathbf{q}.A := \bigwedge\{C \in I_\nu(\mathbf{p}) \mid \mathsf{K} \vdash A \to C\}.$$

Clearly $\exists\mathbf{q}.A$ satisfies (a) and (b). Moreover, $\mathsf{K} \vdash A \to \exists\mathbf{q}.A$. Hence, all we have to prove is that for all B with $\mathsf{PV}(B) \cap \mathbf{q} = \emptyset$:

$$\mathsf{K} \vdash A \to B \Rightarrow \mathsf{K} \vdash \exists\mathbf{q}.A \to B.$$

Suppose, to the contrary, that for some B: $\mathsf{PV}(B) \cap \mathbf{q} = \emptyset$ and $\mathsf{K} \vdash A \to B$ and $\mathsf{K} \nvdash \exists\mathbf{q}.A \to B$. Take $\mathbf{r} := \mathsf{PV}(B) \setminus \mathbf{p}$. Note that $\mathbf{p},\mathbf{q},\mathbf{r}$ are pairwise disjoint, $\mathsf{PV}(A) \subseteq \mathbf{q} \cup \mathbf{p}$ and $\mathsf{PV}(B) \subseteq \mathbf{p} \cup \mathbf{r}$.

Let m be any $\mathbf{p},\mathbf{r}$-node with $m \models \exists \mathbf{q}.A$ and $m \not\models B$. Let $\mathsf{Y} := \mathsf{Y}_{\nu,m[\mathbf{p}]}$ and We claim that: A, Y is consistent. If it were not, we would have: $A \vdash \neg\mathsf{Y}$. And, hence, by definition: $\exists \mathbf{q}.A \vdash \neg\mathsf{Y}$. Quod non, since $m \models \exists \mathbf{q}.A, \mathsf{Y}$ and $\mathfrak{d}(\neg\mathsf{Y}) = \nu$. Let k be any $\mathbf{q},\mathbf{p}$-node such that: $k \models A, \mathsf{Y}$. We find that $k \simeq_{\nu,\mathbf{p}} m$. Apply Lemma 6.1 to find a $\mathbf{q},\mathbf{p},\mathbf{r}$-node n with: $m \simeq_{\mathbf{p},\mathbf{r}} n$ and $m \simeq_{\nu,(\mathbf{p},\mathbf{r})} n$. It follows that $n \not\models B$, but $n \models A$. A contradiction. □

The proof of the following theorem is fully analogous to the the proof of its twin for the case of IPC.

Theorem 6.2. *Consider a node m. Suppose $A \in \mathcal{L}^m$. We have:*

1. $m \models \exists q.A \Leftrightarrow \exists n\ m \simeq_{[q]} n$ *and* $n \models A$.
2. $m \models \forall q.A \Leftrightarrow$ *for all* n *with* $m \simeq_{[q]} n$, $n \models A$.

7. Uniform Interpolation for GL

In this section we prove Uniform Interpolation for GL. It is well known that GL is sound and complete for upward wellfounded Kripke models and that it has the finite model property. Since GL-models are irreflexive we use '$\prec$' for their accessibility relation and '$\preceq$' for the corresponding weak partial order. '$\vdash$' will stand for GL-derivability.

Let X be a finite, adequate set of formulas. *Adequate* means: closed under subformulas. The GL Henkin model $\mathbb{H}_X$ for X is constructed in the following way.

- The nodes are the subsets Δ of X that are *X-saturated*, i.e. if Δ proves some finite disjunction of elements of X then some disjunct is in Δ.
- $\Delta \prec \Delta'$ iff $\Box A \in \Delta \Rightarrow A, \Box A \in \Delta'$

Note that this model may contain *non-trivial loops!* and, thus is not a GL-model. (It is easy to remove these loops, but for the present purposes, we need to keep them.) The height of a model is the maximal depth. The height of the Henkin model is $\leq 2.|\{C \in X \mid C \text{ is boxed}\}|$. To see this, consider $\Delta_0 \prec^+ \Delta_1 \prec^+ \Delta_2$. Clearly, going up the set of boxed formulas in the Δ_i increases. Suppose we had the same boxed formulas in Δ_0, Δ_1 and Δ_2. Suppose $\Box A \in \Delta_2$. Then, ex hypothesi, $\Box A \in \Delta_0$. Hence, $A, \Box A \in \Delta_1$. We may conclude that $\Delta_2 \prec \Delta_1$. Quod non. So, necessarily, the boxed formulas increase by at least one in going from Δ_0 to Δ_2. It follows that if we have a strictly ascending chain of length $2.n$, then there are at least n boxed subformulas.

As in for IPC and K we start with an amalgamation lemma. Consider disjoint sets of propositional variables $\mathcal{Q}$, $\mathbf{p}$ and $\mathcal{R}$. Let $\langle \mathbb{K}, k_0 \rangle \in \mathsf{Pmod}(\mathcal{Q}, \mathbf{p})$ and $\langle \mathbb{M}, m_0 \rangle \in \mathsf{Pmod}(\mathbf{p}, \mathcal{R})$ be pointed GL-models.

Lemma 7.1. *Let $X \subseteq \mathcal{L}^m(\mathcal{Q}, \mathbf{p})$ be a finite adequate set. Let:*

$$\nu := 2.|\{C \in X \mid C \textit{ is boxed}\}|.$$

Suppose that $k_0 \simeq_{2.\nu+1,\mathbf{p}} m_0$. Then there is a $\mathcal{Q}$-extension $\langle \mathbb{N}, n_0\rangle$ of $\langle \mathbb{M}, m_0\rangle$ such that $\mathbb{N}$ is a GL*-model and* $\mathsf{Th}_X(n_0) = \mathsf{Th}_X(k_0)$.

Proof. Let $\mathcal{Z}$ be a downwards closed witness of $k_0 \simeq_{2.\nu+1,\mathbf{p}} m_0$. Define Φ_X from $\mathbb{K}$ to the Henkin model $\mathbb{H} := \mathbb{H}_X$ as follows: $\Phi_X(k) := \Delta(k) := \{B \in X \mid k \models B\}$. Define further for k in $\mathbb{K}$: $d_X(k) := d_{\mathbb{H}}(\Delta(k))$. Note that: $d_X(k) \leq \nu$.

Consider a pair $\langle \Delta, m\rangle$ for Δ in $\mathbb{H}$ and m in $\mathbb{M}$. Consider k', k, m'. Let $\Delta' := \Phi_X(k')$. We say that k', k, m' is a *witnessing triple* for $\langle \Delta, m\rangle$ if:

$$\Delta' \approx \Delta,\ k' \preceq k,\ m' \preceq m,\ k'\mathcal{Z}_{2.d_X(k')+1}m',\ k\mathcal{Z}_{2.d_X(k')}m.$$

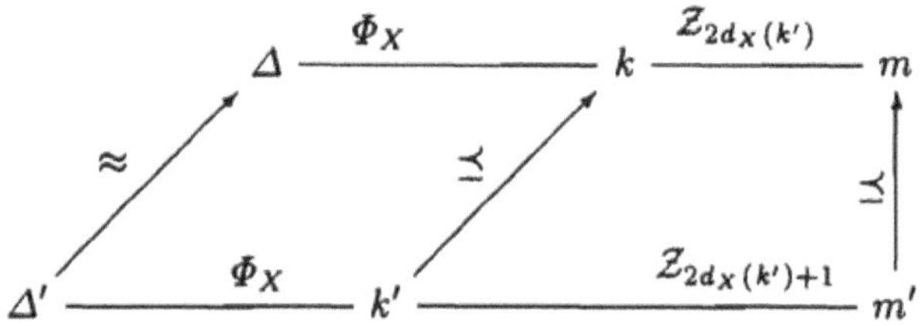

Define:

- $N := \{\langle \Delta, m\rangle \mid$ there is a witnessing triple for $\langle \Delta, m\rangle\}$
- $n_0 := \langle \Delta(k_0), m_0\rangle$
- $\langle \Delta, m\rangle \prec_{\mathbb{N}} \langle \Gamma, n\rangle :\Leftrightarrow \Delta \prec_{\mathbb{H}} \Gamma$ and $m \prec_{\mathbb{M}} n$
- $\langle \Delta, m\rangle \models_{\mathbb{N}} s :\Leftrightarrow \Delta \models_{\mathbb{H}} s$ or $m \models_{\mathbb{M}} s$

Note that by assumption $k_0\mathcal{Z}_{2\nu+1}m_0$. Moreover: $2.d_X(k_0)+1 \leq 2.\nu+1$. Hence: $k_0\mathcal{Z}_{2d_X(k_0)+1}m_0$. So we can take k_0, k_0, m_0 as witnessing triple for n_0. Let k', k, m' be a witnessing triple for $\langle \Delta, m\rangle$. Note that for $p \in \mathbf{p} \cap X$: $\Delta \models p \Leftrightarrow k \models p \Leftrightarrow m \models p$, and hence: $\langle \Delta, m\rangle \models p \Leftrightarrow \Delta \models p \Leftrightarrow m \models p$. It is easy to see that $\mathbb{N}$ is a GL-model (even if $\mathbb{H}_X$ need not be one). We claim:

Claim 1 $n_0 \simeq_{\mathbf{p},\mathcal{R}} m_0$.
Claim 2 For $B \in X : \langle \Delta, m\rangle \models B \Leftrightarrow B \in \Delta$.

Evidently the lemma is immediate from the claims.

We prove Claim 1. Take as bisimulation $\mathcal{B}$ with $\langle \Delta, m\rangle\mathcal{B}m$. It is evident that $\mathsf{Th}_{\mathbf{p},\mathcal{R}}(\langle \Delta, m\rangle) = \mathsf{Th}_{\mathbf{p},\mathcal{R}}(m)$. Moreover, $\mathcal{B}$ has the zig-property. We check that $\mathcal{B}$ has the zag-property. Suppose $\langle \Delta, m\rangle\mathcal{B}m \prec n$. We are looking for a pair $\langle \Gamma, n\rangle$ in N such that $\Delta \prec \Gamma$. Let k', k, m' be a witnessing triple for $\langle \Delta, m\rangle$. We write $\Delta' := \Delta(k')$. Since $k'\mathcal{Z}_{2.d_X(k')+1}m' \prec n$, there is a h such that $k' \prec h\mathcal{Z}_{2.d_X(k')}n$. We take $\Gamma := \Delta(h)$. Clearly $\Delta \prec \Gamma$. We need a witnessing triple k'^*, k^*, m'^* for $\langle \Gamma, n\rangle$ We distinguish two possibilities. First, $\Delta \approx \Gamma$. In this case we can take: $k'^* := k'$, $k^* := h$, $m'^* := m'$.

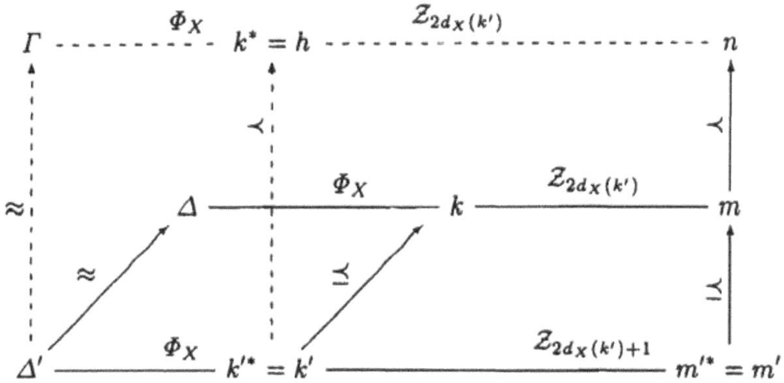

Secondly, $\Delta \not\approx \Gamma$. In this case we can take: $k'^* := h$, $k^* := h$, $m'^* := n$. To see this, note that, since $k' \prec h$, we have: $\Delta \approx \Delta' \prec \Gamma$ and, hence, $\Delta \prec^+ \Gamma$. Ergo $d_X(h) < d_X(k')$. It follows that: $2.d_X(h) + 1 \leq 2.d_X(k')$. So, $h\mathcal{Z}_{2.d_X(k')+1}n$.

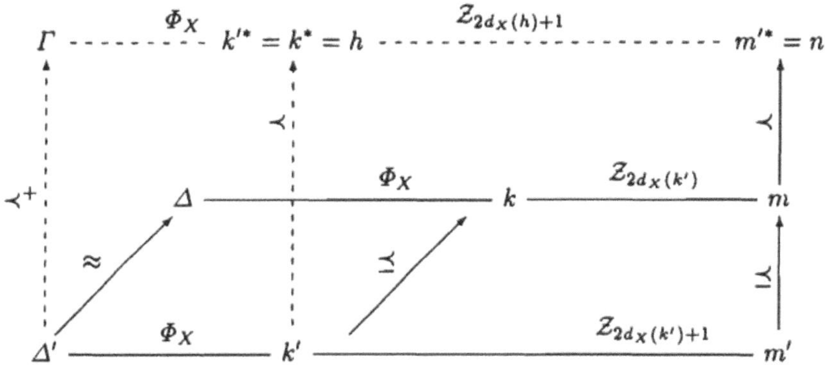

Finally, clearly, $b_{\mathsf{N}}\mathcal{B}b_{\mathsf{M}}$.

We prove Claim 2. The proof is by induction on X. The cases of atoms, conjunction and disjunction are trivial. We treat the only non-trivial case: the left-to-right case of the box. Consider $\Box C \in X$ and consider the node $\langle \Delta, m \rangle$ with witnessing triple k', k, m'. Suppose $\Box C \notin \Delta$. Clearly, $k \not\models \Box C$, so there is an $h' \succ k$ with $h' \not\models C$. Let h be maximal in $\mathbb{K}$ with $h \succ k$ and $h \not\models C$. By maximality, we find: $h \models \Box C$. Let $\Gamma := \Delta(h)$. Since, $\Box C \notin \Delta$ and $\Box C \in \Gamma$, we find: $\Delta \prec^+ \Gamma$. Note that it follows that $d_X(k') \geq 1$. Since, $k\mathcal{Z}_{2.d_X(k')}m$ and $k \prec h$, there is an $n \succ m$ with $h\mathcal{Z}_{2.d_X(k')-1}n$. Moreover: $2.d_X(h) + 1 \leq 2.d_X(k') - 1$. Ergo: $h\mathcal{Z}_{2.d_X(h)+1}n$. So we can take $k'^* := h$, $k^* := h$, $m'^* := n$ to witness $\langle \Gamma, n \rangle$. Clearly, $\langle \Delta, m \rangle \prec \langle \Gamma, n \rangle$. By the Induction Hypothesis: $\langle \Gamma, n \rangle \not\models C$. Hence, $\langle \Delta, m \rangle \not\models \Box C$.

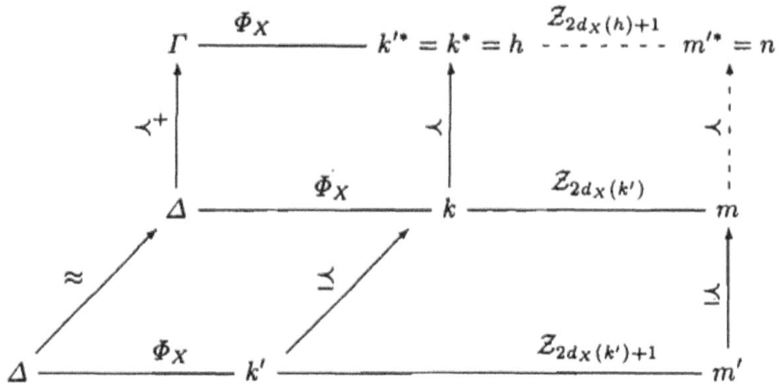

Thus we have proved Claim 2. □

We formulate Uniform Interpolation for GL. Its proof is fully analogous to the one of Uniform Interpolation for K.

Theorem 7.1 (Uniform Interpolation). *We state uniform interpolation for* GL

1. *Consider any formula A and any finite set of variables* $\mathbf{q}$. *Let*

$$\nu := 2.|\{C \in \mathsf{Sub}(A) \mid C \textit{ is boxed}\}|.$$

   There is a formula $\exists \mathbf{q}.A$ such that:
   a) $\mathsf{PV}(\exists \mathbf{q}.A) \subseteq \mathsf{PV}(A) \setminus \mathbf{q}$
   b) $\mathfrak{b}(\exists \mathbf{q}.A) \leq 2.\nu + 1$
   c) For all $B \in \mathcal{L}^m$ with $\mathsf{PV}(B) \cap \mathbf{q} = \emptyset$, we have:

$$\mathsf{GL} \vdash A \to B \Leftrightarrow \mathsf{GL} \vdash \exists \mathbf{q}.A \to B.$$

2. *Consider any formula B and any finite set of variables* $\mathbf{q}$. *Let*

$$\nu := 2.|\{C \in \mathsf{Sub}(B) \mid C \textit{ is boxed}\}|.$$

   There is a formula $\forall \mathbf{q}.B$ such that:
   a) $\mathsf{PV}(\forall \mathbf{q}.B) \subseteq \mathsf{PV}(B) \setminus \mathbf{q}$
   b) $\mathfrak{b}(\forall \mathbf{q}.B) \leq 2.\nu + 1$
   c) For all $A \in \mathcal{L}^m$ with $\mathsf{PV}(A) \cap \mathbf{q} = \emptyset$, we have:

$$\mathsf{GL} \vdash A \to B \Leftrightarrow \mathsf{GL} \vdash A \to \forall \mathbf{q}.B.$$

The semantical interpretation of the propositional quantifiers is fully analogous to the case of K.

8. Uniform Interpolation for S4Grz

S4Grz, a logic called after Andrzej Gregorczyk, is K extended with:

T $\vdash \Box A \to A$
4 $\vdash \Box A \to \Box\Box A$
Grz $\vdash \Box(\Box(A \to \Box A) \to A) \to A$

It is easy to see that T is superfluous. Note also that over KT4 (= S4), *Grz* is equivalent to:

Grz' $\Box(\Box(A \to \Box A) \to A) \to \Box A$

The logic is sound for weak partial orderings such that the associated strict ordering is upward wellfounded. We will show that the completeness of the logic in finite partial orderings. Since we deal with reflexive structures in this section, we will use '$\preceq$' for these relations. In case our relation is a weak partial ordering we write '$\prec$' for the associated strict ordering. For weak partial preorderings we will use $\prec^+$ for the associated strict version to stress the fact that also non-trivial loops are removed. '$\vdash$' will stand for S4Grz-provability.

Let X be a finite adequate set. We construct a Henkin model $\mathbb{J}_X$ as follows. Let

$$X^+ := X \cup \{(B \to \Box B), \Box(B \to \Box B) \mid \Box B \in X\}.$$

Clearly, X is again adequate. Define:

- The domain J is the set of X^+-saturated sets Δ.
- $\Delta \preceq \Delta' :\Leftrightarrow \Delta = \Delta'$ or (for all $\Box C \in \Delta$, $\Box C \in \Delta'$ and for some $\Box D \in \Delta'$, $\Box D \notin \Delta$)
- $\Delta \models p :\Leftrightarrow p \in \Delta$

It is easily seen that $\mathbb{J}_X$ is a finite partial order. We show that for all A in X, $\Delta \models A \Leftrightarrow A \in \Delta$. The crucial feature here is that we do *not* prove this fact for all A in X^+! The proof is by induction on A. We consider the only interesting case. Suppose that A is $\Box B$ and that $\Box B \notin X$. We show $\Delta \not\models \Box B$. We have to produce a Δ' with $\Delta' \succeq \Delta$ and $\Delta' \not\models B$. In case $B \notin \Delta$, and, hence, by the Induction Hypothesis, $\Delta \not\models B$, we are immediately done. So suppose $B \in \Delta$. Note that $\Box(B \to \Box B)$ cannot be in Δ, since, if it were, $\Box B$ would be in Δ. We claim: $\{\Box C \mid \Box C \in \Delta\} \cup \{\Box(B \to \Box B)\} \nvdash B$. If it were otherwise, it would follow by S4-reasoning that: $\{\Box C \mid \Box C \in \Delta\} \vdash \Box(\Box(B \to \Box B) \to B)$. Hence by *Grz'*, $\{\Box C \mid \Box C \in \Delta\} \vdash \Box B$, and, thus $\Delta \vdash \Box B$. Quod non. By the usual methods we can construct an X^+-saturated set Δ' such that $\{\Box C \mid \Box C \in \Delta\} \cup \{\Box(B \to \Box B)\} \subseteq \Delta'$ and $B \notin \Delta'$. It follows that $\Delta \preceq \Delta'$ (with $\Box(B \to \Box B)$ in the role of the D of the definition). Since $B \notin \Delta'$, we have, by the Induction Hypothesis, $\Delta' \not\models B$.

For our proof of Uniform Interpolation we will use a different Henkin model $\mathbb{H}_X$, which is defined like $\mathbb{J}_X$, dropping the clause involving D, which excludes non-trivial loops. The height of $\mathbb{H}_X$ is estimated by the number of boxed formulas in X^+, which is two times the number of boxed formulas in X. We start with an amalgamation lemma. Consider disjoint sets of propositional variables $\mathcal{Q}$, $\mathbf{p}$ and $\mathcal{R}$. Let $\langle \mathbb{K}, k_0 \rangle \in \mathsf{Pmod}(\mathcal{Q}, \mathbf{p})$ and $\langle \mathbb{M}, m_0 \rangle \in \mathsf{Pmod}(\mathbf{p}, \mathcal{R})$ be S4Grz-models.

Lemma 8.1. *Let $X \subseteq \mathcal{L}^m(\mathcal{Q}, \mathbf{p})$ be a finite adequate set. Let:*

$$\nu := 2.|\{C \in X \mid C \text{ is boxed}\}|.$$

Suppose that $k_0 \simeq_{2.\nu+1,\mathbf{p}} m_0$. Then there is a $\mathcal{Q}$-extension $\langle \mathrm{N}, n_0 \rangle$ of $\langle \mathbb{M}, m_0 \rangle$ such that N *is a* S4Grz*-model and* $\mathsf{Th}_X(n_0) = \mathsf{Th}_X(k_0)$.

Proof. Let $\mathcal{Z}$ be a downwards closed witness of $k_0 \simeq_{2.\nu+1,\mathbf{p}} m_0$. Define Φ_X from $\mathbb{K}$ to the Henkin model $\mathbb{H} := \mathbb{H}_X$ as follows: $\Phi_X(k) := \Delta(k) := \{B \in X^+ \mid k \models B\}$. Define further for k in $\mathbb{K}$: $d_X(k) := d_{\mathbb{H}}(\Delta(k))$. Note that: $d_X(k) \leq \nu$.

Consider a pair $\langle \Delta, m \rangle$ for Δ in $\mathbb{H}$ and m in $\mathbb{M}$. Consider k', k, m'. Let $\Delta' := \Phi_X(k')$. We say that k', k, m' is a *witnessing triple* for $\langle \Delta, m \rangle$ if:

$$\Delta' \approx \Delta,\ k' \preceq k,\ m' \preceq m,\ k' \mathcal{Z}_{2.d_X(k')+1} m',\ k \mathcal{Z}_{2.d_X(k')} m.$$

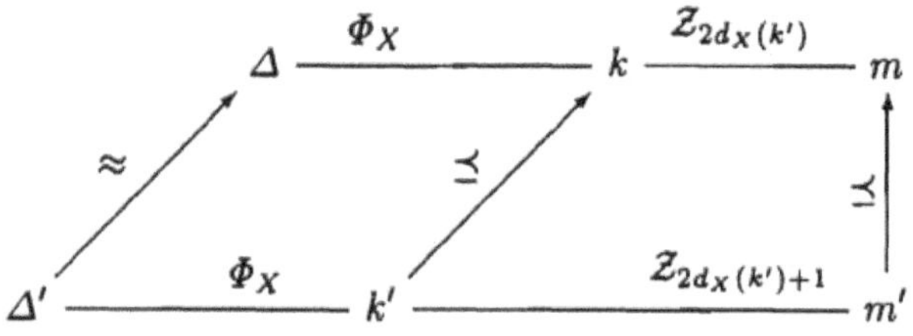

Define:

- $N := \{\langle \Delta, m \rangle \mid$ there is a witnessing triple for $\langle \Delta, m \rangle\}$
- $n_0 := \langle \Delta(k_0), m_0 \rangle$
- $\langle \Delta, m \rangle \preceq_{\mathrm{N}} \langle \Gamma, n \rangle :\Leftrightarrow \langle \Delta, m \rangle = \langle \Gamma, n \rangle$ or ($\Delta \preceq_{\mathbf{H}} \Gamma$ and $m \prec_{\mathbf{M}} n$) or ($\Delta \prec^+_{\mathbf{H}} \Gamma$ and $m \preceq_{\mathbf{M}} n$)
- $\langle \Delta, m \rangle \models_{\mathrm{N}} s :\Leftrightarrow \Delta \models_{\mathbf{H}} s$ or $m \models_{\mathbf{M}} s$

Note that by assumption $k_0 \mathcal{Z}_{2\nu+1} m_0$. Moreover: $2.d_X(k_0)+1 \leq 2.\nu+1$. Hence: $k_0 \mathcal{Z}_{2d_X(k_0)+1} m_0$. So we can take k_0, k_0, m_0 as witnessing triple for n_0. Let k', k, m' be a witnessing triple for $\langle \Delta, m \rangle$. Note that for $p \in \mathbf{p} \cap X$: $\Delta \models p \Leftrightarrow k \models p \Leftrightarrow m \models p$, and hence: $\langle \Delta, m \rangle \models p \Leftrightarrow \Delta \models p \Leftrightarrow m \models p$. It is easy to see that N is a S4Grz-model (even if $\mathbb{H}_X$ need not be one). We claim:

Claim 1 $n_0 \simeq_{\mathbf{p},\mathcal{R}} m_0$.
Claim 2 For $B \in X : \langle \Delta, m \rangle \models B \Leftrightarrow B \in \Delta$.

Evidently the lemma is immediate from the claims.

We prove Claim 1. Take as bisimulation $\mathcal{B}$ with $\langle\Delta, m\rangle\mathcal{B}m$. It is evident that $\mathsf{Th}_{\mathbf{p},\mathcal{R}}(\langle\Delta, m\rangle) = \mathsf{Th}_{\mathbf{p},\mathcal{R}}(m)$. Moreover, $\mathcal{B}$ has the zig-property. We check that $\mathcal{B}$ has the zag-property. Suppose $\langle\Delta, m\rangle\mathcal{B}m \preceq n$. We are looking for a pair $\langle\Gamma, n\rangle$ in N such that $\Delta \preceq \Gamma$. In case $m = n$, we take $\langle\Gamma, n\rangle := \langle\Delta, m\rangle$. Suppose $m \neq n$ and, hence, $m \prec n$. Let k', k, m' be a witnessing triple for $\langle\Delta, m\rangle$. We write $\Delta' := \Delta(k')$. Since $k'\mathcal{Z}_{2.d_X(k')+1}m' \preceq n$, there is a h such that $k' \prec h\mathcal{Z}_{2.d_X(k')}n$. We take $\Gamma := \Delta(h)$. Clearly $\Delta \preceq \Gamma$. We need a witnessing triple k'^*, k^*, m'^* for $\langle\Gamma, n\rangle$ We distinguish two possibilities. First, $\Delta \approx \Gamma$. In this case we can take: $k'^* := k'$, $k^* := h$, $m'^* := m'$.

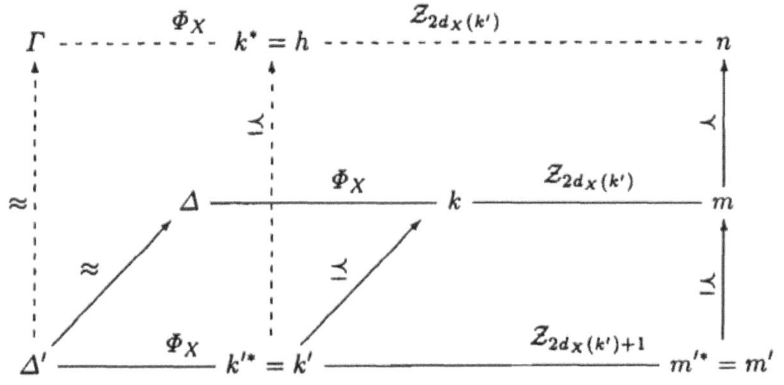

Secondly, $\Delta \not\approx \Gamma$. In this case we can take: $k'^* := h$, $k^* := h$, $m'^* := n$. To see this, note that, since $k' \preceq h$, we have: $\Delta \approx \Delta' \preceq \Gamma$ and, hence, since $\Delta \not\approx \Gamma$, $\Delta \prec^+ \Gamma$. Ergo $d_X(h) < d_X(k')$. It follows that: $2.d_X(h) + 1 \leq 2.d_X(k')$. So, $h\mathcal{Z}_{2.d_X(k')+1}n$.

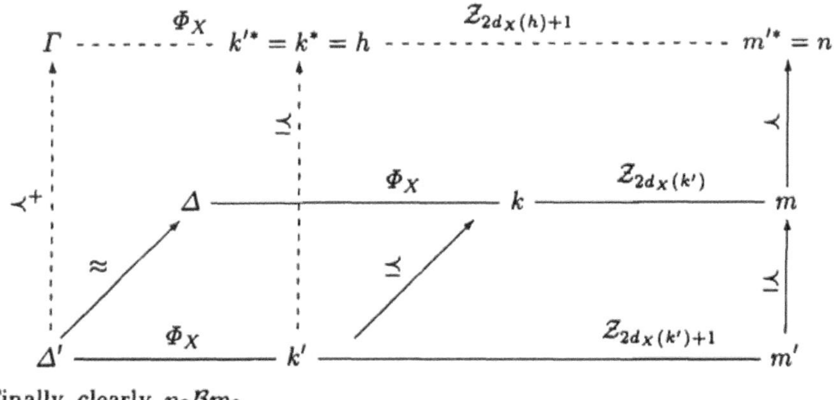

Finally, clearly, $n_0\mathcal{B}m_0$.

We prove Claim 2. The proof is by induction on X. The cases of atoms, conjunction and disjunction are trivial. We treat the only non-trivial case:

the right-to-left case for the box. Consider $\Box C \in X$ and consider the node $\langle \Delta, m\rangle$ with witnessing triple k', k, m'. Suppose $\Box C \notin \Delta$.

In case $C \notin \Delta$, we have, by the Induction Hypothesis, $\langle \Delta, m\rangle \not\models C$ and, hence, $\langle \Delta, m\rangle \not\models \Box C$.

Suppose $C \in \Delta$. It follows that $\Box(C \to \Box C)$ is not in Δ, since, otherwise, $\Box C$ would be in Δ. Clearly, $k \not\models \Box C$, so there is an $h' \succeq k$ with $h' \not\models C$. Let h be maximal in $\mathbb{K}$ with $h \succeq k$ and $h \not\models C$. By maximality, we find: $h \models \Box(C \to \Box C)$. Let $\Gamma := \Delta(h)$. Since, $\Box(C \to \Box C) \notin \Delta$ and $\Box(C \to \Box C) \in \Gamma$, we find: $\Delta \prec^{+} \Gamma$. Note that it follows that $d_X(k') \geq 1$. Since, $k \mathcal{Z}_{2.d_X(k')} m$ and $k \preceq h$, there is an $n \succeq m$ with $h \mathcal{Z}_{2.d_X(k')-1} n$. Moreover: $2.d_X(h) + 1 \leq 2.d_X(k') - 1$. Ergo: $h \mathcal{Z}_{2.d_X(h)+1} n$. So we can take $k'^{*} := h$, $k^{*} := h$, $m'^{*} := n$ to witness $\langle \Gamma, n\rangle$. Clearly, $\langle \Delta, m\rangle \preceq \langle \Gamma, n\rangle$. By the Induction Hypothesis: $\langle \Gamma, n\rangle \not\models C$. Hence, $\langle \Delta, m\rangle \not\models \Box C$.

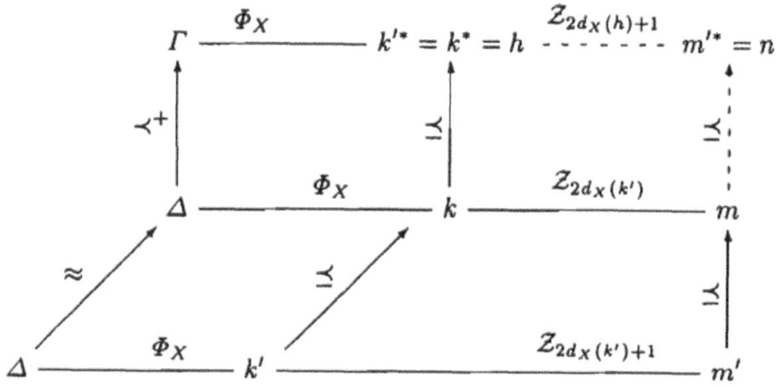

Thus we have proved Claim 2. □

The statement of uniform interpolation and the semantical interpretation of the propositional quantifiers are fully analogous to the case of GL.

We show that Uniform Interpolation for S4Grz implies Uniform Interpolation for IPC. By itself this is not so important, since we proved Uniform Interpolation for IPC directly. I feel, however, that the methodology of such transfers is interesting by itself.

Define $\mathsf{Nec}(A) := \bigwedge\{\Box(p \to \Box p) \mid p \in \mathsf{PV}(A)\}$. The Gödel Translation $(.)^{*}$ from $\mathcal{L}^{i}$ to $\mathcal{L}^{m}$ is specified as follows.

- $(.)^{*}$ commutes with atoms, $\wedge$ and $\vee$
- $(A \to B)^{*} := \Box(A^{*} \to B^{*})$

Lemma 8.2. *1.* $\mathsf{IPC} \vdash A \Leftrightarrow \mathsf{S4Grz} \vdash \mathsf{Nec}(A) \to A^{*}$.
2. $\mathsf{S4Grz} \vdash (\mathsf{Nec}(A) \wedge A) \to \Box A \Rightarrow$ *for some* $A^{i} \in \mathcal{L}^{i}$, $\mathsf{S4Grz} \vdash \mathsf{Nec}(A) \to (A \leftrightarrow A^{i*})$.

Proof. (1) and (2) are a well know facts. (1) is due to Gödel. (2) is probably first due to Rybakov. We prove (2). The proof is by induction on the length of A. Suppose S4Grz $\vdash$ $(\mathsf{Nec}(A) \wedge A) \rightarrow \Box A$. We rewrite A to conjunctive normal form treating the boxed formulas as atoms. Schematically, this form is: $\bigwedge\{\bigvee\{\Box B, \neg\Box C, p, \neg q\}\}$. We find, in S4Grz + Nec(A):

$$\begin{aligned} A &\leftrightarrow \bigwedge\{\bigvee\{\Box B, \neg\Box C, p, \neg q\}\} \\ &\leftrightarrow \Box\bigwedge\{\bigvee\{\Box B, \neg\Box C, p, \neg q\}\} \\ &\leftrightarrow \bigwedge\{\Box\bigvee\{\Box B, \neg\Box C, p, \neg q\}\} \\ &\leftrightarrow \bigwedge\{\Box(\bigwedge\{\Box C, q\} \rightarrow \bigvee\{\Box B, p\})\} \\ &\leftrightarrow \bigwedge\{\Box(\bigwedge\{(\Box C)^{i*}, q\} \rightarrow \bigvee\{(\Box B)^{i*}, p\})\} \end{aligned}$$

So we can take $A^i := \bigwedge\{(\bigwedge\{(\Box C)^i, q\} \rightarrow \bigvee\{(\Box B)^i, p\})\}$. □

Theorem 8.1. *Uniform Interpolation for* S4Grz *implies Uniform Interpolation for* IPC

Proof. Consider A in $\mathcal{L}^i$. Let $\mathbf{q}$ be some subset of PV(A). Let $\tilde{A}$ be the post-interpolant w.r.t. $\mathbf{q}$ of $\mathsf{Nec}(A) \wedge A^*$ in S4Grz. Note that: S4Grz $\vdash (\mathsf{Nec}(A) \wedge A^*) \rightarrow \Box\tilde{A}$. Hence, by the properties of the post-interpolant: S4Grz $\vdash \tilde{A} \rightarrow \Box\tilde{A}$. Thus, we can find an $\mathcal{L}^i$-formula $\tilde{A}^i$, such that S4Grz $\vdash \mathsf{Nec}(\tilde{A}) \rightarrow (\tilde{A} \leftrightarrow \tilde{A}^{i*})$. We show that $\tilde{A}^i$ is the desired post-interpolant. Note that, S4Grz $\vdash (\mathsf{Nec}(A) \wedge A^*) \rightarrow \tilde{A}^{i*}$. We may conclude: IPC $\vdash A \rightarrow \tilde{A}^i$.

Suppose IPC $\vdash A \rightarrow B$, where the shared variables of A and B are in $\mathbf{q}$. It follows that: S4Grz $\vdash \mathsf{Nec}(A \rightarrow B) \rightarrow (A^* \rightarrow B^*)$. Hence, S4Grz $\vdash (\mathsf{Nec}(A) \wedge A^*) \rightarrow (\mathsf{Nec}(B) \rightarrow B^*)$. Thus: S4Grz $\vdash \tilde{A}^{i*} \rightarrow (\mathsf{Nec}(B) \rightarrow B^*)$. And so, S4Grz $\vdash (\mathsf{Nec}(\tilde{A}^i \rightarrow B) \wedge \tilde{A}^{i*}) \rightarrow B^*$. Ergo, IPC $\vdash \tilde{A}^i \rightarrow B$.

We turn to pre-interpolants. Consider B in $\mathcal{L}^i$. Let $\mathbf{q}$ be some subset of PV(B). Let B' be the pre-interpolant w.r.t. $\mathbf{q}$ of $\mathsf{Nec}(B) \rightarrow B^*$ in S4Grz. Take $\breve{B} := \Box B'$. We can find an $\mathcal{L}^i$-formula $\breve{B}^i$, such that S4Grz $\vdash \mathsf{Nec}(\breve{B}) \rightarrow (\breve{B} \leftrightarrow \breve{B}^{i*})$. We show that $\breve{B}^i$ is the desired pre-interpolant. Note that, S4Grz $\vdash (\mathsf{Nec}(B) \wedge \breve{B}^{i*}) \rightarrow B^*$. We may conclude: IPC $\vdash \breve{B}^i \rightarrow B$.

Suppose IPC $\vdash A \rightarrow B$, where the shared variables of A and B are in $\mathbf{q}$. It follows that: S4Grz $\vdash \mathsf{Nec}(A \rightarrow B) \rightarrow (A^* \rightarrow B^*)$. Hence, S4Grz $\vdash (\mathsf{Nec}(A) \wedge A^*) \rightarrow (\mathsf{Nec}(B) \rightarrow B^*)$. Thus: S4Grz $\vdash (\mathsf{Nec}(A) \wedge A^*) \rightarrow B'$. And so, S4Grz $\vdash (\mathsf{Nec}(A) \wedge A^*) \rightarrow \breve{B}$ (since $(\mathsf{Nec}(A) \wedge A^*)$ is self-necessitating). So, finally, S4Grz $\vdash (\mathsf{Nec}(A) \wedge A^*) \rightarrow \breve{B}^{i*}$. Ergo, IPC $\vdash A \rightarrow \breve{B}^i$. □

It would be interesting to find a similar argument to prove Uniform Interpolation for S4Grz from Uniform Interpolation for GL. In their paper [4] Ghilardi and Zawadowski show that S4 does not satisfy uniform interpolation. In fact, the following formula $A(p, q, r)$,

$$p \wedge \Box(p \rightarrow \Diamond q) \wedge \Box(q \rightarrow \Diamond p) \wedge \Box(p \rightarrow r) \wedge \Box(q \rightarrow \neg r)$$

does not have a post-interpolant w.r.t. r.

Acknowledgement. I thank Marco Hollenberg and Volodya Shavrukov for reading earlier versions of the manuscript. I thank Giovanna d' Agostino, Dick de Jongh, Jelle Gerbrandy and Domenico Zambella for their comments on the material and their interest during a course on IPC. I am grateful to Paul Taylor for the use of his macros to make the diagrams.

References

1. R. Dyckhoff. Contraction-free sequent calculi for intuitionistic logic. *Journal of Symbolic Logic* **78** (1992) 795–807.
2. S. Ghilardi. An algebraic theory of normal forms. *Annals of Pure and Applied Logic* **71** (1995) 189–245.
3. S. Ghilardi and M. Zawadowski. A sheaf representation and duality for finitely presented Heyting algebras. *Journal of Symbolic Logic* **60** (1995) 911–939.
4. S. Ghilardi and M. Zawadowski. Undefinability of Propositional Quantifiers in the Modal System S4. *Studia Logica* **55** (1995) 259–271.
5. Z. Gleit and W. Goldfarb. Characters and Fixed Points in Provability Logic. *Notre Dame Journal of Formal Logic* **31** (1990) 26–36.
6. R. Goldblatt. Saturation and the Hennessy-Milner Property. In: [11] 107–129.
7. W. Hodges. *Model theory. Encyclopedia of Mathematics and its Applications* **42** (1993), Cambridge University Press, Cambridge.
8. M. Hollenberg. Hennessy-Milner Classes and Process Algebra. In: [11] 187–216.
9. J. Hudelmaier. *Bounds for cut elimination in intuitionistic propositional logic.* Ph.D. Thesis (1989), University of Tübingen, Tübingen.
10. A. Pitts. On an interpretation of second order quantification in first order intuitionistic propositional logic. *Journal of Symbolic Logic* **57** (1992) 33–52.
11. A. Ponse, M. de Rijke and Y. Venema (eds.). *Modal Logic and Process Algebra, a Bisimulation Perspective. CSLI Lecture Notes* **53** (1995), Center for the Study of Language and Information, Stanford.
12. V. Yu. Shavrukov. Subalgebras of diagonalizable algebras of theories containing arithmetic. *Dissertationes Mathematicae* **323** (1993), Polska Akademia Nauk, Mathematical Institute, Warszawa
13. A. Visser, J. van Benthem, D. de Jongh and G. Renardel de Lavalette. NNIL, a Study in Intuitionistic Propositional Logic. In: [11] 289–326.

Part II

Contributed Papers

Gödel's Ontological Proof Revisited *

C. Anthony Anderson and Michael Gettings

University of California, Santa Barbara
Department of Philosophy

Gödel's version of the modal ontological argument for the existence of God has been criticized by J. Howard Sobel [5] and modified by C. Anthony Anderson [1]. In the present paper we consider the extent to which Anderson's emendation is defeated by the type of objection first offered by the Monk Gaunilo to St. Anselm's original Ontological Argument. And we try to push the analysis of this Gödelian argument a bit further to bring it into closer agreement with the details of Gödel's own formulation. Finally, we indicate what seems to be the main weakness of this emendation of Gödel's attempted proof.

1.

Gaunilo observed against St. Anselm that his form of argument, if cogent, could be used to "prove" all sorts of unwelcome conclusions — for example, that there is somewhere a perfect island. It would seem to even follow that there are near-perfect, but defective, demi-gods and all matter of other theologically repugnant entities. Gaunilo concluded, reasonably enough, that something must be wrong with the argument.

Kurt Gödel's modern version of the Ontological Argument [12] involves an attempt to complete the details of Leibniz's proof that it is possible that there is a perfect being or a being with all and only "positive" attributes. Given this conclusion, other assumptions about positive properties, and, well, a second-order extension of the modal logic S5, Gödel successfully deduced the actual existence, indeed the necessary existence, of the being having all and only positive attributes. Alas, or "Oh, joy!", depending on ones' theological prejudices, J. Howard Sobel showed that Gödel's assumptions lead also to the conclusion that whatever is true is necessarily true. Followers of Spinoza aside, this casts quite considerable doubt on the premisses of the argument. We shall consider here Anderson's emendation which does not suffer from the mentioned defect and which is still recognizably closely related to Gödel's argument. [1]

---

* This paper is in its final form and no similar paper has been or is being submitted elsewhere.

[1] Petr Hájek [13] has argued that Anderson's version of the argument has superfluous premisses, but the truth of this claim depends on the details of the underlying second-order modal logic adopted. In the context of the (quite reasonable) version of that logic formulated by Nino Cocchiarella [2] and cited by

Here are the assumptions and definitions — the notion of a positive attribute is taken as a primitive by Gödel and in the present version. We hasten to add that the idea is not crystal clear; Gödel's own explanations are extremely terse and somewhat cryptic. A property's being positive is supposed to be a good thing, such properties being characteristic of a completely and necessarily non-defective being.

(A1*) If Φ is positive, then its complement non-Φ is not positive.
(A2*) If Φ is positive and necessarily all Φ's are Ψ's, then Ψ is positive — that is, properties entailed by positive properties are themselves positive.
Definition: "x is Godlike" means by definition that x has a property Φ necessarily, if and only if the property Φ is positive.
(A3*) Godlikeness is a positive property.
(A4*) If a property Φ is positive, then it is necessarily positive.
Definition: "Property Φ is an essence of an entity x" means that Φ entails all and only those properties which x has necessarily.
Definition: aa$x$ is necessarily existent" means that every essence of x is necessarily instantiated.
(A5*) Necessary existence is a positive property.

From these it follows (in the mentioned logic) that it is necessary that there exists a Godlike being — indeed that such a being is unique. (Details of the argument may be found in Anderson [1].)

What would Gaunilo say? Following a suggestion of Patrick Grim's, he might argue thus: Let's say that a property is "restricted-positive" if it is positive, but does *not* entail some particular positive property — say moral goodness. Now define x to be "nearly-Godlike" if it has a property necessarily, if and only if that property is *restricted*-positive. Can we not argue, heretically, that there necessarily exists a nearly Godlike being — a being otherwise perfect, but lacking moral perfection?

Curiously, the objection fails. We find this a bit surprising since Gaunilo-type objections seem to apply powerfully and persuasively to virtually every other version of the Ontological Argument with which we are acquainted.

To produce a persuasive reductio, the Gaunilist parallel argument must make the corresponding assumptions in his Near-Ontological Argument. In the present case, he must assume, in addition to the analogues of the other axioms, the analogue of axiom (A3*), namely:

Anderson as the underlying logic, Hájek's point does *not* hold. It does indeed follow from the diminished set of premisses that: if THERE IS a Godlike being, then necessarily THERE IS such a being. From this one can easily deduce that THERE IS a Godlike being, as Hajek observes. But the sense of the quantifier indicated by the capitalized phrase here (i.e., as formalized in Cocchiarella's logic) is that of "possible existence" or "subsistence". To express actual existence requires a separate quantifier. So without using the other premisses, nothing yet follows about the actual existence of a Godlike being.

(N3*) Near-godlikeness is a restricted-positive property.

But this won't do. Use "G'" for the property of near-Godlikeness and attend to the definition of this property and of restricted-positivity. Since moral goodness, call it M, is positive and entails the disjunctive property of being either non-G'-or-M, this latter property must be positive. [2] Now this disjunctive property does not itself entail M (assuming, what is evident, that the property of being non-G' is possibly exemplified). Hence the disjunctive property non-G'-or-M is restricted-positive. Given the definition of near-Godlikeness (viz., having all and only restricted-positive properties necessarily), it follows that near-Godlikeness, G', entails the disjunctive property non-G'-or-M. Hence G' entails M — and so is *not* itself restricted-positive, contrary to (N3*).

Of course the Gaunilist might find some defect in the axiom we have used against him — that properties entailed by a positive property are themselves positive. But this is very plausibly construed as a crucial feature of the very idea of positiveness to which, obscure as it is, the Ontological Arguer is entitled. There is also the further possibility that some other Gaunilo-type argument will succeed where this one has failed, but our best efforts to produce such have so far been fruitless. We conclude, tentatively, that this emendation of Gödel's Ontological Proof may well be immune to the Gaunilo-type of objection.

2.

We suggest further revisions of Anderson's modifications of Gödel's Ontological Argument which bring the assumptions and reasoning closer to Gödel's original intent. In the place of the Axiom A1 (and Anderson's modified A1*), we propose:

(A1') $\neg[Pos(\Box F) \equiv Pos(\neg \Box F)]$,

that is, exactly one of the two properties *being necessarily F* and *not being necessarily F* is positive. (We use obvious abbreviations here. For example, '$\Box F$' abbreviates '$\lambda x \Box F(x)$' — the property which anything x has when it is necessarily F.) Gödel's corresponding axiom asserts that of each property and its negation, or complement, exactly one is positive. This can be seen as in some measure responsible for the "modal collapse" noted by Sobel — that is, from Gödel's assumptions and using a natural second-order modal logic, it follows that every proposition which is true is necessary. Anderson's emendation replaced Gödel's premiss by its weaker consequence: if a property is positive, then its negation is not. The present axiom seems closer in spirit to

[2] The Gaunilist has offered no objection to axiom (A2*) and indeed may use it to prove his analogue thereof.

the original, since it just modalizes the property before asserting the exclusive disjunction.

A second alteration we propose is to eliminate what is apparently Dana Scott's gloss on one of Gödel's premisses in favor of that exact assumption (See Sobel [5]). Scott simplified Gödel's argument by assuming that *being Godlike* is positive and Anderson paralleled this with the premiss that *being Godlike** is positive. But in the original note, Gödel had instead that the conjunction of any number of positive properties is positive, including the "infinite conjunction" formed by universal quantification. Since to be Godlike is to have all and only positive properties, Scott's premiss follows easily from Gödel's definition and the Conjunction Assumption (as we shall call it). So we adopt:

(A3') If F is any set of positive properties, then the property obtained by taking the conjunction of the properties in F is positive. (One needs third-order logic to state this formally. In that context, the assumption is:

(A3') $\forall F[\Phi(F) \supset Pos(F)] \supset \forall G \Box \forall x\{G(x) \equiv \forall F[\Phi(F) \supset F(x)]\} \supset Pos(G)$.)

Finally, we adopt an assumption which Gödel explicitly endorses elsewhere in his notes on the Ontological Argument (Cf. [12]): that the necessitation of a positive property is positive:

(A6') $Pos(F) \supset Pos(\Box F)$.

Gödel observes that with this addition the assumption (A5) that Necessary Existence is a positive property can be replaced by the weaker premiss that Existence itself is a positive property. (Here, to avoid needless controversy, we can take Existence to be defined as follows: x exists if and only if some essence F of x is exemplified, i.e. there is a y such that $F(y)$).

It is interesting to observe that with these assumptions and utilizing Anderson's emended definitions of "essence" and "Godlikeness", we can deduce the conclusion that necessarily there exists a Godlike being. As presently advised, then, we propose as a rational reconstruction of Gödel's reasoning, the assumptions stated herein, together with A2, and A4. In addition we still maintain that the emended definitions of essence and Godlikeness are preferable for the present purposes. In an appendix, and on the handout, we sketch proofs of the technical claims we have made.

3.

Does Gödel's Ontological Argument then rest secure? We think not. Anderson and, especially, Robert Adams have emphasized that taking the conjunction of two or more positive properties to be positive or else just assuming that

Godlikeness is positive, together with the assumption that positive properties entail only positive properties comes quite close, epistemically, to just assuming that the existence of God (defined in Gödel's or in Anderson's way) is possible. We can of course demand that a positive property be *purely* positive — it must entail no negation, and in particular, no contradiction. But then how can we be sure that taking conjunctions of such will not yield an impossibility (given that the incompatibility involved need not be purely "formal").

Leibniz thought that the premiss that God's existence is possible needs to be proved and he attempted such a proof using the idea that a perfection (which corresponds roughly to Gödel's notion of a positive property) must be "simple". And he thought that it is then reasonable to conclude that the conjunction of two or more perfections cannot be impossible. But it just isn't clear that this is so and if we make assumptions that guarantee that it is so, along the lines of Gödel's axioms, then it is difficult to see that any epistemic progress has been made. We suggest that the Gödelian Ontological Arguer should simply admit that neither the possibility of God nor the truth of the axioms used to "prove" that possibility are self-evident. And he might just maintain that the less evident axioms, for example that a conjunction of positive properties is positive, is an assumption which he adopts on grounds of mere plausibility and is entitled to accept until some incompatibility between clearly positive properties is discovered.

Appendix

Assuming A1', we show that
(1) $\neg Pos(F) \supset Pos(\neg \Box F)$.

Proof. Assume that F is not positive. Then necessarily F cannot be positive, for necessarily F entails F — and the properties entailed by a positive property are positive (A2). Hence the second alternative of A1' must hold: $Pos(\neg \Box F)$.

Now contemplate the definition of Godlikeness* — something is Godlike* if and only if it has all and only positive properties essentially. We show, using Gödel's general "conjunction axiom" (A3'), that Godlikeness* is positive. To do so, it is evidently sufficient to show that any property of the form $\lambda x\{\Box H(x) \equiv Pos(H)\}$ is positive — since Godlikeness* is the "infinite conjunction" of them. In turn, to show this, it suffices to show that each property of the form: $\lambda x\{\Box H(x) \supset Pos(H)\}$ and each property of the form $\lambda x\{Pos(H) \supset \Box H(x)\}$ is positive.
(2) $\lambda x\{\Box H(x) \supset Pos(H)\}$ is always a positive property.

Proof. Either H is positive or it is not. If so, then the indicated property is positive, being entailed by the (vacuous) and necessary property $\lambda x Pos(H)$

((A4), (A2)). If it is not, then by (1), $Pos(\neg\Box H)$ — and the indicated property is entailed by $\neg\Box H$ (Again use A2).

(3) $\lambda x\{Pos(H) \supset \Box H(x)\}$ is always a positive property.

Proof. If $Pos(H)$, then $Pos(\Box H)$ by A6 — and the indicated property is entailed by $\Box H$ (A2).

We may conclude from (2), (3) and the conjunction axiom (A3') that Godlikeness* is positive — Anderson's original Axiom A3*.

We note further that Anderson's original modification of Gödel's first axiom is a consequence of the present set of assumptions:
(4) $Pos(F) \supset \neg Pos(\neg F)$.

Proof. Assume that F is positive and, for a reductio, that $\neg F$ is positive as well. Then by (A6), $Pos(\Box F)$. Furthermore, if $\neg F$ is positive, then so is $\Diamond\neg F$, being entailed by it. But this latter is just the property $\neg\Box F$, so that we have both $Pos(\Box F)$ and $Pos(\neg\Box F)$ — contradicting A1'.

The line of reasoning Gödel suggests for proving that Necessary Existence is positive, from the premiss that Existence is positive may be reproduced here if we take as our definition of Existence: $E(x) =_d f\forall F[FEssx \supset \exists xF(x)]$. [If this reasoning is formalized using Cocchiarella's system, the existential quantifier here is actualist.]

Thus the reasoning of Anderson's version of the Gödel's Ontological Argument goes through, given the present premisses.

References

1. C. A. Anderson. Some emendations of Gödel's ontological argument. *Faith and Philosophy*, 7:291–303, 1990.
2. N. B. Cocchiarella. A completeness theorem in second order modal logic. *Theoria*, 35:81–103, 1969.
3. K. Gödel. In S. Feferman and all, editors, *Collected Works. Volume 3.* Oxford University Press, New York, 1995.
4. P. Hájek. Magari and others on Gödel's ontological proof. In Ursini and Agliano, editors, *Logic and Algebra*, pages 125–136. Marcel Dekker, Inc., 1996.
5. J. H. Sobel. Gödel's ontological proof. In J. J. Thomson, editor, *On Being and Saying. Essays for Richard Cartwright.* The MIT Press, Cambridge, Mass. & London, England, 1987.

A Uniform Theorem Proving Tableau Method for Modal Logic*

Tadashi Araragi

NTT Communication Science Laboratories
2-2, Hikaridai, Seika-cho, Souraku-gun, Kyoto 619-02 Japan
araragi@cslab.kecl.ntt.jp

Summary. In this paper, we propose a uniform theorem proving tableau method for a wide class of systems in propositional modal logic. The class is wide enough to include most well-known systems. In this method, for a given natural number μ, a modal formula θ is effectively transformed to a first-order formula $\Delta(\theta)_\mu$ without depending on the system addressed. The transformation is based on the idea of tableau methods. Now, if S is a system that is complete for a class of Kripke frames characterized by a first-order formula Σ, then $S \vdash \theta$ iff $\Sigma \supset \Delta(\theta)_\mu$ is provable in first-order logic for some μ. This method also raises questions that are interesting from a theoretical viewpoint.

1. Introduction

In this paper, we propose a uniform tableau method for a wide class of systems in propositional modal logic. Tableau methods are efficient ways of theorem proving, based on the idea of model elimination [4], [6]. These methods are known to be especially useful for modal logic. Tableau methods have been proposed for a number of well-known systems: K, D, T, B, S4, and S5. However, there are many systems to which tableau methods have not been proposed yet. For example, S4.1 and S4.2 are not covered. Moreover, each of the proposed methods requires a system dependent individual device to achieve a complete theorem proving procedure. At present, there is no general strategy for obtaining tableau methods for all systems.

The aim of this paper is to propose a uniform tableau method applicable to a wide class of systems. The class consists of complete normal systems which are complete for a class of Kripke frames characterized by a first-order formula. The class is natural and wide enough to include most well-known systems. In our method, a given modal formula θ is effectively transformed to a first-order formula $\Delta(\theta)_\mu$ for a given natural number μ. This transformation is independent of the system being addressed. We have the following correctness theorem.

If S is a system, complete for a class of Kripke frames characterized by a first-order formula Σ, then $S \vdash \theta$ iff $\Sigma \supset \Delta(\theta)_\mu$ is provable in first-order logic for some μ.

* This paper is in its final form and no similar paper has been or is being submitted elsewhere.

Our method is an extension of prefixed tableau methods [4], [5], [11]. Tableau methods are based on the idea of model elimination. In these methods, we assume an input formula θ is false in some world of a Kripke structure of a system and, according to the definition of Kripke structures, we create worlds, decide the truth of subformulas in those worlds and derive a contradiction: some formula is both true and false at the same time in some world. Hence, we conclude that θ is valid in all the Kripke structures; that is, θ is a theorem of the system. In a prefixed tableau method, in contrast, we do not create an actual world, but attribute world indexes to each formula generated during the proving procedure. Then, we select each pair of prefixed formulas whose formula parts are the same, but whose truths are different, and check if the indexes can be the same under special unifications derived from the accessibility relation of Kripke frames for the system. In [11], the unifications are given for systems K, D, T, K4, D4, S4, and S5. The unification method for these systems is efficient, but is restricted to only a few number of systems and requires a special device depending on systems. In our method, on the other hand, we introduce a more general notion for indexes by using Skolem functions and transform the condition of the contradictions to a first-order formula. As a result, we obtain a system independent theorem proving method for a wide class of systems in modal logic.

2. Preliminaries for Modal Logic

2.1 Syntax of modal logic

Definition 2.1. The alphabet used in this paper for a language $\mathcal{L}$ of modal logic is as follows.
logical connectives: $\wedge$, $\vee$, $\neg$, $\supset$, L
countably many propositional variables: p, q, r, . . .

Definition 2.2. *Well-formed formulas* (*wffs*) of the language $\mathcal{L}$ are defined recursively in the following.
1) Propositional variables are wffs.
2) If ϕ and ψ are wffs, then $\phi\wedge\psi$, $\phi\vee\psi$ and $\phi\supset\psi$ are all wffs.
3) If ϕ is a wff, then $\neg\phi$ and $L\phi$ are wffs.

As usual, we use brackets (and) for convenience.

2.2 Semantics of modal logic

Definition 2.3. Let W be a non-empty set and R be a binary relation on W $(R\subset W \times W)$. We call a pair $\langle W, R\rangle$ a *frame*. W and R are called a set of worlds and an accessibility relation on worlds, respectively. Let V be a binary relation on W and the set of all propositional variables of $\mathcal{L}$ (denoted by PV). That is $V \subset W \times PV$. We call a triple $\langle W, R, V\rangle$ a *Kripke structure* of $\mathcal{L}$ and V a *valuation*. We define models by using the Kripke structure below. In the following, we define the relation $M,w \models \phi$, where M is a Kripke structure $\langle W, R, V\rangle$, $w \in W$ and ϕ is a wff.

1) If p is a propositional variable , $M,w \models p$ iff $(w,p) \in V$.
2) $M,w \models \phi \vee \psi$ iff $M,w \models \phi$ or $M,w \models \psi$.
3) $M,w \models \phi \wedge \psi$ iff $M,w \models \phi$ and $M,w \models \psi$.
4) $M,w \models \neg\phi$ iff not $M,w \models \phi$ (We denote this by $M,w \not\models \phi$).
5) $M,w \models \phi \supset \psi$ iff $M,w \not\models \phi$ or $M,w \models \psi$.
6) $M,w \models L\phi$ iff for all v such that $(w,v) \in R$, $M,v \models \phi$.

When $M,w \models \phi$ for all $w \in W$, we write $M \models \phi$ and call the Kripke structure M a *(Kripke) model* of ϕ.

For convenience, we may write $M,w \models T\phi$ and $M,w \models F\phi$ for $M,w \models \phi$ and $M,w \not\models \phi$, respectively. Here, T and F denote "true" and "false", respectively.

A system S is called a *complete normal* system if there is a class of Kripke frames C such that the set of all the theorems of S coincides with the set $\{\phi \mid$ for all $\langle W, R\rangle \in C$ and for all V, $\langle W, R, V\rangle \models \phi\}$. Most well-known systems are complete normal.

2.3 First-order definable systems

We will define a class of systems to which our proving procedure is uniformly applicable.

Definition 2.4. Let Σ be a closed wff of FOL whose predicate symbols are among R and $=$. The class of frames $C(\Sigma)$ is defined to be $\{$frame $\langle W, R\rangle \mid \langle W, R\rangle \models \Sigma\}$. A system is called a *complete normal system defined by Σ* (we will call it *Σ-system* for short), if the set of all theorems of the system coincides with $\{\phi \mid$ for any $\langle W, R\rangle \in C(\Sigma)$ and any valuation V, $\langle W, R, V\rangle \models \phi\}$. A system is called *first-order definable* if it is a complete normal system defined by Σ for some Σ.

Examples:
If Σ is a tautology, then Σ-system is the system K.
If Σ is $\forall x R(x,x)$, then Σ-system is the system T.
If Σ is $\forall x R(x,x) \wedge \forall x \forall y \forall z (R(x,y) \wedge R(y,z) \supset R(x,z))$, then Σ-system is the system S4.
If Σ is $\forall x R(x,x) \wedge \forall x \forall y \forall z (R(x,y) \wedge R(y,z) \supset R(x,z)) \wedge \forall x \forall y \forall z (R(x,y) \wedge R(x,z) \supset \exists w R(y,w) \wedge R(z,w))$, then Σ-system is the system S4.2.
If Σ is $\forall x R(x,x) \wedge \forall x \forall y \forall z (R(x,y) \wedge R(y,z) \supset R(x,z)) \wedge \forall x \forall y \forall z (R(x,y) \wedge R(x,z) \supset R(y,z) \vee R(z,y))$, then Σ-system is the system S4.3.

3. A Uniform Theorem Proving Method

A *prefix* is a list of terms constructed from variables and functional symbols. A *signature* is a symbol of either T or F. A *prefixed formula* is a sequence of three elements in the following order: a prefix, a signature and a wff. For example, $[f, x, y, g(x, y), z]Fp \vee L(p \wedge q)$ is a prefixed formula.

[Procedure] A wff θ and a natural number μ are given and perform the following operations in the given order.

step 1. First, associate a new Skolem function symbol with each subformula of θ that has form $L\phi$ and occurs positively in θ; in other words, negatively in $F\theta$. We say that the Skolem function symbol *corresponds* to the subformula $L\phi$. This notion is used in the π-rule below.
Next, let $\mu = 1$ and E be the prefixed formula $[0]F\theta$. Here, 0 is a Skolem function with 0-ary.

step 2. Apply the following rules to the subformulas of E repeatedly until no rule is applicable to them. Then, let E' be the resulting expression.

⟨α- rule⟩
$sT\phi\wedge\psi \longrightarrow sT\phi \times sT\psi$
$sF\phi\vee\psi \longrightarrow sF\phi \times sF\psi$
$sF\phi\supset\psi \longrightarrow sT\phi \times sF\psi$
$sT\neg\phi \longrightarrow sF\phi$
$sF\neg\phi \longrightarrow sT\phi$

⟨β- rule⟩
$sT\phi\vee\psi \longrightarrow (sT\phi + sT\psi)$
$sF\phi\wedge\psi \longrightarrow (sF\phi + sF\psi)$
$sT\phi\supset\psi \longrightarrow (sF\phi + sT\psi)$

(Here, s denotes a prefix.)

⟨ν- rule⟩
$sTL\phi \longrightarrow s^*[x_1]T\phi\times\ldots\times s^*[x_\mu]T\phi$
Here, $x_1, .., x_\mu$ are μ pieces of new variables and * denotes a concatenation of lists.

⟨π- rule⟩
$sFL\phi \longrightarrow s^*[f(t)]F\phi$
Here, f is the Skolem function symbol corresponding to subformula $L\phi$ and t is the last element of list s.

step 3. Apply the following rules to the subexpressions of E' repeatedly until no rule is applicable to them.
$S \times (P + Q) \longrightarrow S \times P + S \times Q$
$(P + Q) \times S \longrightarrow P \times S + Q \times S$

Then, we get the resulting expression $S_{1,1} \times \ldots \times S_{1,k_1} + \ldots + S_{n,1} \times \ldots \times S_{n,k_n}$. Here, S, P, Q, and $S_{i,j}$ denote prefixed formulas.

[Notations] We call each $S_{j,1} \times \ldots \times S_{j,kj}$ above a *branch* of θ and denote the set of all branches of θ by $tab(\theta)_\mu$. A pair of prefixed formulas whose formulas are the same, but whose signatures are different, is called a *connection*. For a branch br, $co(br)$ denotes the set of all connections occurring in br. For a connection co of form $([g_1,..,g_m]Tp, [h_1,..,h_n]Fp)$, $rel(co)$ is defined to be the first-order formula $R(g_1,g_2)\wedge\ldots\wedge R(g_{m-1},g_m)\wedge R(h_1,h_2)\wedge\ldots\wedge R(h_{n-1},h_n)\wedge(g_m=h_n)$, where R is a binary predicate symbol.

step 4. Let $\nu(\theta)_\mu$ be $\bigwedge_{br\in tab(\theta)}\bigvee_{co\in co(br)} rel(co)$.
Let $\pi-set(\theta)_\mu$ be $\{R(g_1,g_2)\wedge\ldots\wedge R(g_{p-2},g_{p-1})\supset R(g_{p-1},g_p)\mid$ prefix $[g_1,..,g_p,..,g_t]$ appears in some prefixed formula of some branch in $tab(\theta)$ and g_p is a function term$\}$.
Let $\pi(\theta)_\mu$ be the conjunction of all elements of $\pi-set(\theta)_\mu$.
Lastly, let $\Delta(\theta)_\mu$ be $\exists x_1..\exists x_p[\pi(\theta)_\mu\supset\nu(\theta)_\mu]$. Here, $x_1,..,x_p$ are all the variables in $\pi(\theta)\supset\nu(\theta)$.

step 5. If $\Sigma\supset\Delta(\theta)_\mu$ is a theorem of first-order logic, then output the sentence "θ is a theorem of Σ-system" and halt.
Otherwise, let $\mu=\mu+1$ and go to step 2.

step 2 and **step 3** are based on the usual transformations of prefixed tableau methods. $rel(co)$ expresses that the formulas in the connection co are in the same world with opposite signatures. $\nu(\theta)$ expresses that every branch has some connection whose formulas belong to the same world. $\pi(\theta)$ expresses, by using function symbols, some information included in the given wff that is concerned with the existence of worlds. μ plays the same role as in [11]. That is, in the interpretation of the formula $L\phi$, we consider only μ worlds, instead of all worlds. $\Delta(\theta)_\mu$ can be written as $\forall x_1..\forall x_p\pi(\theta)_\mu\supset\exists x_1..\exists x_p\nu(\theta)_\mu$. We may abbreviate it as $\forall x\pi(\theta)_\mu$ for $\forall x_1..\forall x_p\pi(\theta)_\mu$.

We have the following correctness theorem.

Theorem 3.1 ((Correctness Theorem)). *θ is a theorem of the Σ-system iff $\Sigma\supset\Delta(\theta)_\mu$ is a theorem of first-order logic for some μ.*

4. Examples

By using examples, we will show how our procedure transforms any given wff θ of modal logic to the wff $\Delta(\theta)_\mu$ of first-order logic.

Example 1.
Input formula: $\theta=L(Lp\supset q)\vee L(Lq\supset p)$
$\mu=1$

The process of transformation is as follows.
$[0]FL(Lp\supset q)\vee L(Lq\supset p)$
$[0]FL(Lp\supset q)\times[0]FL(Lq\supset p)$
$[0,1]FL(Lp\supset q)\times[0]FL(Lq\supset p)$

$[0,1]F(Lp \supset q) \times [0,2]F(Lq \supset p)$
$[0,1]TLp \times [0,1]Fq \times [0,2]F(Lq \supset p)$
$[0,1]TLp \times [0,1]Fq \times [0,2]TLq \times [0,2]Fp$
$[0,1,w1]Tp \times [0,1]Fq \times [0,2]TLq \times [0,2]Fp$
$[0,1,w1]Tp \times [0,1]Fq \times [0,2,w2]Tq \times [0,2]Fp$

$\pi(\theta)_1 = R(0,1) \wedge R(0,2)$
$\nu(\theta)_1 = [R(0,1) \wedge R(1,w1) \wedge \underline{R(0,2)} \wedge (w1 = 2)] \vee [\underline{R(0,2)} \wedge R(2,w2) \wedge \underline{R(0,1)} \wedge (w2 = 1)]$

The underlined atoms are redundant because they appear in $\pi(\theta)_1$ as a consequence.
$\Sigma \supset [R(0,1) \wedge R(0,2) \supset R(1,2) \vee R(2,1)]$
is the resulting transformed formula.

θ is an axiom of the system S4.3 and S4.3 is the complete normal system defined by $\forall x \, \forall y \, \forall z (R(x,y) \wedge R(x,z) \supset R(y,z) \vee R(z,y)) \wedge$ reflexiveness $\wedge$ transitivity.

Example 2.
Input formula: $\theta = L(p \vee \neg L \neg q) \supset (Lp \vee \neg L \neg q)$
$\mu = 1$

The process of transformation is as follows.
$Lp[0]FL(p \vee \neg L \neg q) \supset (Lp \vee \neg L \neg q)$
$[0]TL(p \vee \neg L \neg q) \times [0]F(Lp \vee \neg L \neg q)$
$[0,w1]T(p \vee \neg L \neg q) \times [0]F(Lp \vee \neg L \neg q)$
$[0,w1]T(p \vee \neg L \neg q) \times [0]FLp \times [0]F \neg L \neg q$
$([0,w1]Tp + [0,w1]T \neg L \neg q) \times [0]FLp \times [0]F \neg L \neg q$
$([0,w1]Tp + [0,w1]FL \neg q) \times [0]FLp \times [0]F \neg L \neg q$
$([0,w1]Tp + [0,w1]FL \neg q) \times [0]FLp \times [0]TL \neg q$
$([0,w1]Tp + [0,w1,f(w1)]F \neg q) \times [0]FLp \times [0]TL \neg q$
$([0,w1]Tp + [0,w1,f(w1)]F \neg q) \times [0,1]Fp \times [0]TL \neg q$
$([0,w1]Tp + [0,w1,f(w1)]F \neg q) \times [0,1]Fp \times [0,w2]T \neg q$
$([0,w1]Tp + [0,w1,f(w1)]Tq) \times [0,1]Fp \times [0,w2]T \neg q$
$([0,w1]Tp + [0,w1,f(w1)]Tq) \times [0,1]Fp \times [0,w2]Fq$
$[0,w1]Tp \times [0,1]Fp \times [0,w2]Fq + [0,w1,f(w1)]Tq \times [0,1]Fp \times [0,w2]Fq$

$\pi(\theta)_1 = R(0,1) \wedge (R(0,w1) \supset R(w1,f(w1)))$

$\nu(\theta)_1 = R(0,w1) \wedge \underline{R(0,1)} \wedge (w1{=}1) \wedge R(0,w1) \wedge \underline{R(w1,f(w1))} \wedge R(0,w2) \wedge (f(w1) = w2)$

The underlined atoms are redundant.
$\Sigma \supset \exists w[(R(0,1) \wedge (R(0,w) \supset R(w,f(w)))) \supset R(0,f(1))]$

is the resulting transformed formula.

It is known that θ is not a theorem of the system T, but of S4. Actually, for $\Sigma = \forall x R(x,x)$, $\Sigma \supset \exists w[(R(0,1) \wedge (R(0,w) \supset R(w,f(w)))) \supset R(0,f(1))]$ is not proved, but for Σ that includes transitivity, it is proved. Of course, the former does not mean that θ is not a theorem of T.

Example 3.
Input formula: $\theta = L(p \wedge q) \supset Lp \wedge Lq$
$\mu = 2$

The process of transformation is as follows.
$[0]FL(p \wedge q) \supset Lp \wedge Lq$
$[0]TL(p \wedge q) \times [0]F(Lp \wedge Lq)$
$[0,w1]T(p \wedge q) \times [0,w2]T(p \wedge q) \times [0]F(Lp \wedge Lq)$
$\vdots$
$[0,w1]Tp \times [0,w1]Tq \times [0,w2]Tp \times [0,w2]Tq \times ([0,1]Fp + [0,2]Fq)$

$\pi(\theta)_2 = R(0,1) \wedge R(0,2)$
$\nu(\theta)_2 = ((R(0,w1) \wedge R(0,1) \wedge (w1 = 1)) \vee (R(0,w2) \wedge R(0,1) \wedge (w2 = 1)))$
$\quad \wedge ((R(0,w1) \wedge \underline{R(0,2)} \wedge (w1 = 2)) \vee (R(0,w2) \wedge \underline{R(0,2)} \wedge (w2 = 2)))$
The underlined atoms are redundant.
$\Sigma \supset \exists w1 \exists w2\ (R(0,1) \wedge R(0,2) \supset ((R(0,w1) \wedge (w1 = 1)) \vee (R(0,w2) \wedge (w2 = 1)))$
$\quad \wedge ((R(0,w1) \wedge (w1 = 2)) \vee (R(0,w2) \wedge (w2 = 2))))$
is the resulting transformed formula.

θ is a theorem of any normal system. In fact, $\Delta(\theta)_2$ is proved, while $\Delta(\theta)_1$ is not a theorem.

Remark. We can simplify the proving procedure in some cases. First, we can prove that if some branch does not include any connection for some input μ, then, for any μ' larger than μ, there is always a branch not including any connection. Therefore, if we find a branch not including a connection, we can stop the calculation and conclude that θ is not a theorem. Secondly, as seen in example 1 above, by the completeness of paramodulation [3], if Σ does not include the predicate = positively, we can remove each predicate of the form $t_1 = t_2$ in the formula ν by applying $mgu(t_1, t_2)$ to ν, when they are unifiable and their variables appear only in the conjunct. If t_1 and t_2 are not unifiable, then we can remove the conjunct that includes the predicate $t_1 = t_2$ from ν.

5. Soundness and Completeness of the Procedure

The formal proof of soundness and completeness of the procedure is long. We give only the outline of the proof.

As for the soundness, we must show that if $\Sigma \supset \Delta(\theta)_\mu$ is a theorem of first-order logic for some μ, then θ is a theorem of Σ-system. This is achieved by purely using the definition of models of modal logic and first-order logic.

As for the completeness, we must show the reverse. We will show the contraposition of it. That is, if $\Sigma \supset \Delta(\theta)_\mu$ is not a theorem for any μ, then θ is not a theorem of Σ-system. Hence, all we have to do is construct a Kripke structure M of the system and locate a world w such that $M,w \models F\theta$ from the precondition.

First, we prove that $\Delta(\theta)_\mu \supset \Delta(\theta)_{\mu+1}$ is valid. Then, by the compactness theorem, the precondition means that the set of formulas $\{\Sigma,\neg\Delta(\theta)_1,\neg\Delta(\theta)_2, \neg\Delta(\theta)_3,...\}$ has a model. Moreover, by Löwenheim Skolem theorem, we can assume that the model is countable. Let the model be (W,I), where W is the underlying set and I is the interpretation of the predicate symbol R and the Skolem functions appearing in $\{\Delta(\theta)_\mu\}_{1\leq\mu}$. Then, we can employ $\langle W, R^I\rangle$ as the Kripke frame of the model we are seeking, where R^I is the interpretation of R by I. We index the worlds with natural numbers.

In the next step, we give a valuation to the frame. We construct partial valuations, step by step, depending on μ. First, we place the signed formula $F\theta$ at the world 0^I. We consider that partial valuations make θ false at 0^I. Next, we decompose the formula $F\theta$ just as done in **step 2** of the proving procedure. For example, if the formula has the form $F\phi\vee\psi$, then we place the two formulas $F\phi$ and $F\psi$ at the world. If the formula has the form $F\phi \wedge \psi$, then we split the partial valuation and place $F\phi$ at the world for one of the partial valuations and $F\psi$ for the other. If a placed formula has the form $TL\psi$, we place $T\psi$ to the first μ pieces of the accessible worlds according to their indexes. If a placed formula has the form $FL\psi$, we place $F\psi$ at the accessible world determined by the interpretation of Skolem functions. We then iterate this procedure.

As a result, we obtain maps (partial valuations) V^i_μ from W to the power set of the set of signed subformulas of θ ($1 \leq i \leq n_\mu$), where $V^i_\mu(w)$ is the set of all signed formulas placed at the world w in the branch corresponding to V^i_μ. Here, we can prove that for any μ there is a partial valuation V^i_μ without contradiction: for any world w and any formula ϕ, $F\phi \in V^i_\mu$ and $T\phi \in V^i_\mu$ cannot be simultaneously valid. This is due to the fact that (W,I) is a model of $\{\Sigma,\neg\Delta(\theta)_1,\neg\Delta(\theta)_2, \neg\Delta(\theta)_3,...\}$. The proof is straightforward from the construction of the partial valuations, but long and tedious. We define the extension relation $V^i_{\mu1} \lhd V^j_{\mu2}$ ($\mu1 \leq \mu2$) as $V^i_{\mu1}(w) \subset V^j_{\mu2}(w)$ for any world w. Note that if $V^i_{\mu1} \lhd V^j_{\mu2}$ and $V^j_{\mu2}$ has no contradiction, then $V^i_{\mu1}$ also has none. Therefore, by this relation, the partial valuations without contradiction form an infinite tree with a finite number of branches at each node. Then, by

König's lemma, we can find an infinite chain of partial valuations $V_1 \lhd V_2 \lhd \ldots$, such that each V_μ has no contradiction. Let V_∞ be the inductive limit of this sequence. In other words, $S\psi \in V_\infty(w)$ iff $S\psi \in V_\mu(w)$ for some μ, where S is either T or F. We define the valuation V as $\langle W, R^I, V\rangle, w \models p$ if $Tp \in V_\infty(w)$ for any propositional variable p. Then we can prove that if $S\phi \in V_\infty(w)$ then $\langle W, R^I, V\rangle, w \models S\phi$ for any formula ϕ, where S is as above. This is shown by formula induction. The key point is that $T\phi$ is now placed at every world accessible from one where $TL\phi$ is placed, by taking the limit of the chain. Therefore, in particular, $\langle W, R^I, V\rangle, 0^I \models F\theta$, and we have thus obtained a desired model.

6. Termination of the Procedure

The proving method for modal logic proposed here transforms the statement "a given wff of modal logic is a theorem of a certain system" to the statement "one of some countably many wffs of first-order logic is a theorem". Therefore, this procedure is semi-decidable, but not decidable in general. This corresponds to the fact that there is a first-order definable system that is not decidable [7]. On the other hand, many well-known first-order definable systems are known to be decidable. It is desirable that our proving method guarantees termination for those systems. A simple way to realize this is to find two computable functions ω and d from the set of all wffs to the set of natural numbers, which have the following properties.

(1) For a given θ, $\Sigma \supset \Delta(\theta)_\mu$ is a theorem for some μ iff $\Sigma \supset \Delta(\phi)_\mu$ is a theorem for some μ less then $\omega(\theta)$.
(2) The proof search of $\Sigma \supset \Delta(\theta)_\mu$ $(\mu \leq \omega(\theta))$ can be restricted to the bounded Herbrand space, the depth of whose terms are less then $d(\theta)$.

These functions have been obtained for K, D, T, S4, and S5, by analyzing their tableaux. The basic idea is to find the effectively calculated finite part of their tableaux such that if all branches have connections, we can find a connection in the finite part for all branches. To find the finite part, we capture the periodicity of their tableaux. The method depends on each system and we have not yet found a general framework for it.

7. Discussion

There is another approach to a uniform proving method called the translation method [8], which is applicable to the same class of systems. In this method, a modal formula is directly translated to a first-order formula by introducing predicate symbols with 1-ary for each propositional variable and an accessibility relation symbol. Our method, in contrast, uses only two predicate symbols:

an accessibility relation symbol and =. It is also known that in the translation method, a transformed formula grows exponentially. Recently, in [9] and [10], this method was improved by introducing special Skolem functions and axioms on them. Nonetheless, the method requires a special modification for systems not containing D and then loses efficiency and generality.

Example 1 in section 4 shows that the procedure has some relationship with Correspondence Theory [2]. For Sahlqvist's formulas, the procedure seems to give the condition on accessibility relations corresponding to an input formula. But, at present, we have no idea about for what class of systems this method gives the corresponding conditions mechanically.

References

1. T. Araragi: *A Uniform Prefixed Tableau Method for Positive First-Order Definable Systems*, Workshop of Theorem Proving with Analytic Tableaux and Related Methods, Technical Report 8/92, University of Karlsruhe, Institut für Logik, Komplexität und Deduktionssysteme, pp.4-6, 1992.
2. J. van Benthem: *Correspondence Theory*, in Handbook of Philosophical Logic II, pp.167-242, D.Gabbay and F.Guenthner eds., Dordrecht, Reidel, 1984.
3. C.L. Chang and R.C. Lee: *Symbolic Logic and Mechanical Theorem Proving*, Academic Press, 1973.
4. M.C. Fitting: *Proof Methods for Modal and Intuitionistic Logics*, vol. 169 of Synthese library, Dordrecht, Reidel, 1983.
5. P. Jackson and H. Reichgelt: *Logic-Based Knowledge Representation*, The MIT Press, 1989.
6. G.H. Hughes and M.J. Cresswell: *An Introduction to Modal Logic*, London, Methuen, 1968.
7. M. Kracht: *Highway to the Danger Zone*, Journal of Logic and Computation, to appear.
8. R. Moore. *Reasoning About Knowledge and Action*, PhD Thesis, MIT, Cambridge, 1980.
9. A. Nonnengart: *First-Order Modal Logic Theorem Proving and Functional Simulation*, 13th IJCAI, pp.80-85, 1993.
10. H.J. Ohlbach. *Semantics-based translation methods for modal logics*, Journal of Logic and Computation, vol. 1, pp.691-746, 1991.
11. L. Wallen: *Automated Proof Search in Non-Classical Logics*, The MIT Press, 1989.

Decidability of the $\exists^*\forall^*$-Class in the Membership Theory NWL *

Dorella Bellè and Franco Parlamento

Department of Mathematics and Computer Science
University of Udine, via Delle Scienze 206, 33100 Udine, Italy.
fax: +39 432 558499, e-mail: {*belle,parlamen*}@*dimi.uniud.it*

Summary. Let NWL be the theory having the obvious axioms for the existence of the empty set and of the result of adding or removing an element from a set. The problem of establishing whether sentences of the form $\exists x_1 \ldots \exists x_n \forall y_1 \ldots y_m F$, with F quantifier free, are satisfiable with respect to NWL is decidable.

1. Introduction

The basic role set theoretic notions play in mathematics, especially in its foundation, makes it quite natural to enquire which 'fragment' of set theory Gödel's incompleteness and undecidability results apply to. Already in [22], Tarski addressed this problem stating that the *small axiomatic fragment* - to use Tarski's wording - formulated in the language $=, \in$, endowed with classical first order logic, whose axioms are

N:	$(\forall x)(x \notin \emptyset)$	(Null-set Axiom),
W:	$(\forall x)(\forall y)(\forall z)(x \in y \,\mathbf{w}\, z \leftrightarrow x \in y \vee x = z)$	(With Axiom),
E:	$(\forall x)(\forall y)((\forall z)(z \in x \leftrightarrow z \in y) \rightarrow x = y))$	(Extensionality Axiom),

was sufficiently strong to interpret Robinson's Arithmetic **Q** (see also [6], [23]); a result recently improved by dropping the axiom E [11]. Later on R. Vaught establishes with a different method the essential undecidability of NW [21].

On the other hand, already in the early time of automated deduction, the *problem of handling the membership relation in an efficient way, so that the problems involving set theoretic notions can be treated* was brought to evidence by J. Robinson in [20]. Awareness of the importance of that problem has steadily increased and it is at present particularly evident in connection with the enhancement of declarative programming [1], [7], [8], [10]. Earlier, throughout the eighties, a project led by J.Schwartz at the Courant Institute (NYU), aiming at the development of a proof verifier for 'elementary set theory', conceived as a particularly important subdomain of mathematics, led to the discovery of various decision procedure to establish whether in the 'intended' model there are sets satisfying given constraints expressed with (mainly) unquantified formulae over some of the most basic set theoretic constructors [5].

* This work has been supported by funds 40% and 60% MURST; it is in its final form and will not be published elsewhere.

First limitations to what could be accomplished in that direction were established in [19]. As a matter of fact by a (standard) development of Gödel's incompleteness results combined with suitable coding, [14] establishes the essential undecidability of the theory NWLE obtained from NWL by adding the axiom

$$\mathrm{L} : (\forall x)(\forall y)(\forall z)(z \in x \,\mathrm{l}\, y \leftrightarrow z \in x \wedge z \neq y) \quad \text{(Less Axiom)}$$

with respect to existential closures of a restricted subclass of the formulae on the language $=, \in$, involving only the restricted quantifiers $\forall x \in y, \exists x \in y$ (Δ_0-formulae). Following [14] significant improvements of the limitative results, concerning both the theories and the class of sentences involved, have been obtained, noticeable the undecidability of the (logical) satisfiability of $\forall^*\exists$ sentences with respect to NWLE [3].

On the positive side some decision results (for membership theories) had been obtained in [25] as well as [9]. In particular [9] establishes by a model theoretic argument the completeness of ZFC with respect to $\exists^*\forall$ sentences. [13] improves this result by showing that the theory NWLER, where

$$\mathrm{R} : (\forall x)(x \neq \emptyset \rightarrow (\exists y)(y \in x \wedge (\forall z)(z \in y \rightarrow z \notin x))) \quad \text{(Regularity Axiom)},$$

is already complete with respect to $\exists^*\forall$ sentences. Furthermore [13] yields a number of positive results for such a class of sentences providing decision procedures (for their satisfiability) with respect to the theories NWLE,NWLR, and [18] extends such decidability results to the class of $\exists^*\forall\forall$-sentences with respect to the theory NWL.

In this work we establish the decidability of the full Bernays-Schönfinkel class, namely the class of $\exists^*\forall^*$-sentences with respect to the theory NWL. In the special case of the satisfiability of $\forall$ sentences the decision procedure we obtain does not differ significantly from the one already obtained in [13]. On the contrary for the special case of the $\forall\forall$-formulae our decision procedure is quite different from the one provided in [18] as it reduces to a set-(of finite graphs)-inclusion test, while the procedure presented in [18] consists in a number of attempts to build an Herbrand's model, searching for a construction which proceeds for sufficiently many steps. More importantly the method we present here works for the full Bernays-Schönfinkel class of which the $\forall\forall$ case is merely a special one, and it can be applied also to the Theory NWLR. The addition of the Extensionality Axiom makes matters combinatorially much harder, as it is suggested also by the existence of $\forall\forall$-formulae which are satisfiable but not finitely satisfiable when both the Extensionality and the Regularity Axiom are assumed and of $\forall\forall\forall$ formulae of this kind when the Extensionality Axiom alone is assumed (see [15], [16], [17]). As a matter of fact the decision problem for the full Bernays-Schönfienkel-class when the Extensionality Axiom is assumed is still open.

2. Reduction

A $\forall^m$-formula with free variables $x_1, \ldots, x_\ell$,

$$\forall y_1, \ldots, y_m A(x_1, \ldots, x_\ell, y_1, \ldots, y_m)$$

with A quantifier free in the language $\{\in, =\}$, is logically equivalent to a disjunction of formulae of the following form

$$B(x_1, \ldots, x_n) \wedge \forall y_1, \ldots, y_m (y_1 \neq x_1 \wedge \ldots \wedge y_m \neq x_n \rightarrow C(x_1, \ldots, x_n, y_1, \ldots, y_m))$$

to be abbreviated as $B(\mathbf{x}) \wedge \forall \mathbf{y} \neq \mathbf{x}\, C(\mathbf{x}, \mathbf{y})$, where $B(x_1, \ldots, x_n)$ has the form

$$\bigwedge_{1 \leq i \neq j \leq n} x_i \neq x_j \wedge \bigwedge_{1 \leq i,j \leq n} x_i \in_{i,j} x_j$$

with $\in_{i,j}$ either $\in$ or $\notin$, and all the literals appearing in C contain at least one of the variables $y_1, \ldots, y_m$.

Therefore the satisfiability problem for $\exists^*\forall^*$-sentences in the language $\in, =$ is reducible to the satisfiability problem for sentences of the form

$$B(\mathbf{c}) \wedge \forall \mathbf{y} \neq \mathbf{c}\, C(\mathbf{c}, \mathbf{y}) \tag{2.1}$$

where $\mathbf{c} = (c_1, \ldots c_n)$ is an n-tuple of distinct new constants replacing the free variables $x_1, \ldots, x_n$. $B(\mathbf{c})$ naturally induces a graph G over $\{1, \ldots, n\}$ by letting $(i,j) \in G$ iff $c_i \in c_j$ is a conjunct in $B(\mathbf{c})$.

In the sequel we will let F denote a sentence over $c_1, \ldots, c_n, \in, =$ of the form 2.1 above. We now reduce the satisfiability of such a sentence F to the set inclusion between two finite collections of graphs over $\{1, \ldots, n, n+1, \ldots, n+m\}$.

Definition 2.1. *$\Gamma_m(F)$ is the collection of $n+m$-graphs, i.e. graphs over $\{1, \ldots, n+m\}$ which satisfy F when $c_1, \ldots, c_n$ are interpreted with $1, \ldots, n$ respectively, $\in$ with the graph's relation and $=$ with the identity relation.*

Definition 2.2. *Let M be an interpretation of $\{c_1, \ldots, c_n, \in, =\}$ such that $=_M$ is an equivalence relation congruent with respect to $\in_M$ having more than $n+m$ $=_M$-equivalence classes and the interpretations $e_1, \ldots, e_n$ of $c_1, \ldots, c_n$ are not $=_M$ related. Then $\Gamma_m(M)$ is the collection of the $n+m$-graphs induced over $\{1, \ldots, n+1, \ldots, n+m\}$ by $\in_M$ restricted to $\{e_1, \ldots, e_n, a_1, \ldots, a_m\}$ where $a_1, \ldots, a_m$ is an m-tuple of elements of M not $=_M$-related to each other nor with any of $e_1, \ldots, e_n$.*

Proposition 2.1. *Let M be an interpretation of $\{c_1, \ldots, c_n, \in, =\}$ such that $=_M$ is an equivalence relation congruent with respect to $\in_M$ having more than $n+m$ $=_M$-equivalence classes and the interpretations of $c_1, \ldots, c_n$ are distinct. Then*

$$M \models F \quad \textit{iff} \quad \Gamma_m(M) \subseteq \Gamma_m(F) .$$

Proof. Let $M/=_M$ be the quotient structure of M with respect to $=_M$. Then $M \models F$ iff $M/=_M \models F$. Since $|M/=_M| \geq n+m$ $\Gamma_m(M)$ and $\Gamma_m(M/=_M)$ are both defined and $\Gamma_m(M) = \Gamma_m(M/=_M)$. Furthermore for any normal interpretation N of $\{c_1, \ldots, c_n, \in, =\}$, as it is straightforward to check, $N \models F$ iff $\Gamma_m(N) \subseteq \Gamma_m(F)$. □

Proposition 2.2. *If M is an interpretation of $c_1, \ldots, c_n, \in, =$ which is a model of NWL then*

$$M \models F \quad \textit{iff} \quad \Gamma_m(M) \subseteq \Gamma_m(F) \ .$$

Proof. It follows immediately from the previous proposition since if $M \models NWL$ then $=_M$ determines infinitely many $=_M$-equivalence classes. □

From now on we will refer to the skolemized version of NWL.

Definition 2.3. *H_n is the Herbrand's preinterpretation of the language $\emptyset, c_1, \ldots, c_n, \mathbf{w}, \mathbf{l}$, namely the collection of closed terms of $\emptyset, c_1, \ldots, c_n, \mathbf{w}, \mathbf{l}$, with the canonical interpretation of the constants and of the functional symbols* $\mathbf{w}, \mathbf{l}$.

Since F is a universal sentence, F is satisfiable with respect to NWL iff there is an Herbrand's model of NWL over H_n in which F is true. Therefore, by the previous Proposition 2.2, the satisfiability problem for F with respect to NWL is reducible to the problem of determining whether there exists an Herbrand's model M, over H_n, of NWL such that $\Gamma_m(M) \subseteq \Gamma_m(F)$. The latter problem is reducible to the problem of determining whether a family of finite set of graphs $\mathit{\Gamma}$ contains an element Γ such that $\Gamma \subseteq \Gamma_m(F)$ provided $\mathit{\Gamma}$ fulfills the following two conditions:

1. for every $\Gamma \in \mathit{\Gamma}$ there exists an Herbrand's model M of NWL such that $\Gamma_m(M) \subseteq \Gamma$;
2. for every Herbrand's model M of NWL there exists $\Gamma \in \mathit{\Gamma}$ such that $\Gamma_m(M) \subseteq \Gamma$.

In fact from the previous proposition, given such a π, F is satisfiable in a model of NWL iff there is $\Gamma \in \mathit{\Gamma}$ such that $\Gamma \subseteq \Gamma_m(F)$. Thus our decision problem is reduced to the problem of effectively determining a family $\mathit{\Gamma}$ which fulfills the above conditions.

3. The Decision Test

3.1 Defining $\mathit{\Gamma}$

Let $M = (H_n, \in_M, =_M)$ be an Herbrand's model of NWL. A term t in M has the form $c_h \bullet_1 t_1 \ldots \bullet_\ell t_\ell$, where $\ell \in \aleph$, c_h is a constant in $\{c_0 = \emptyset, c_1, \ldots c_n\}$

and $\bullet_i \in \{\mathbf{w}, \mathbf{l}\}$. We will call c_h the 'seed' of t. Let t', t'' be terms in H_n; we define by induction on the construction of terms the binary relations 'to be added in' and 'to be removed in' (with respect to $=_M$) as follows:

if t'' is a constant then t' is not added nor removed in t''
if $t'' = t_1 \bullet t_2$ then t' is added in t'' iff t' is added in t_1 and $t' \neq_M t_2$ or $\bullet = \mathbf{w}$ and $t' =_M t_2$,
t' is removed in t'' iff t' is removed in t_1 and $t' \neq_M t_2$ or $\bullet = \mathbf{l}$ and $t' =_M t_2$.

We say that t' is added (removed) syntactically in t'' if t' is added (removed) in t'' with respect to the syntactical identity.

Due to the form of F we can restrict our attention to Herbrand's models M over H_n such that $c_i \neq_M c_j$, for $i \neq j$. Let S_M be the function that maps a term t in the set

$$S_M(t) = \{i : t \in_M c_i, 1 \leq i \leq n\} .$$

Since $M \models \mathrm{NWL}$, given an m-tuple of terms $\mathbf{t} = (t_1, \ldots, t_m)$ in M not $=_M$ related to each other nor with any of the c_i's, the $n+m$-graph induced over $\{1, \ldots, n+m\}$ is uniquely determined by:

1. the restriction G of $\in_M$ to $\{c_1, \ldots, c_n\}$;
2. the m-tuple $\sigma_M(\mathbf{t}) = (sit_M(t_1), \ldots, sit_M(t_m))$ where $sit_M(t)$ is the triple (i, J, I) where c_i is the seed of t, $J = \{j : c_j \in_M t\}$, $I = S_M(t)$;
3. the map $B_M(\mathbf{t}) : \{1, \ldots, m\}^2 \to \{\mathbf{w}, \mathbf{l}, \times\}$ defined by letting

$$B_M(\mathbf{t})(i,j) \quad = \quad \left[\begin{array}{ll} \mathbf{w} & \text{if } t_i \text{ is added in } M \text{ to } t_j \\ \mathbf{l} & \text{if } t_i \text{ is removed in } M \text{ from } t_j \\ \times & \text{otherwise} \end{array} \right.$$

We will call m-situation of $\mathbf{t} = (t_1, \ldots, t_m)$ in M the tern $(G, \sigma_M(\mathbf{t}), B_M(\mathbf{t}))$.

Abstracting from any given interpretation we give the following further definition:

Definition 3.1. *Let $\mathcal{G}$ be the family of n-graph. An m-situation is a tern of the form (G, σ, B) where*
$G \in \mathcal{G}$;
σ is an m-tuple $s_1, \ldots, s_m$, with $s_h = (i_h, I_h, J_h)$, $1 \leq i_h \leq n$, $I_h \subseteq \{0, 1, \ldots, n\}$, $J_h \subseteq \{1, \ldots, n\}$;
$B : \{1, \ldots, m\}^2 \to \{\mathbf{w}, \mathbf{l}, \times\}$.

Given an m-situation (G, σ, B) and an m-tuple of terms $\mathbf{t}$ in a model M which induces the graph G over $\{1, \ldots, n\}$, we say that $\mathbf{t}$ realizes σ, B if $\sigma = (sit_M(t_1), \ldots, sit_M(t_m))$ and B is the map from $\{1, \ldots, m\}^2$ into $\{\mathbf{w}, \mathbf{l}, \times\}$ induced by $(t_1, \ldots, t_m)$.

An m-situation (G, σ, B), with $\sigma = (s_1, \ldots, s_m)$, $s_h = (i_h, I_h, J_h)$ determines an $n+m$-graph $Ghp(G, \sigma, B) = (\{1, \ldots, n, n+1, \ldots, n+m\}, \in_E)$ by letting

$i \in_E j$ if $i, j \leq n$ and $i \in_G j$,
$i \in_E h$ if $i \leq n < h \leq n+m$ and $i \in J_h$,
$h \in_E i$ if $i \leq n < h \leq n+m$ and $i \in I_h$,
$h \in_E k$ if $n < h, k \leq n+m$ and $B_{h-n,k-n} = \mathbf{w}$ or $c_{i_k} \in I_h$ and $B_{h-n,k-n} \neq \mathbf{l}$.

It is immediate that the $n+m$-graph induced by $t_1, \ldots, t_m$ is precisely the graph induced in that way by its situation. The m-situation (G, σ, B) which induces the $n+m$-graph $(\{1, \ldots, n+m\}, \in_E)$, unless B takes only values in $\{\mathbf{w}, \mathbf{l}\}$ is by no means unique, as a matter of fact it is immediate to verify that the following holds:

Proposition 3.1. *If the m-situation (G, σ, B) induces the relation E over $\{1, \ldots, n+m\}$, and B' is obtained from B by changing $(i, j, \times)$ with $(i, j, \mathbf{w})$, if $(i \in_E j)$ or with $(i, j, \mathbf{l})$, if $(i \notin_E j)$, then $Ghp(G, \sigma, B') = Ghp(G, \sigma, B)$.*

Definition 3.2. *Let $\mathcal{G}$ be the family of n-graph;*
$\mathcal{I}$ the family of maps $I : \{0, 1, \ldots, n\} \times \mathcal{P}(\{1, \ldots, n\}) \to \mathcal{P}(\{1, \ldots, n\})$;
$\mathcal{B}_m$ the family of maps $B : \{1, \ldots, m\}^2 \to \{\mathbf{w}, \mathbf{l}, \times\}$ such that $B(i, j) \in \{\mathbf{w}, \mathbf{l}\}$ if $i < j$, $B(i, j) = \times$ otherwise;
For π a permutation of $\{1, \ldots, m\}$ and $B \in \mathcal{B}_m$, B^π-the permutation of B under π- is defined by letting, for $1 \leq i, j \leq m$, $B^\pi(i, j) = B(\pi(i), \pi(j))$.
For G an n-graph and $I \in \mathcal{I}$,
$\Gamma_m(G, I) = \{Ghp(G, \sigma, B^\pi) : \sigma \in I^m, B \in \mathcal{B}_m, \pi$ permutation of $\{1, \ldots, m\}\}$.

The family Γ having the desired properties, as we will show, is the collection of the finite set of $n+m$-graphs $\{\Gamma_m(G, I) : G \in \mathcal{G}, I \in \mathcal{I}\}$.

Example 3.1. By way of illustration we offer a few examples of how the method could be used to determine the satisfiability/unsatisfiability of specific formulae.

Let $m = n = 1$ and

$$F_1 = \forall y(y \notin c) \Leftrightarrow c \notin c \wedge \forall y \neq c(y \notin c)$$
$$F_2 = c \notin c \wedge \forall y \neq c(y \in c)$$
$$F_3 = c \notin c \wedge \forall y \neq c(y \in c \leftrightarrow c \in y).$$

Let I_1 and I_2 be the constant maps on $\{0, 1\} \times \mathcal{P}(\{1\})$ with value $\emptyset$ and $\{1\}$ respectively, and

$$I_3 = \{(0, \emptyset, \emptyset), (0, \{1\}, \{1\}), (1, \emptyset, \emptyset), (1, \{1\}, \{1\})\} .$$

Since the family $\mathcal{B}_1$ consists of the single map $B = \{((1, 1), \times)\}$, $\Gamma_1(G, I_i)$ is determined by four 1-situations of the form: $\{(G, \sigma, B) : \sigma \in I_i\}$. We have that

$\Gamma_1(G, I_1) = \{\emptyset, \{(1, 2)\}\}$
$\Gamma_1(G, I_2) = \{\{(2, 1)\}, \{(1, 2), (2, 1)\}, \{(2, 1), (2, 2)\}, \{(1, 2), (2, 1), (2, 2)\}\}$;
$\Gamma_1(G, I_3) = \{\emptyset, \{(1, 2), (2, 1)\}, \{(1, 2), (2, 1), (2, 2)\}\}$;

where we have omitted to explicitly indicate the common domain $\{1,2\}$. As it is easy to see, for $i = 1,2,3$, $\Gamma_1(G,I) \subseteq \Gamma_1(F_i)$ iff $I = I_i$. That makes it easy to establish the unsatisfiability of the following formula

$$F_3' = F_3 \wedge \forall x \neq c(c \notin x \vee x \notin c \vee x \notin x) .$$

In fact, since $F_3' \to F_3$ the only $\Gamma \in \Gamma_1$ that contains models of F_3' is $\Gamma_1(G,I_3)$, but $\Gamma_1(G,I_3) \not\subseteq \Gamma(F_3')$ since $(\{1,2\},\{(1,2),(2,1),(2,2)\}) \not\models F_3'$.

For $n=1$ $m=2$, we do not explicitly list any $\Gamma_2(G,I)$, since 128 different 2-situation have to be taken into account to determine $\Gamma_2(G,I)$. Consider however the formula:

$$F_4 = c \notin c \wedge \forall x_2, x_3 \neq c(x_2 \in c \wedge x_3 \in c) .$$

Clearly the only $\Gamma(G,I)$ that is contained in $\Gamma(F_4)$ is $\Gamma_2(G,I_4)$, where I_4 is the constant map with value $\{1\}$. In particular F_4 is satisfiable.

$\Gamma_2(G,I_4)$ must contains, for instance, the graphs induced by the following two 2-situations:

$$(G,((0,\emptyset,\{1\}),(1,\{1\},\{1\})),B^w) \quad \text{and}$$
$$(G,((0,\emptyset,\{1\}),(1,\{1\},\{1\})),B_l),$$

where $B^w = \begin{array}{c|cc} & 1 & 2 \\ \hline 1 & \times & w \\ 2 & \times & \times \end{array}$ and $B_l = \begin{array}{c|cc} & 1 & 2 \\ \hline 1 & \times & \times \\ 2 & \ell & \times \end{array}$.

They are respectively the following:

$$(\{1,2,3\},\{(1,2),(2,1),(1,1),(3,1),(2,3),(3,2)\})$$
$$(\{1,2,3\},\{(1,2),(2,1),(1,1),(3,1)\})$$

That suffices to establish the unsatisfiability of the following formula:

$$F_4' = F_4 \wedge \forall x_2, x_3 \neq c((c \in x_2 \wedge c \notin x_3) \to (x_2 \notin x_2 \vee x_2 \in x_3 \vee x_3 \in x_2)) .$$

3.2 Proving the Conditions on Γ

Proposition 3.2. *For every graph G over $\{1,\dots,n\}$ and function $I \in \mathcal{I}$ there is a normal Herbrand's model $H(G,I)$ over H_n such that*

$$\Gamma_m(H(G,I)) \subseteq \Gamma_m(G,I) .$$

Proof. Let $H(G,I)$ be the Herbrand's model over H_n, defined by letting:

$$c_i \in_{H(G,I)} c_j \quad \text{iff} \quad (i,j) \in G$$

For $r,s \in H_n$ having seeds c_h and c_k respectively:

$$r \in_{H(G,I)} s \quad \text{iff} \quad r \text{ is added sintactically to } s \text{ or } k \in I(h,J_r) ,$$

where $J_r = \{i : c_i \in_{H(G,I)} c_h$ or (c_i is added sintactically to r)$\}$.
$H(G,I)$ is a normal Herbrand model of NWL such that for every term $t \notin \{c_1,\ldots,c_n\}$, if c_h is the seed of t and $J_t = \{i : c_i \in_{H(G,I)} t\}$ then $S_{H(G,I)}(t) = I(h, J_t)$.

Given $(t_1,\ldots,t_m)$, m-tuple of distinct terms in $H(G,I)$ all distinct from $c_1,\ldots,c_n$, we have to show that $Ghp_{H(G,I)}(t_1,\ldots,t_m) \in \Gamma_m(G,I)$, namely to find $\sigma \in I^m, B \in \mathcal{B}_m, \pi$ permutation of $\{1,\ldots,n\}$ such that the m-situation (G,σ,B^π) induces $Ghp_{H(G,I)}(t_1,\ldots,t_m)$.
Clearly $\sigma = (sit_{H(G,I)}(t_1),\ldots,sit_{H(G,I)}(t_m)) \in I^m$.

Let $B' : \{1,\ldots,m\}^2 \to \{\mathrm{w},\mathrm{l},\times\}$ be defined as follows:

$$B' = \begin{bmatrix} \mathrm{w} & \text{if } t_i \text{ is added syntactically to } t_j \\ \mathrm{l} & \text{if } t_i \text{ is removed syntactically from } t_j \\ \times & \text{otherwise} \end{bmatrix}$$

Since B' is defined with reference to the notion of being added or removed syntactically, the relation R defined by letting $R(i,j)$ iff $B(i,j) \in \{\mathrm{w},\mathrm{l}\}$ is well-founded and by R-recursion we can define a permutation λ of $\{1,\ldots,m\}$ such that if $B^\lambda(i,j) \in \{\mathrm{w},\mathrm{l}\}$ then $i < j$.

Let B be the uniquely determined - according to Proposition 3.1 - function $B : \{1,\ldots,m\}^2 \to \{\mathrm{w},\mathrm{l},\times\}$ such that:

if $B^\lambda(i,j) \in \{\mathrm{w},\mathrm{l}\}$ then $B(i,j) = B^\lambda(i,j)$,
$B(i,j) \in \{\mathrm{w},\mathrm{l}\}$ iff $i < j$

and that induces the same $n+m$-graph as $(G,\lambda\sigma,B")$,
where $\lambda\sigma = (sit(t_{\lambda(1)}),\ldots,sit(t_{\lambda(m)}))$. Letting $\pi = \lambda^{-1}$it is easy to verify that
$Ghp(G,(t_1,\ldots,t_m),B') = Ghp(G,\sigma,B^\pi)$. Since $B \in \mathcal{B}_m$, that proves that $Ghp(G,(t_1,\ldots,t_m),B') \in \Gamma_m(G,I)$. □

Lemma 3.1. *Let M be an Herbrand's model of NWL. Then for every $m \geq 1$ there exists $I_m \in \mathcal{I}$ such that*
for every $1 \leq h \leq m$, $\sigma \in (I_m)^h$ and $B \in \mathcal{B}_h$, there are infinitely many disjoint h-tuples of distinct elements of M which realize (σ, B).

Proof. Let $\{h_k\}_{k\in\kappa}$ be an increasing sequence on numerals distinct from c_0, $c_1,\ldots,c_n$. For every $i \in \{0,1,\ldots,n\}$ and $J \subset \{0,1,\ldots,n\}$ let:

$$\begin{aligned} t^1_{i,J}(r) \quad = \quad & c_i \cup J^c \setminus (\{c_1,\ldots c_n\} \setminus J^c) \\ & \bullet_0 h_0 \ldots \bullet_n h_n \\ & \mathrm{l}\ h_{n+1} \mathrm{w}\ h_{n+2} \\ & \bullet_{n+1} h_{n+3+r} \end{aligned}$$

where $J^c = \{c_j : j \in J\}$ and for every $0 \leq j \leq n$, $\bullet_j$ is w if $h_j \notin_M c_j$, l otherwise and $\bullet_{n+1}$ is w if $h_{n+3+r} \notin_M c_i$, l otherwise.

Assuming $m > 1$, for $0 \leq h < m$, σ h-tuple of terns in $SIT_n = \{0,\ldots,n\} \times \mathcal{P}(\{0,\ldots,n\}) \times \mathcal{P}(\{0,\ldots,n\})$, $B \in \mathcal{B}_h$ such that there are infinitely many disjoint h-tuple of distinct terms in M which realize (σ, B), let en_σ^B be an enumeration of infinitely many such h-tuples. Without loss of generality we may suppose that as h, σ and B vary, the ranges of the enumerations en_σ^B are disjoint from each other, from $\{c_0, c_1, \ldots, c_n\}$, from $\{h_k\}_{k\in\kappa}$ and from a further increasing sequences of numerals $\{h'_k\}_{k\in\kappa}$ in its turn disjoint from $\{h_k\}_{k\in\kappa}$.

Let

$$\Lambda_m(r) = \bigcup\{range(en_\sigma^B(r+j)) : 1\leq j\leq 2^h,\ 1\leq h< m,\ B \in \mathcal{B}_h,\ \sigma \in (SIT_n)^h\}$$

and

$$\Delta_m(r)=\Lambda_m(r)\cup\{t : t \text{ is added or removed syntactically in some term in } \Lambda_m(r)\}$$

$\Delta_m(r)$ is a finite set and let $\delta_m(r)$ be its cardinality; furthermore let $\delta_m^{<r} = \Sigma_{i<r}\delta_m(i)$.

Let O_j^h be the operation that assigns to a h-tuple $(t_1, \ldots, t_h)$ of terms the string $\bullet_1 t_1 \ldots \bullet_h\, t_h$ where $(\bullet_1 \ldots \bullet_h)$ is the j-th h-tuple in $\{\mathrm{w}, \mathrm{l}\}^h$, in any fixed ordering whatsoever, and $\circ X$, where X is a set of strings, be the concatenation of the strings in X in any fixed ordering. Let

$$\begin{aligned} t^m_{i,J}(r) \quad = \quad & t^1_{i,J}(r) \\ & \circ\{O_j^h en_\sigma^B(r+j) : 1\leq j\leq 2^h,\ 1\leq h<m,\ \sigma \in (SIT_n)^m,\ B \in \mathcal{B}_h\} \\ & \bullet'_1 h'_{\delta_m^{<r}+1} \cdots \bullet'_{\delta_m(r)}\, h'_{\delta_m^{<r}+\delta_m(r)} \end{aligned}$$

where for a given enumeration $a_1, \ldots, a_{\delta_m(r)}$ of $\Delta_m(r)$, $\bullet'_i$ is w if $h'_{\delta_m^{<r}+i} \notin_M a_i$, l otherwise. Then the following hold:

1. every term added or removed in $t^m_{i,J}(r)$ is added or removed only once; as a consequence it can't happen that a term added is successively removed in $t^m_{iJ}(k)$ or vice versa;
2. the seed of $t^m_{i,J}(r)$ is c_i and the set of predecessor of $t^m_{iJ}(r)$ among the constants $c_0, c_1, \ldots, c_n$ is J^c.
3. $t^m_{iJ}(r)$ is distinct from any constant c_j since $h_j \in_M t^m_{iJ}(r)$ iff $h_j \notin_M c_j$; furthermore it is distinct from any numeral since $h_{n+2} \in_M t^m_{iJ}(r)$ while $h_{n+1} \notin_M t^m_{iJ}(r)$ and we have assumed $h_{n+1} < h_{n+2}$;
4. If $r \neq r'$ then $t^m_{iJ}(r) \neq t^m_{iJ}(r')$: in fact $h_{n+3+r} \in_M t^m_{iJ}(r)$ iff $h_{n+3+r} \notin_M c_i$, furthermore h_{n+3+r} is not added nor removed in $t^m_{iJ}(r')$, hence $h_{n+3+r} \in_M t^m_{iJ}(r)$ iff $h_{n+3+r} \notin_M t^m_{iJ}(r')$;
5. $t^m_{i,J}(r)$ is distinct from any term in $\Delta_m(r)$ since for every $1 \leq j \leq \delta_m(r)$ $h'_j \in_M t^m_{i,J}(r)$ iff $h'_j \notin_M a_j$
6. $t^m_{i,J}(r)$ is not added or removed in itself: it follows from 3 and 4 since the elements added or removed in $t^m_{iJ}(r)$ are either constants in $\{c_0, \ldots, c_n\}$ or numerals or terms in $\Delta_m(r)$;
7. $t^m_{i,J}(r)$ is not added or removed in any of the terms in $\Lambda_m(r)$ otherwise we would have that $t^m_{i,J}(r) =_M a_j$ for some $a_j \in \Delta_m(r)$ which is impossible by 5 above.

Since the terms $t^m_{i,J}(r)$ are all distinct form each other and there are only 2^n subsets of $\{1, \ldots, n\}$ there must be at least one such subset say $I_{i,J}$ such that for infinitely many $r's$, $S_M(t^m_{iJ}(r)) = I_{i,J}$. Let

$$I_m(i, J) = S_M(t^m_{i,J}(r_0))$$

where r_0 is the least natural number such that for infinitely many $r's$ $S_M(t^m_{iJ}(r)) = S_M(t^m_{i,J}(r_0))$.

By induction on h we now prove that if $1 \leq h \leq m, \sigma \in (I_m)^h$ and $B \in \mathcal{B}_h$ then there are infinitely many disjoint h-tuples of distinct terms in M which realize σ, B.

Base case: $h = 1$. Because of 6 above the infinitely many terms of the form $t^m_{i,J}(r)$ such that $S_M(t^m_{i,J}(r)) = S_M(t^m_{i,J}(r_0))$ realize $(i, J, I_m(i, J))$, $B_\times$.

Inductive step: assume the property holds for h and that $h + 1 \leq m$. Assume $(s_1, \ldots, s_h, s_{h+1}) \in (I_m)^{h+1}$ and $B \in \mathcal{B}_{h+1}$. Let $s_{h+1} = (i, J, S_M(t^m_{i,J}(r_0))$ and $(B(1, h+1), \ldots, B(h, h+1))$ be the j-th h-tuple in $\{\mathbf{w}, \mathbf{l}\}^h$, namely $O^h_j = (B(1, h+1), \ldots, B(h, h+1))$. Let B' be the restriction of B to $\{1, \ldots, h\}^2$. By inductive hypothesis there are infinitely many disjoint h-tuples of distinct terms in M which realize $(s_1, \ldots, s_h), B'$. Therefore $O^h_j en^{B'}_{(s_1,\ldots,s_n)}(r + j)$ is a substring of $t^m_{i,J}(r)$. If $en^{B'}_{(s_1,\ldots,s_n)}(r + j) = (t_{r,1}, \ldots, t_{r,h})$ and $S_M(t^m_{i,J}(r)) = S_M(t^m_{i,J}(r_0))$ it is easy to check by using 5 above that $(t_{r,1}, \ldots, t_{r,h}, t^m_{i,J}(r))$ realizes $(s_1, \ldots, s_h, s_{h+1}), B$.

As r varies the h-tuples $(t_{r,1}, \ldots, t_{r,h})$ are disjoint from each other by the assumption on $en^{B'}_{(s_1,\ldots,s_n)}$, furthermore the terms $t^m_{i,J}(r)$ are distinct from each other and from the terms $t_{r,1}, \ldots, t_{r,h}$ as well, because of 4 and 5 above. Therefore the $h+1$-tuples $(t_{r,1}, \ldots, t_{r,h}, t^m_{i,J}(r))$ are infinitely many disjoint $h+1$-tuples of distinct terms of M which realize $(s_1, \ldots, s_h, s_{h+1})$, B and we are done. □

Proposition 3.3. *Let M be an Herbrand's model of NWL. Then for every $m \geq 1$ there exists $I_m \in \mathcal{I}$ such that for every $\sigma \in (I_m)^m$, $B \in \mathcal{B}_m$ and permutation π of $\{1, \ldots, m\}$ σ, B^π is realized in M, therefore $\Gamma_m(G, I) \subseteq \Gamma_m(M)$*

Proof. When π is the identity the proposition follows immediately from the previous lemma. Otherwise let σ' be defined by $\sigma'(i) = \sigma(\pi^{-1}(i))$. By the previous lemma 3.1 there is an m-tuple $(t'_1, \ldots, t'_m)$ of distinct terms in M which realizes σ', B. It is immediate that the m-tuple $(t_1, \ldots, t_m)$ such that $t_i = t'_{\pi(i)}$ realizes σ, B^π. □

Theorem 3.1. *Let F be a $\forall^m$-sentence in normal form over $\{c_1, \ldots, c_n, \in, =\}$. Then the following are equivalent:*

1. *F is satisfiable with respect to NWL*
2. *F is true in a normal Herbrand's model of NWL of the form $H(G, I)$ with $G \in \mathcal{G}_n$ and $I \in \mathcal{I}$*

3. for some $G \in \Gamma$ and $I \in \mathcal{I}$ $\quad \Gamma_m(G,I) \subseteq \Gamma_m(F)$

Proof. If F is satisfiable with respect to NWL then there is an Herbrand's model M over H_n of NWL such that $M \models F$. Then by Proposition 2.2, $\Gamma_m(M) \subseteq \Gamma_m(F_0)$. By Proposition 3.3, letting G be the graph induced on $\{1,\ldots,m\}$ by $\in_M$ on $\{c_1,\ldots,c_n\}$ there is a function $I \in \mathcal{I}$ such that $\Gamma_m(G,I) \subseteq \Gamma_m(G,I)$. Thus $\Gamma_m(G,I) \subseteq \Gamma_m(F)$. Hence 1) entails 3). Assuming 3), let $G \in \mathcal{G}$ and $I \in \mathcal{I}$ be such that $\Gamma_m(G,I) \subseteq \Gamma_m(F)$. By Proposition 3.2 $\Gamma_m(H(G,I)) \subseteq \Gamma_m(G,I)$, therefore $\Gamma_m(H(G,I)) \subseteq \Gamma_m(F)$. Thus, by Proposition 2.2, $H(G,I)' \models F$. Hence 3) entails 2). Since, obviously 2) entails 1), the proof is completed. □

As an immediate consequence of the reduction established in Section 2. and of the previous theorem we have our main decidability result:

Theorem 3.2. *The satisfiability problem for $\exists^*\forall^*$-sentences in $\{\in,=\}$ with respect to NWL is decidable.*

References

1. A. Aiken. Set Constraints: Results, Applications and Future Directions. Technical Report, University of California, Berkeley, 1994.
2. D. Bellè and F. Parlamento. Decidability and Completeness of Open Formulae in Membership Theories. *Notre Dame Journal of Formal Logic*, 36(2):304–318, 1995.
3. D. Bellè and F. Parlamento. The decidability of the $\forall^*\exists$-class and the axiom of foundation. Technical Report, Università di Udine, 1995.(Submitted)
4. D. Bellè and F. Parlamento. Undecidability of Weak Membership Theories. In *Logic and Algebra*:327–338. MDI(New York), 1995.
5. D. Cantone, A. Ferro and E.G. Omodeo. *Computable Set Theory. Vol. 1.* Oxford University Press, 1989. Int. Series of Monographs on Computer Science.
6. G. E. Collins and J. D. Halpern. On the Interpretability of Arithmetic in Set Theory. *Notre Dame Jour. Formal Logic*, XI(4):477–483, 1970.
7. A .Dovier. Computable Set Theory and Logic Programming. PhD Thesis, TD196 Universit'a di Pisa, 1996.
8. A. Dovier, E. G. Omodeo, E. Pontelli and G. Rossi. {log}:A Language for Programming in Logic with Finite Sets. *Journal of Logic Programming*, to appear.
9. D. Gogol. The $\forall_n\exists$-completeness of Zermelo-Fraenkel set theory. *Zeitschr. fur Math. Logik und Grundlagen d. Math.*, 1978.
10. P.M. Hill and J.W. Lloyd. The Gödel Programming Language. The MIT Press, Cambridge, Mass., 1994.
11. A. Mancini and F. Montagna. A Minimal Predicative Set Theory. *Notre Dame Jour. Formal Logic*, 35:186–203, 1994.
12. E. Omodeo, F. Parlamento and A. Policriti. A derived algorithm for evaluating ϵ-expressions over sets. *J.Symbolic Computation*, 15:673-704, 1993.

13. E. Omodeo, F. Parlamento, and A.. Policriti. Decidability of $\exists^*\forall$-sentences in Membership Theories. *Mathematical Logic Quarterly*, 1(1), 1996.
14. F. Parlamento and A. Policriti. Decision Procedures for Elementary Sublanguages of Set Theory. IX. Unsolvability of the Decision Problem for a Restricted Subclass of the Δ_0-Formulas in Set Theory. *Communications on Pure and Applied Mathematics*, XLI:221–251, 1988.
15. F. Parlamento and A. Policriti. The Logically Simplest Form of the Infinity Axiom. *Proceedings of the American Mathematical Society*, 103(1):274–276, May 1988.
16. F. Parlamento and A. Policriti. Note on: The Logically Simplest Form of the Infinity Axiom. *Proceedings of the American Mathematical Society*, 108(1), 1990.
17. F. Parlamento and A. Policriti. Expressing Infinity without Foundation. *Journal of Symbolic Logic*, 56(4):1230–1235, 1991.
18. F. Parlamento and A. Policriti. The satisfiability problem for the class $\exists^*\forall\forall$ in a set-theoretic setting. Technical Report 13/92, Universitá di Udine, October 1992.
19. F. Parlamento and A. Policriti. Undecidability Results for Restricted Universally Quantified Formulae of Set Theory. *Communications on Pure and Applied Mathematics*, XLVI:57–73, 1993.
20. J.A. Robinson A Review of Automatic Theorem-Proving. *Proc. Symp. Appl. Math.* Amer. Math. Soc., Providence, R.I. volume 19:1–18, 1967.
21. R.L.Vaught. On a Theorem of Cobham Concerning Undecidable Theories. In *Proceedings of the 1960 international Congress on Logic,Methodology, and Philosophy of Science*, 1962.
22. A. Tarski, A. Mostowsky and R.M. Robinson. *Undecidable Theories*. North-Holland, 1953.
23. A. Tarski and W. Szmielew. Mutual Interpretability of Some Essentially Undecidable Theories. In *Proc. of Intl. Cong. of Mathematicians. Cambridge.*, volume 1. Cambrige University Press, 1950.
24. S. Givant Tarski. *A formalization of Set Theory without variables*. American Mathematical Society, 1987.
25. F. Ville. Decidabilite' des Formules Existentielles en Theorie des Ensembles. *C. R. Acad. Sc. Paris*, pages 513–516, 1971.

A Logical Approach to Complexity Bounds for Subtype Inequalities

Marcin Benke*

Institute of Informatics
Warsaw University
ul. Banacha 2
02-097 Warsaw, POLAND
e-mail: Marcin.Benke@mimuw.edu.pl

Summary. We study complexity of type reconstruction with subtypes. As proved recently, this problem is polynomially equivalent to checking satisfiability of systems of inequalities. Therefore we concentrate on the latter problem and show how a variant of the transitive closure logic can be used to find an interesting class of posets for which this problem can be solved in polynomial time. Further we propose alternation as a framework suitable for presenting and explaining the aforementioned complexity for various classes of underlying subtype relation.

Introduction

Recent results of Hoang and Mitchell [3] show that the problem of Type Reconstruction with subtyping (TRS) is polynomial-time equivalent to the problem of Satisfiability of Subtype inequalities (SSI). So now the latter problem, as the only known algebraic equivalent of the former, gains importance in the study of foundations of programming languages involving subtyping.

In connection with SSI problem, its special case called FLAT-SSI was considered by many authors [10, 7, 8, 4, 2]. The latter is equivalent to the retractability problem, known from the theory of partial orders [6]. The purpose of the research was to provide some kind of 'taxonomy' amongst posets, having in mind the complexity of satisfiability-checking. The problem of FLAT-SSI attracted research interests mainly as an 'attack route' towards the general SSI problem, and thus towards the problem of type reconstruction with subtyping. The aim of this paper is to establish further links between SSI and FLAT-SSI. Sections 2. through 4. show that for posets for which feasibility of FLAT-SSI is witnessed by formulae of transitive closure logic, SSI is feasible too. Section 5. shows that for posets for which FLAT-SSI is NP-complete (wrt some class of reductions), SSI is PSPACE complete. It also proposes alternation as the framework within which relations between complexity of FLAT-SSI and SSI can be explained.

---

* This work has been partially supported by Polish KBN grant 2 P301 031 06 and ESPRIT BRA "Gentzen".

1. Preliminaries

1.1 Subtype inequalities

Let Q be a finite poset. The elements of Q are constant symbols of the signature which in addition contains a binary operation symbol $\to$. Let $\mathcal{T}_Q$ be the term algebra over this signature. The carrier of $\mathcal{T}_Q$ is partially ordered by extending the order from Q to all terms by the rule

$$\frac{r_1 \leq t_1 \quad t_2 \leq r_2}{(t_1 \to t_2) \leq (r_1 \to r_2)}$$

A *system* Σ of *inequalities* is a finite set of formulas of the form

$$\Sigma = \{\tau_1 \leq \rho_1, \ldots, \tau_n \leq \rho_n\},$$

where τ's and ρ's are terms over the above signature with variables from set V. Σ is said to be *flat* if every term in Σ is of size 1, i.e. it is either a constant symbol or a variable. Σ is said to be *satisfiable* in $\mathcal{T}_C$ if there is a valuation $v : V \to \mathcal{T}_C$ such that $\tau_i[v] \leq \rho_i[v]$ holds in $\mathcal{T}_C$ for all i.

1.2 Shapes and weak satisfiability

The set $\mathcal{T}_\star$ of *shapes* is the set of terms without variables over the signature $\Sigma = \langle 0, \to \rangle$.

The shape of a term $t \in \mathcal{T}_Q$ (without variables) is defined as follows:

$$(c)_\star = 0 \text{ for } c \in Q, \qquad (t \to u)_\star = (t)_\star \to (u)_\star$$

Note that the subtype order on $\mathcal{T}_Q$ is stratified, i.e. only terms of the same shape are comparable. In the sequel we shall operate on strata of this ordering, defined as follows:

$$\begin{array}{rcl} Q_0 & = & Q \\ Q_{\sigma \to \tau} & = & \{t \to u : t \in Q_\sigma, u \in Q_\tau\} \end{array}$$

A system of inequalities $\Sigma = \{\tau_1 \leq \rho_1, \ldots, \tau_n \leq \rho_n\}$ is said to be *weakly satisfiable* if $\Sigma_\star = \{(\tau_1)_\star = (\rho_1)_\star, \ldots, (\tau_n)_\star = (\rho_n)_\star\}$ is satisfiable in $\mathcal{T}_\star$. The most general unifier of $\Sigma_\star$ will be denoted by $mgu(\Sigma_\star)$

Weak satisfiability is clearly a necessary condition for satisfiability. It is decidable in (and in fact complete for) polynomial time since it is an instance of the unification problem.

In the sequel, we shall deal only with weakly satisfiable sytems. In some places we shall assume (for the sake of proofs, not algorithms) that all inequalities of the system are annotated with proper shape and use the notation

$$t \leq_\sigma u$$

for an inequality in shape σ.

1.3 Retractions and obstacles

We say that $R \supseteq Q$ retracts to Q ($R \rhd Q$) if there exists an order preserving and idempotent (i.e such that $f \circ f = f$) map $f : R \to Q$.

The problem of *Q-retractability* is defined as follows: given $R \supseteq Q$, does R retract to Q. For every Q, Q-FLAT-SSI is logspace-equivalent to Q-retractability. Henceforth we shall identify flat systems of inequalities over Q with corresponding extensions of Q.

V. Pratt and J. Tiuryn [7] introduce the notion of an *obstacle* to retractability — a property of a larger poset which prevents it from retracting onto another one. An obstacle is called complete for Q if R retracts to Q whenever R does not satisfy it. The reader is referred to this paper for an in-depth explanation of this concept.

1.4 Intractable posets

An n-crown is a poset with $2n$ elements $0, 1, \ldots, 2n-1$ ordered in such a way that $2i \leq (2i \pm 1) \bmod 2n$.

V. Pratt and J. Tiuryn [7] show that for n-crowns ($n \geq 2$), FLAT-SSI is NP-complete. Moreover, in [Tiu92] it is shown that for these posets SSI is PSPACE-complete. In section 5. we show how this result can be generalized.

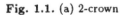

Fig. 1.1. (a) 2-crown (b) 3-crown

2. Transitive closure logic for subtype inequalities

2.1 Syntax

Let $\sigma, \sigma_1, \sigma_2 \ldots$ be shapes. The set of *annotated TC-formulas over Q* is the least set ATC_Q such that

- Every atomic formula $t_1 \leq_\sigma t_2$, where t_1, t_2 are terms from $\mathcal{T}_Q(X)$ is in ATC_Q.

- If φ and ψ are in TC_Q, and every variable x free in φ and ψ has identical annotations in both formulae, then

$$(\varphi \vee \psi),\ (\varphi \wedge \psi)$$

- If φ is in TC_Q, and every free occurrence of x is annotated by σ then

$$(\exists x^\sigma.\varphi)$$

is in ATC_Q.

- if φ is in ATC_Q, $\sigma = \sigma_1, \ldots, \sigma_n$, then

$$TC(\lambda\, \mathbf{x}^\sigma, \mathbf{y}^\sigma.\varphi)(\mathbf{t_1}, \mathbf{t_2})$$

is in ATC_Q, where $\mathbf{x}, \mathbf{y}$ are n-vectors of individual variables, $\mathbf{t_1}, \mathbf{t_2}$ are n-vectors of Q-terms, and $\mathbf{t}$ denotes the vector $t_1, \ldots, t_n$

We shall say that a formula is *flat* if it contains no occurrences of an arrow and all its variables are annotated with 0. In such a case the annotations are of no consequence and we can safely omit them.

2.2 Projections

First we define projections on shapes:

$$0 \downarrow i = 0, \quad i = 1,2 \qquad (\sigma_1 \to \sigma_2) \downarrow i = \sigma_i$$

Next we define projections on terms:

$$c \downarrow i = c \quad x^\sigma \downarrow i = x^{\sigma \downarrow i}, \ i = 1,2$$

$$(t_1 \to t_2) \downarrow i = t_i, \ i = 1,2$$

Now we define projections of ATC-formulae: $(\cdot)\downarrow 1, (\cdot)\downarrow 2 : ATC_Q \to ATC_Q$

$$\begin{array}{rcl}
(t \le_0 u) \downarrow i & = & t \le_0 u \\
(t \le_{\sigma_1 \to \sigma_2} u) \downarrow 1 & = & (u \downarrow 1) \le_{\sigma_1} (t \downarrow 1) \\
(t \le_{\sigma_1 \to \sigma_2} u) \downarrow 2 & = & (t \downarrow 2) \le_{\sigma_2} (t \downarrow 2) \\
(\exists x^\sigma.\varphi) \downarrow i & = & \exists x^{\sigma\downarrow i}.\varphi[x^{\sigma\downarrow i}/x^\sigma],\ i = 1,2 \\
(TC(\lambda\, \mathbf{x}^\sigma, \mathbf{y}^\sigma.\varphi)(t,u)) \downarrow i & = & TC(\lambda x^{\sigma\downarrow i}, y^{\sigma\downarrow i}.(\varphi \downarrow i))(t \downarrow i, u \downarrow i)
\end{array}$$

3. The proof system

3.1 Lonely Variables

Given an ATC-formula φ (or a term t), we define the set of its *lonely variables*, $LV(\varphi)$ as follows:

$$\begin{aligned}
LV(x) &= \{x\} \\
LV(t \to u) &= \emptyset \\
LV(t \leq u) &= LV(t) \cup LV(u) \\
LV(\varphi \wedge \psi) &= LV(\varphi) \cup LV(\psi) \\
LV(\varphi \vee \psi) &= LV(\varphi) \cup LV(\psi) \\
LV(\exists x.\varphi) &= LV(\varphi) \setminus \{x\} \\
LV(TC(\lambda \mathbf{x}^\sigma, \mathbf{y}^\sigma.\varphi)(\mathbf{t}, \mathbf{u}) &= (LV(\varphi) \setminus \{\mathbf{x}, \mathbf{y}\}) \cup LV(\mathbf{t}) \cup LV(\mathbf{u})
\end{aligned}$$

3.2 Closures

Let $t \preceq u$ denote the formula $TC(\lambda x^\sigma, y^\sigma.x \leq y)(t, u)$, The closure of a formula φ (denoted $\overline{\varphi}$) is defined as follows:

$$\begin{aligned}
\overline{t \leq u} &= t \preceq u \\
\overline{\varphi \wedge \psi} &= \overline{\varphi} \wedge \overline{\psi} \\
\overline{\varphi \vee \psi} &= \overline{\varphi} \vee \overline{\psi} \\
\overline{\exists x^\sigma.\varphi} &= \exists x^\sigma.\overline{\varphi}
\end{aligned}$$

3.3 Inference rules

Let Σ be weakly satisfiable and all its variables be annotated according to $mgu(\Sigma_\star)$. Consider the inference system depicted in Fig. 3.1

3.4 Normal derivations

We shall say that a derivation is in *normal form* if all the applications of the rule ($\downarrow$) are made as early as possible. Now it is easy to observe, that

In the normal derivation of	the last rule is
$\varphi_1 \wedge \varphi_2$	$(\wedge)$
$\varphi_1 \vee \varphi_2$	$(\vee)$
$TC(\lambda \mathbf{x}^\sigma, \mathbf{y}^\sigma.\varphi)(\mathbf{t}, \mathbf{u})$	TC_0 or TC_S

Proposition 3.1. *Any derivation from a flat system Σ is normal, and the last rule is always an introduction of the main connective.*

$$\Sigma \vdash t \leq u \text{ for } t \leq u \in \Sigma$$

$$\frac{\Sigma \vdash \varphi}{\Sigma \vdash \varphi \downarrow i}\ (\downarrow) \qquad LV(\varphi) = \emptyset$$

$$\frac{\Sigma \vdash \varphi \quad \Sigma \vdash \psi}{\Sigma \vdash \varphi \wedge \psi}\ (\wedge)$$

$$\frac{\Sigma \vdash \varphi_i}{\Sigma \vdash \varphi_1 \vee \varphi_2}\ (\vee)$$

$$\frac{\Sigma \vdash \varphi[t/x^\tau] \quad (t)_* = \tau}{\Sigma \vdash \exists x^\tau . \varphi}\ (\exists)$$

$$\frac{\Sigma \vdash \varphi[\mathbf{t}/\mathbf{x}^\tau, \mathbf{u}/\mathbf{y}^\tau]}{\Sigma \vdash TC(\lambda \mathbf{x}^\sigma, \mathbf{y}^\sigma . \varphi)(\mathbf{t}, \mathbf{u})}\ (TC_0)$$

$$\frac{\Sigma \vdash TC(\lambda \mathbf{x}^\sigma, \mathbf{y}^\sigma . \varphi)(\mathbf{t}, \mathbf{s}) \quad \Sigma \vdash TC(\lambda \mathbf{x}^\sigma, \mathbf{y}^\sigma . \varphi)(\mathbf{s}, \mathbf{u})}{\Sigma \vdash TC(\lambda \mathbf{x}^\sigma, \mathbf{y}^\sigma . \varphi)(\mathbf{t}, \mathbf{u})}\ (TC_S)$$

Fig. 3.1. An inference sytem for ATC-formulae

Proposition 3.2. *Any normal derivation of* $TC(\lambda \mathbf{x}^\sigma, \mathbf{y}^\sigma . \varphi)(\mathbf{t}, \mathbf{u})$ *always ends either with single use of* TC_0 *or like*

$$\frac{\Sigma \vdash \varphi[\mathbf{t}/\mathbf{x}, \mathbf{r_1}/\mathbf{y}]}{\Sigma \vdash TC(\lambda \mathbf{x}^\sigma, \mathbf{y}^\sigma . \varphi)(\mathbf{t}, \mathbf{r_1})} \quad \cdots \quad \frac{\Sigma \vdash \varphi[\mathbf{r_k}/\mathbf{x}, \mathbf{u}/\mathbf{y}]}{\Sigma \vdash TC(\lambda \mathbf{x}^\sigma, \mathbf{y}^\sigma . \varphi)(\mathbf{r_k}, \mathbf{u})}$$

$$\vdots$$

$$\frac{\Sigma \vdash TC(\lambda \mathbf{x}^\sigma, \mathbf{y}^\sigma . \varphi)(\mathbf{t}, \mathbf{r_i}) \quad \Sigma \vdash TC(\lambda \mathbf{x}^\sigma, \mathbf{y}^\sigma . \varphi)(\mathbf{r_i}, \mathbf{u})}{TC(\lambda \mathbf{x}^\sigma, \mathbf{y}^\sigma . \varphi)(\mathbf{t}, \mathbf{u})}$$

Proposition 3.3. *For fixed* φ*, one can check in time polynomial in* $|\Sigma|$*, whether* $\Sigma \vdash \varphi$*.*

4. Results

Lemma 4.1. *Let* φ *be the complete obstacle for* Q*. For every flat system of inequalities* Σ*,* Σ *is satisfiable iff*

$$Q \cup \Sigma \nvdash \bar{\varphi} \vee NGC(Q)$$

Theorem 4.1. *Let* φ *be the complete obstacle for* Q*. For every system of inequalities* Σ*,* Σ *is satisfiable iff it is weakly satisfiable and*

$$Q \cup \Sigma \nvdash \overline{\varphi} \vee NGC(Q)$$

Proof. The ($\Rightarrow$) implication is obvious. The opposite implication is proved by induction on the number of equivalence classes of $\sim$ defined on $var(\Sigma)$ as follows

$$x \sim y \quad \text{iff} \quad \Sigma_* \models x = y$$

where the induction basis follows from the lemma 4.1.

Corollary 4.1. *For any TC-feasible Q and Σ — a system of inequalities over Q one can check in time polynomial in $|\Sigma|$, whether Σ is satisfiable.*

5. Subtyping and Alternation

The aim of this chapter is to establish further links between SSI and FLAT-SSI, claiming the following:

Conjecture 5.1. Given a poset Q such that Q-FLAT-SSI is complete for $NTM(s,t)$, Q-SSI is complete for $ATM(s,t)$.

In our opinion, the 'nondeterminism vs alternation' concept constitutes a framework within which various complexity phenomena bound with subtyping can be explained. Sure enough, there is still a lot of open questions and gaps to be filled, but we present it with hope that it will encourage further research in this area. One example would be the apparent 'gap' in the poset hierarchy. So far we know no posets for which SSI is NP-complete or FLAT-SSI — P-complete. Within our framework, the explanation for this gap is provided by the fact that (unless P=NP or NP=PSPACE) NP is not an alternating complexity class and (unless P=NLOGSPACE or P=NP), P is not a nondeterministic complexity class.

5.1 Motivating examples

First let us look at several examples known so far that supporting the thesis that arrows in the systems of inequalities correspond on the complexity level exactly to the transition from nondeterministic classes to corresponding alternating classes. This is at the same time a resume of current knowledge about the complexity of SSI:

1. If P is discrete, then
   - P-FLAT-SSI is in NLOGSPACE[1];
   - P-SSI is equivalent to the unification, and hence ALOGSPACE-complete.
2. If P is a disjoint union of lattices (but not discrete), then
   - P-FLAT-SSI is NLOGSPACE-complete [2];

[1] the problem whether it is NLOGSPACE-hard is equivalent to a known open problem in complexity, whether SYMLOGSPACE=NLOGSPACE

   - P-SSI is ALOGSPACE-complete [8].
3. If P is a non-discrete Helly poset, then
   - P-FLAT-SSI is NLOGSPACE-complete [2];
   - P-SSI is ALOGSPACE-complete [1].
4. If P is a non-discrete TC-feasible poset, then
   - P-FLAT-SSI is NLOGSPACE-complete [7];
   - P-SSI is ALOGSPACE-complete (Corollary 4.1).
5. If P is an n-crown ($n > 1$), then
   - P-FLAT-SSI is NP-complete [7];
   - P-SSI is AP-hard [8].

5.2 Encoding alternation

In this section we show that the result of Tiuryn can be generalized stating that for all posets for which FLAT-SSI is NP-hard, SSI is AP-hard. To this end, we construct an encoding for QBF as an SSI, given encoding of SAT as FLAT-SSI.

First let us make some assumptions about encodings of instances of SAT as systems of inequalities. Later we show how these assumptions can be either removed or replaced. Intuitively, these assumptions express the requirement that whenever there exists a simulation of NTM, there exists one which is "regular" enough to be transformed to a simulation of an ATM. This intuition is formalized in the following

Definition 5.1. *Let $\varphi = \varphi(\mathbf{x})$ be a 3-CNF propositional formula with variables $\mathbf{x} = x_1, \dots x_n$ (and no other)*

We say that Σ_φ, a flat system of inequalities encodes φ if there exist variables $z_1, \dots, z_n$ and constants such that for every $p_1, \dots, p_n \in \{0, 1\}$

$$\models \varphi[\mathbf{p}/\mathbf{x}] \iff \Sigma_\varphi[\mathbf{c}/\mathbf{z}] \text{ is satisfiable}$$

We say the encoding is symmetric, *if there exists an antimonotonic bijection $f : P \to P$ that extends to an antimonotonic bijection of (the poset corresponding to) Σ_φ onto itself and such that $c_i^1 = f(c_i^0)$ for $i = 1, \dots, n$.*

Theorem 5.1. *Let P be a poset such that P-FLAT-SSI is complete for NP under* symmetric *reductions. Then P-SSI is hard for AP.*

Proof. Let

$$\forall x_1 \exists y_1 \dots \forall x_{p(n)} \exists y_{p(n)} \varphi$$

be an instance of QBF, φ contains no quantifiers

Let Σ_φ be a symmetric encoding of φ. We show how to construct a system of inequalities Σ_k such that

$$\psi_k \text{ holds} \iff \Sigma_k \text{ is satisfiable}$$

where

$$\psi_k = \exists x_n \exists y_n \ldots \exists x_{k+1} \exists y_{k+1} \forall x_k \exists y_k \ldots \forall x_1 \exists y_1 \; \varphi$$

The construction of Σ_k is by induction on k, the number of quantifier alternations in ψ_k.

Let q be the smallest positive integer such that $f^q = id$ (such q must exist since Σ is finite, moreover it can't be greater than $|\Sigma|$).

In what follows we use a with sub- or super-scripts. These are new variables. We will also use new variables $[u]_k^{i,j}$, where $0 \leq k \leq n$, $i,j \in P$ and u is a propositional variable of φ. The variable $[u]_k^{i,j}$ is a version of $[u]^{i,j}$, lifted to level k. The variable a_k^i, which we use below, represents constant i lifted to level k.

Let us first define sets Δ_k, for $0 \leq k \leq n$.

$$\Delta_0 = \{\, a_{0,0}^{i,j} = a_0^j \mid i,j \in P \,\} \cup \{\, a_0^i = i \mid i \in P \,\}$$

For $k < n$, Δ_{k+1} is Δ_k plus the equations (5.1–5.4), with i,j ranging over P.

$$a_{k+1}^i = a_k^{f(i)} \to a_k^i \tag{5.1}$$

For $k+1 < p \leq n$ and $z_p \in \{x_p, y_p\}$,

$$f^i(z_{p,k+1}) = f^{i+1}(z_{p,k}) \to f^i(z_{p,k}) \quad \text{for } i = 0, \ldots q-1 \tag{5.2}$$

For $1 \leq p \leq k$,

$$a_{p,k+1}^{i,j} = a_{p,k}^{f(j),f(i)} \to a_{p,k}^{i,j} \tag{5.3}$$

$$a_{k+1,k+1}^{i,j} = a_{k+1}^j \tag{5.4}$$

For every $k \geq 0$, let $\hat{\Sigma}_k$ be the system of inequalities obtained from $\hat{\Sigma}$ by replacing every variable $[u]^{i,j}$ of $\hat{\Sigma}$ by $[u]_k^{i,j}$, and replacing the constant $i \in P$ by a (new) variable a_k^i. Hence, there are no constants in $\hat{\Sigma}_k$.

Finally we set $\Sigma_{k+1} = \Delta_{k+1} \cup \hat{\Sigma}_{k+1}$ plus the equation (5.5) with i,j ranging over P and $1 \leq p \leq k+1$.

$$z_{p,k+1} = a_{p,k+1}^{c_p^0, c_p^1} \tag{5.5}$$

The thesis follows from the following lemmas:

Lemma 5.1. *Let $V_k = \{x_{k+1}, y_{k+1}, \ldots, x_n, y_n\}$. For all $k \geq 0$, and for every function $\xi : V_k \to \{0,1\}$, $\Sigma_{k+1} \cup \{\, z_k = a_k^{c_k^{\xi(v)}} \mid v \in V_{k+1} \,\}$ is satisfiable iff for every $i \in \{0,1\}$, $\Sigma_k \cup \{\, z_k = a_k^{c_k^{\xi(v)}} \mid v \in V_k \,\} \cup \{z_{k+1,k} = a_k^{c_i}\}$ is satisfiable.*

For $0 \leq k \leq n$ let

$$\varphi_k = \forall x_k \exists y_k \ldots \forall x_1 \exists y_1 \; \varphi$$

Hence, free variables of φ_k are among $V_k = \{x_{k+1}, y_{k+1}, \ldots, x_n, y_n\}$. The following result shows correctness of the choice of Σ_k.

Lemma 5.2. *For every $0 \leq k \leq n$ and for every valuation $\xi : V_k \to \{0,1\}$, ξ satisfies φ_k iff $\Sigma_k \cup \{z_j = a_k^{c_j^{\xi(z_j)}} \mid z_j \in V_k\}$ is satisfiable.*

Acknowledgments

This paper would be never written without the continous advice, encouragement and patience of Professor Jerzy Tiuryn. Many thanks go also to Damian Niwiński for for the fruitful discussions and suggestions. I am also grateful to Jakob Rehof for careful lecture of drafts for this paper and helpful remarks.

References

1. M. Benke. Efficient type reconstruction in the presence of inheritance (extended abstract). In *Proc. Int. Symp. MFCS 1993*, 1993.
2. M. Benke. Efficient type reconstruction in the presence of inheritance. Technical Report TR94-10(199), Institute of Informatics, Warsaw University, Dec. 1994.
3. M. Hoang and J. C. Mitchell. Lower bounds on type inference with subtypes. In *Conf. Rec. ACM Symp. Principles of Programming Languages*, 1995.
4. P. Lincoln and J. C. Mitchell. Algorithmic aspects of type inference with subtypes. In *Conf. Rec. ACM Symp. Principles of Programming Languages*, pages 293–304, 1992.
5. J. C. Mitchell. Coercion and type inference. In *Conf. Rec. ACM Symp. Principles of Programming Languages*, pages 175–185, 1984.
6. P. Nevermann and I. Rival. Holes in ordered sets. *Graphs and Combinatorics*, (1):339–350, 1985.
7. V. Pratt and J. Tiuryn. Satisfiability of inequalities in a poset. Technical Report TR 95-15(215), Institute of Informatics, Warsaw University, 1995.
8. J. Tiuryn. Subtype inequalities. In *Proc. 7th IEEE Symp. Logic in Computer Science*, pages 308–315, 1992.
9. J. Tiuryn and M. Wand. Type reconstruction with recursive types and atomic subtyping. In *M.-C. Gaudel and J.-P. Jouannaud (Eds.) TAPSOFT'93: Theory and Practice of Software Development, Proc. 4th Intern. Joint Conf. CAAP/FASE, Springer-Verlag LNCS 668*, pages 686–701, 1993.
10. M. Wand and P. O'Keefe. On the complexity of type inference with coercion. In *Proc. ACM Conf. Functional Programming and Computer Architecture*, 1989.

How to characterize provably total functions by the Buchholz operator method *

Benjamin Blankertz and Andreas Weiermann

Institut für mathematische Logik und Grundlagenforschung
der Westfälischen Wilhelms-Universität Münster
Einsteinstr. 62, D-48149 Münster, Germany

Summary. Inspired by Buchholz' technique of operator controlled derivations (which were introduced for simplifying Pohlers' method of local predicativity) a straightforward, perspicuous and conceptually simple method for characterizing the provably recursive functions of Peano arithmetic in terms of Kreisel's ordinal recursive functions is given. Since only amazingly little proof and hierarchy theory is used, the paper is intended to make the field of ordinally informative proof theory accessible even to non-prooftheorists whose knowledge in mathematical logic does not exceed a first graduate level course.

1. Introduction and Motivation

A fascinating result of ordinally informative proof theory due to Kreisel (1952) is as follows:

Theorem: (*)
The provably recursive functions of Peano arithmetic are exactly the ordinal recursive functions.

Folklore (proof-theoretic) proofs for (*) [cf., for example, Schwichtenberg (1977), Takeuti (1987), Buchholz (1991) or Friedman and Sheard (1995) for such proofs] rely on non trivial metamathematical evaluations of the Gentzen- or Schütte-style proof-theoretic analyses of Peano arithmetic. Alternatively a proof- and recursion-theoretic analysis of Gödel's 1958 functional interpretation of Heyting arithmetic can be employed for proving (*), cf. for example [Tait (1965), Buchholz (1980), Weiermann (1995)]. A proof of (*) which does not rely on metamathematical considerations – like primitive recursive stipulations of codes of infinite proof-trees – has been given in [Buchholz (1987), Buchholz and Wainer (1987)]. A proof of (*) using the slow growing hierarchy is given in [Arai (1991)]. A local predicativity style proof – which generalizes uniformly to theories of proof-theoretic strength less than or equal to KPM, cf. [Rathjen (1991)] – of (*) has been given in [Weiermann (1993), Blankertz and Weiermann (1995)]. Other proofs for (*) which are based on model theory can be found, for example, in [Hájek and Pudlák (1993)]. Buchholz (1992) introduced the technique of operator controlled derivations which allows a

* This paper is in its final form and no similar paper has been or is being submitted elsewhere.

simplified and conceptually improved exposition of Pohlers' local predicativity. One aim of the present paper is to give a contribution to the following question (Buchholz, private communication, 1993): Is it possible to use operator controlled derivations to give a proof for (*) – and generalizations of (*) – which is technically smooth? In this paper appropriate operators on subsets of the natural numbers are introduced via the Buchholz-Cichon-Weiermann (1994) approach to subrecursive hierarchies. It turns out that these operators work smoothly – i.e. virtually no auxiliary computations are needed – during the embedding and collapsing procedure. In the critical step of the argument (Reduction lemma and Cut-elimination) it is shown up how cut reduction directly corresponds to composition and diagonalization of the majorization functions involved. Only here an operator analysis is needed but nevertheless the critical arguments can be carried out in some few lines, cf. lemma 2.1 (vi) and (viii). Another aim of this paper is to present a method which can presumably be employed for giving as a direct corollary from Rathjen's proof-theoretic analyses – in which adaptations of operator controlled derivations are used – a classification of the provably recursive functions of $KP + \Pi^0_3 - Reflection$, $\Pi^1_2 - (CA)$ and related systems, cf. [Rathjen (1994),(1995)] and also [Arai (1995)].
The paper is self-contained. It only requires knowledge of basic facts about the ordinals up to ε_0 and elementary level facts about cut elimination in Tait's calculus for predicate logic.

2. Proof of the main Theorem

The set of non logical constants of PA includes the set of function symbols for primitive recursive functions and the relation symbol $=$.
(In the sequel $\underline{0}$ denotes the constant symbol for zero and S denotes the successor function symbol.) The logical operations include $\wedge, \vee, \forall, \exists$. We have an infinite list of variables $x_0, x_1, \ldots$ The set of *PA-terms* (which are denoted in the sequel by $r, s, t \ldots$) is the smallest set which contains the variables and constants and is closed under function application, i.e. if f is a k-ary function symbol and $t_1, \ldots, t_k$ are terms, so is $ft_1 \ldots t_k$. If $t(\mathbf{x})$ is a PA-term with $FV(t) \subseteq \{\mathbf{x}\}$ then $t^{\mathbb{N}}$ denotes the represented function in the standard structure $\mathbb{N}$. The set of *PA-formulas* (which are in the sequel denoted by A, B, F) is the smallest set which includes $s = t$, $\neg s = t$ (*prime formulas*) and is closed under conjunction, disjunction and quantification. The notation $\neg A$, for A arbitrary, is an abbreviation for the formula obtained via the de Morgan laws; $A \to B$ abbreviates $\neg A \vee B$. We denote finite sets of PA-formulas by $\Gamma, \Delta, \ldots$ As usual Γ, A stands for $\Gamma \cup \{A\}$ and Γ, Δ stands for $\Gamma \cup \Delta$.
The formal system PA is presented in a Tait calculus. PA includes the *logical axioms* $\Gamma, \neg A, A$, the *equality axioms* $\Gamma, t = t$ and $\Gamma, t \neq s, \neg A(t), A(s)$, the *successor function axioms* $\Gamma, \forall x(\neg \underline{0} = Sx)$ and $\Gamma, \forall x \forall y(Sx = Sy \to x = y)$,

the defining equations for primitive recursive function symbols (cf. [Pohlers (1989)]) and the *induction scheme* $\Gamma, A(\underline{0}) \wedge \forall x(A(x) \to A(Sx)) \to \forall x A(x)$. The *derivation relation* $\vdash$ for PA is defined as follows:

(Ax) $\vdash \Gamma$ if Γ is an axiom of PA.
($\wedge$) $\vdash \Gamma, A_i$ for all $i \in \{0,1\}$ imply $\vdash \Gamma, A_0 \wedge A_1$.
($\vee$) $\vdash \Gamma, A_i$ for some $i \in \{0,1\}$ implies $\vdash \Gamma, A_0 \vee A_1$.
($\forall$) $\vdash \Gamma, A(y)$ implies $\vdash \Gamma, \forall x A(x)$ if y does not occur in $\Gamma, \forall x A(x)$.
($\exists$) $\vdash \Gamma, A(t)$ implies $\vdash \Gamma, \exists x A(x)$.
(cut) $\vdash \Gamma, A$ and $\vdash \Gamma, \neg A$ imply $\vdash \Gamma$.

Definition 2.1 (Rank of a formula).

- $\mathrm{rk}(A) := 0$, *if A is a prime formula*
- $\mathrm{rk}(A \vee B) := \mathrm{rk}(A \wedge B) := \max\{\mathrm{rk}(A), \mathrm{rk}(B)\} + 1$
- $\mathrm{rk}(\forall x A(x)) := \mathrm{rk}(\exists x A(x)) := \mathrm{rk}(A) + 1$

We consider only ordinals less than ε_0. These ordinals are denoted by $\alpha, \beta, \gamma, \xi, \eta$. Finite ordinals are denoted by $k, m, n, \ldots$ $\#$ denotes the natural sum of ordinals (cf. [Schütte 1977] or [Pohlers 1989]). For each $\alpha < \varepsilon_0$ let $\mathrm{N}(\alpha)$ be the number of occurences of ω in the Cantor normal form representation of α. Thus $\mathrm{N}(0) = 0$, and $\mathrm{N}(\alpha) = n + \mathrm{N}(\alpha_1) + \cdots + \mathrm{N}(\alpha_n)$ if $\alpha = \omega^{\alpha_1} + \cdots + \omega^{\alpha_n} > \alpha_1 \geq \ldots \geq \alpha_n$. N satisfies the following conditions:

(N1) $\mathrm{N}(0) = 0$
(N2) $\mathrm{N}(\alpha \# \beta) = \mathrm{N}(\alpha) + \mathrm{N}(\beta)$
(N3) $\mathrm{N}(\omega^\alpha) = \mathrm{N}(\alpha) + 1$
(N4) $\mathrm{card}\, \{\alpha < \varepsilon_0 : \mathrm{N}(\alpha) < k\} < \omega$ for every $k \in \mathbb{N}$

Furthermore we see that $\mathrm{N}(\alpha + n) = \mathrm{N}(\alpha) + n$ for $n < \omega$.
Abbreviation: $\mathrm{N}\alpha := \mathrm{N}(\alpha)$.

Definition 2.2. *Let $\Phi(x) := 3^{x+1}$ and $f^{(2)}(x) := f(f(x))$ denote iteration.*

$$\mathrm{F}_\alpha(x) := \max\left(\{2^x\} \cup \{\mathrm{F}_\gamma{}^{(2)}(x) \mid \gamma < \alpha \ \& \ \mathrm{N}\gamma \leq \Phi(\mathrm{N}\alpha + x)\}\right)$$

This expression is well defined due to (N4).

Remark: It follows immediately from [Wainer (1970)] and [Buchholz, Cichon et Weiermann (1994)] that this hierarchy is equivalent to the ordinal recursive functions.

Definition 2.3. *Throughout the whole article we assume* $\mathrm{F} : \mathbb{N} \to \mathbb{N}$ *to be a monotone function. Given $\Theta \in \mathbb{N}$ we denote by* $\mathrm{F}[\Theta]$ *the function defined by*

$$\mathrm{F}[\Theta] : \mathbb{N} \to \mathbb{N};\ x \mapsto \mathrm{F}(\max\{x, \Theta\})$$

We write $\mathrm{F}[k_1, \ldots, k_n]$ *instead of* $\mathrm{F}[\max\{k_1, \ldots, k_n\}]$ *and use the abbreviation* $\mathrm{F} \leq \mathrm{F}' \quad :\Leftrightarrow \quad \forall x \in \mathbb{N} \quad \mathrm{F}(x) \leq \mathrm{F}'(x)$.

Remark: $\mathrm{F}[n,m] = \mathrm{F}[n][m]$ and $\mathrm{F}[n][n] = \mathrm{F}[n]$.

Lemma 2.1. *(i)* $x < \mathrm{F}_\alpha(x) < \mathrm{F}_\alpha(x+1)$
(ii) $\mathrm{N}\alpha < \mathrm{F}_\alpha(x) < \mathrm{F}_\alpha{}^{(2)}(x) \leq \mathrm{F}_{\alpha+1}(x)$, *in particular* $n < \mathrm{F}_n(x)$
(iii) $\alpha < \gamma \;\&\; \mathrm{N}\alpha \leq \Phi(\mathrm{N}\gamma + \Theta) \quad \Rightarrow \quad \mathrm{F}_\alpha[\Theta] \leq \mathrm{F}_\gamma[\Theta]$
(iv) *For every* n*-ary primitive recursive function* f *there exists a* $p \in \mathbb{N}$ *such that* $\forall \mathbf{k} \in \mathbb{N}^n \quad f(\mathbf{k}) < \mathrm{F}_{\omega \cdot p}(\max \mathbf{k})$.
(v) $\mathrm{F}_\gamma[\Theta] \leq \mathrm{F}_{\gamma \# \beta}[\Theta]$
(vi) $k < \mathrm{F}_\gamma(\Theta) \quad \Rightarrow \quad \mathrm{F}_\gamma[\Theta][k] \leq \mathrm{F}_{\gamma+1}[\Theta]$
(vii) $\mathrm{N}\alpha, \mathrm{N}\beta < \mathrm{F}_\gamma(\Theta) \quad \Rightarrow \quad \mathrm{N}(\alpha+\beta), \mathrm{N}\omega^\alpha < \mathrm{F}_{\gamma+1}(\Theta)$
(viii) $\alpha_0 < \alpha \;\&\; \mathrm{N}\alpha_0 < \mathrm{F}_\gamma(\Theta) \quad \Rightarrow \quad \mathrm{F}_{\gamma \# \alpha_0+2}[\Theta] \leq \mathrm{F}_{\gamma \# \alpha+1}[\Theta]$

Proof. (i)-(vii) are simple. In (vi) for example $k < \mathrm{F}_\gamma(\Theta)$ implies for all $x \in \mathbb{N}$ $\mathrm{F}_\gamma[\Theta][k](x) = \mathrm{F}_\gamma(\max\{\Theta, k, x\}) \leq \mathrm{F}_\gamma{}^{(2)}(\max\{\Theta, x\}) \leq \mathrm{F}_{\gamma+1}(\max\{\Theta, x\})$.
(viii) If $\alpha = \alpha_0 + n$ for some $n \geq 1$ the claim follows from (v). So we can assume

$$(1) \qquad \gamma \# \alpha_0 + 2 < \gamma \# \alpha$$

and the premise $\mathrm{N}\alpha_0 < \mathrm{F}_\gamma(\Theta)$ implies

$$(2) \qquad \mathrm{N}(\gamma \# \alpha_0 + 2) \leq \mathrm{N}\gamma + \mathrm{F}_\gamma(\Theta) + 1 \leq \Phi(\mathrm{N}(\gamma \# \alpha) + \mathrm{F}_{\gamma \# \alpha}(\Theta))$$

For all $x \in \mathbb{N}$ we obtain from (1) and (2) by part (iii) and (ii)

$$\mathrm{F}_{\gamma \# \alpha_0+2}[\Theta](x) = \mathrm{F}_{\gamma \# \alpha_0+2}(\max\{\Theta, x\}) <^{(iii)} \mathrm{F}^{(2)}_{\gamma \# \alpha}(\max\{\Theta, x\})$$
$$\leq^{(ii)} \mathrm{F}_{\gamma \# \alpha+1}[\Theta](x).$$

Definition 2.4 (F-controlled derivations).
$\mathrm{F}\overset{\alpha}{\underset{r}{\vdash}} \Gamma$ *holds iff* $\mathrm{N}\alpha < \mathrm{F}(0)$ *and one of the following is true:*

(Ax)	$\Gamma \cap \Delta(\mathbb{N}) \neq \emptyset$			
$(\vee)$	$A_0 \vee A_1 \in \Gamma$	$\& \; \exists i \in \{0,1\}$	$\exists \alpha_0 < \alpha$	$\mathrm{F}\overset{\alpha_0}{\underset{r}{\vdash}} \Gamma, A_i$
$(\wedge)$	$A_0 \wedge A_1 \in \Gamma$	$\& \; \forall i \in \{0,1\}$	$\exists \alpha_0 < \alpha$	$\mathrm{F}\overset{\alpha_0}{\underset{r}{\vdash}} \Gamma, A_i$
$(\exists)$	$\exists x A(x) \in \Gamma$	$\& \; \exists n < \mathrm{F}(0)$	$\exists \alpha_0 < \alpha$	$\mathrm{F}\overset{\alpha_0}{\underset{r}{\vdash}} \Gamma, A(\underline{n})$
$(\forall)$	$\forall x A(x) \in \Gamma$	$\& \; \forall n \in \mathbb{N}$	$\exists \alpha_n < \alpha$	$\mathrm{F}[n]\overset{\alpha_n}{\underset{r}{\vdash}} \Gamma, A(\underline{n})$
(cut)	$\mathrm{rk}(A) < r$	$\&$	$\exists \alpha_0 < \alpha$	$\mathrm{F}\overset{\alpha_0}{\underset{r}{\vdash}} \Gamma, (\neg)A$

The abbreviations used in (Ax) and (cut) are the following:

- $\Delta(\mathbb{N}) := \{A \mid A \text{ is prime formula and } \mathbb{N} \models A\}$
- $\mathrm{F}\overset{\alpha_0}{\underset{r}{\vdash}} \Gamma, (\neg)A \quad :\Leftrightarrow \quad \mathrm{F}\overset{\alpha_0}{\underset{r}{\vdash}} \Gamma, A \;\&\; \mathrm{F}\overset{\alpha_0}{\underset{r}{\vdash}} \Gamma, \neg A$

So the F-controlled derivations are just like usual PA-derivations but with ω-rule and some information about the $\exists$-witnesses and the derivation length. The first one is the essential aid for collapsing (lemma 2.6) while the latter is used to apply lemma 2.1 (viii) in the cut-elimination procedure 2.5.

Lemma 2.2 (Monotonicity).
For $r \leq s$ & $\alpha \leq \beta$ & $\mathrm{N}\beta < \mathrm{F}'(0)$ & $\mathrm{F} \leq \mathrm{F}'$ *the following holds*

$$\mathrm{F}\Big|\frac{\alpha}{r}\ \Gamma \quad \Rightarrow \quad \mathrm{F}'\Big|\frac{\beta}{s}\ \Gamma, \Gamma'$$

Lemma 2.3 (Inversion).

(i) $\mathrm{F}\Big|\frac{\alpha}{r}\ \Gamma, A \ \&\ \neg A \in \Delta(\mathbb{N}) \quad \Rightarrow \quad \mathrm{F}\Big|\frac{\alpha}{r}\ \Gamma$

(ii) $\mathrm{F}\Big|\frac{\alpha}{r}\ \Gamma, A_0 \wedge A_1 \quad \Rightarrow \quad \forall i \in \{0,1\} \quad \mathrm{F}\Big|\frac{\alpha}{r}\ \Gamma, A_i$

(iii) $\mathrm{F}\Big|\frac{\alpha}{r}\ \Gamma, \forall x A(x) \quad \Rightarrow \quad \forall n \in \mathbb{N} \quad \mathrm{F}[n]\Big|\frac{\alpha}{r}\ \Gamma, A(\underline{n})$

Lemma 2.4 (Reduction). *Let* A *be a formula* $A_0 \vee A_1$ *or* $\exists x B(x)$ *of rank* $\leq r$.

$$\mathrm{F}_\gamma[\Theta]\Big|\frac{\alpha}{r}\ \Gamma, \neg A \ \&\ \mathrm{F}_\gamma[\Theta]\Big|\frac{\beta}{r}\ \Delta, A \quad \Rightarrow \quad \mathrm{F}_{\gamma+1}[\Theta]\Big|\frac{\alpha+\beta}{r}\ \Gamma, \Delta$$

Proof by induction on β. The interesting case is that $A \equiv \exists x B(x)$ is the main formula of the last deduction step. For some $k < \mathrm{F}_\gamma(\Theta)$ and $\beta_0 < \beta$ we have the premise

$$\mathrm{F}_\gamma[\Theta]\Big|\frac{\beta_0}{r}\ \Delta, A, B(\underline{k})$$

The induction hypothesis yields

$$\mathrm{F}_{\gamma+1}[\Theta]\Big|\frac{\alpha+\beta_0}{r}\ \Delta, B(\underline{k})$$

By the inversion lemma 2.3 (iii) and the fact $\mathrm{F}_\gamma[\Theta][k] \leq \mathrm{F}_{\gamma+1}[\Theta]$ (lemma 2.1 (vi)) we can transform the first derivation into

$$\mathrm{F}_{\gamma+1}[\Theta]\Big|\frac{\alpha}{r}\ \Gamma, \neg B(\underline{k})$$

$\mathrm{N}(\alpha+\beta) < \mathrm{F}_{\gamma+1}(\Theta)$ is true by lemma 2.1 (vii). Furthermore $\mathrm{rk}(B(\underline{k})) < \mathrm{rk}(\forall x B(x))$, so the claim can be obtained by a cut (using monotonicity before).

Lemma 2.5 (Cut-elimination). $\mathrm{F}_\gamma[\Theta]\Big|\frac{\alpha}{r+1}\ \Gamma \quad \Rightarrow \quad \mathrm{F}_{\gamma\#\alpha+1}[\Theta]\Big|\frac{\omega^\alpha}{r}\ \Gamma$

Proof by induction on α. In the interesting case of a cut we have for some $\alpha_0 < \alpha$

$$\mathrm{F}_\gamma[\Theta]\Big|\frac{\alpha_0}{r+1}\ \Gamma, (\neg)A$$

for some A of rank $\leq r$. The induction hypothesis yields

$$\mathrm{F}_{\gamma\#\alpha_0+1}[\Theta]\Big|\frac{\omega^{\alpha_0}}{r}\ \Gamma, (\neg)A$$

By reduction 2.4 (or lemma 2.3 (i) for prime formulas $(\neg)A$) we obtain

$$\mathrm{F}_{\gamma\#\alpha_0+2}[\Theta]\Big|\frac{\omega^{\alpha_0}\cdot 2}{r}\ \Gamma$$

and the claim follows from lemma 2.1 (vii), (viii), $\omega^{\alpha_0}\cdot 2 < \omega^\alpha$ and monotonicity.

Lemma 2.6 (Collapsing).

$$\mathrm{F}\overset{\alpha}{\underset{0}{\vdash}} \exists y A(y) \;\&\; \mathrm{rk}(A) = 0 \quad \Rightarrow \quad \exists n < \mathrm{F}(0) \;\; \mathbb{N} \models A(\underline{n})$$

Proof by induction on α. Since we have a cut-free derivation the last deduction step was $(\exists)$ so there is an $n < \mathrm{F}(0)$ and an $\alpha_0 < \alpha$ such that

$$\mathrm{F}_\gamma[\Theta]\overset{\alpha_0}{\underset{0}{\vdash}} A(\underline{n}), \exists y A(y)$$

If $\mathbb{N} \models A(\underline{n})$ we are done otherwise we can use lemma 2.3 (i) and the induction hypothesis to proof the claim.

Definition 2.5. $A \sim A' \quad :\Leftrightarrow$ *There are a PA-formula B, pairwise distinct variables $x_1, ..., x_n$ and closed PA-terms $t_1, s_1, \ldots, t_n, s_n$ such that $t_i^{\mathbb{N}} = s_i^{\mathbb{N}}$ $(i = 1, ..., n)$ and $A \equiv B_{x_1,...,x_n}(t_1, ..., t_n)$, $A' \equiv B_{x_1,...,x_n}(s_1, ..., s_n)$.*

Lemma 2.7 (Tautology and embedding of mathematical axioms).

(i) $A \sim A' \quad \Rightarrow \quad \mathrm{F}_k\overset{k}{\underset{0}{\vdash}} A, \neg A'$ *where* $k := 2 \cdot \mathrm{rk}(A)$

(ii) For every formula A with $FV(A) \subseteq \{x\}$ there is a $k \in \mathbb{N}$ such that

$$\mathrm{F}_k\overset{\omega+3}{\underset{0}{\vdash}} A(\underline{0}) \wedge \forall x(A(x) \to A(\mathrm{S}x)) \to \forall x A(x)$$

(iii) For every other math. axiom of PA A there is a $k < \omega$ such that $\mathrm{F}_k\overset{k}{\underset{0}{\vdash}} A$

Proof. (i) as usual using $\mathrm{N}k < \mathrm{F}_k(0)$. (iii) If Γ contains only Π-formulas and $\overset{k}{\underset{0}{\vdash}} \Gamma$ denotes a usual cut-free PA-deriviation (without F-controlling) we can easily conclude $\mathrm{F}_k\overset{k}{\underset{0}{\vdash}} \Gamma$ by induction on k. Since the result $\overset{k}{\underset{0}{\vdash}} A$ for mathematical axioms A except induction is folklore the claim follows.
(ii) Let $k := \mathrm{rk}(A(\underline{0}))$. We show by induction on n

$$(*) \quad \mathrm{F}_{2k}[n]\overset{2k+2n}{\underset{0}{\vdash}} \neg A(\underline{0}), \neg\forall x(A(x) \to A(\mathrm{S}x)), A(\underline{n})$$

$n = 0$: The tautology lemma 2.7 (i) yields $\mathrm{F}_{2k}\overset{2k}{\underset{0}{\vdash}} \neg A(\underline{0}), A(\underline{0})$. So the claim follows by monotonicity.
$n \mapsto n+1$: By tautology lemma we have $\mathrm{F}_{2k}\overset{2k}{\underset{0}{\vdash}} \neg A(\mathrm{S}\underline{n}), A(\underline{n+1})$. Connecting this with the derivation given by induction hypothesis by $(\wedge)$ yields
$\mathrm{F}_{2k}[n{+}1]\overset{2k+2n+1}{\underset{0}{\vdash}} \neg A(\underline{0}), \neg\forall x(A(x) \to A(\mathrm{S}x)), A(\underline{n}) \wedge \neg A(\mathrm{S}\underline{n}), A(\underline{n+1})$
So $(*)$ follows by an application of $(\exists)$. The claim follows from $(*)$ by an application of $(\forall)$ and three applications of $(\vee)$.

Lemma 2.8 (Embedding). *For every Γ satisfying $FV(\Gamma) \subseteq \{x_1, \ldots, x_m\}$ and $PA \vdash \Gamma$ there exist $\gamma < \omega^2, \alpha < \omega \cdot 2$, and $r < \omega$ such that*

$$\forall \mathbf{n} \in \mathbb{N}^m \quad \mathrm{F}_\gamma[\mathbf{n}]\overset{\alpha}{\underset{r}{\vdash}} \Gamma(\underline{\mathbf{n}})$$

Proof by induction on the derivation of Γ.
1. $A, \neg A \in \Gamma$ is an axiom. The tautology lemma yields the claim.
2. $\forall x A(x, \mathrm{x}) \in \Gamma(\mathrm{x})$ and Γ results from $\Gamma(\mathrm{x}), A(y, \mathrm{x})$ and $y \notin \{x_1, \ldots, x_m\}$ holds true. By induction hypothesis there are $\gamma < \omega^2, \alpha < \omega \cdot 2, r < \omega$ such that

$$\forall n \in \mathbb{N}\, \forall \mathrm{n} \in \mathbb{N}^m \quad \mathrm{F}_\gamma[n, \mathrm{n}] \overset{\alpha}{\underset{r}{\vdash}} \Gamma(\underline{\mathrm{n}}), A(\underline{n}, \underline{\mathrm{n}})$$

We obtain the claim by an application of $(\forall)$ since $\mathrm{F}_\gamma[n, \mathrm{n}] = \mathrm{F}_\gamma[\mathrm{n}][n]$.
3. $\exists x A(x, \mathrm{x}) \in \Gamma(\mathrm{x})$ and Γ results from $\Gamma(\mathrm{x}), A(t(\mathrm{x}), \mathrm{x})$. We can assume that $x \notin \{x_1, \ldots x_m\}$ and $FV(t) \subseteq \{x_1, \ldots x_m\}$. By induction hypothesis there are $\gamma < \omega^2, \alpha_0 < \omega \cdot 2$ and $r_0 < \omega$ such that

$$(3) \qquad \forall \mathrm{n} \in \mathbb{N}^m \quad \mathrm{F}_\gamma[\mathrm{n}] \overset{\alpha_0}{\underset{r_0}{\vdash}} \Gamma(\underline{\mathrm{n}}), A(t(\underline{\mathrm{n}}), \underline{\mathrm{n}})$$

The tautology lemma 2.7 (i) yields a $k < \omega$ such that

$$(4) \qquad \forall \mathrm{n} \in \mathbb{N}^m \quad \mathrm{F}_k \overset{k}{\underset{0}{\vdash}} A(\underline{t^{\mathbb{N}}(\mathrm{n})}, \underline{\mathrm{n}}), \neg A(t(\underline{\mathrm{n}}), \underline{\mathrm{n}})$$

Since $\lambda \mathrm{x}.t^{\mathbb{N}}(\mathrm{x})$ is a primitive recursive function due to lemma 2.1 (iv) there is a $p < \omega$ satisfying $\forall \mathrm{x} \in \mathbb{N}^m\ t^{\mathbb{N}}(\mathrm{x}) < \mathrm{F}_{\omega \cdot p}(\max \mathrm{x})$. Choosing $p > k$ implies $\mathrm{F}_k \leq \mathrm{F}_k[\mathrm{n}] \leq \mathrm{F}_{\omega \cdot p}[\mathrm{n}]$ with aid of lemma 2.1 (iii). Letting $\alpha := \max\{\alpha_0, k\}$ and $r > \max\{r_0, \mathrm{rk}(A)\}$ we obtain

$$\forall \mathrm{n} \in \mathbb{N}^m \quad \mathrm{F}_{\gamma \# \omega \cdot p}[\mathrm{n}] \overset{\alpha}{\underset{r}{\vdash}} \Gamma(\mathrm{n}), A(\underline{t^{\mathbb{N}}(\mathrm{n})}, \underline{\mathrm{n}}), (\neg)A(t(\underline{\mathrm{n}}), \underline{\mathrm{n}}).$$

from (3) and (4) by monotonicity 2.2. Applying a cut we get

$$\forall \mathrm{n} \in \mathbb{N}^m \quad \mathrm{F}_{\gamma \# \omega \cdot p}[\mathrm{n}] \overset{\alpha+1}{\underset{r}{\vdash}} \Gamma(\mathrm{n}), A(\underline{t^{\mathbb{N}}(\mathrm{n})}, \underline{\mathrm{n}})$$

and $(\exists)$ proves the claim since $t^{\mathbb{N}}(\mathrm{n}) < \mathrm{F}_{\gamma \# \omega \cdot p}(\mathrm{n})$.
4. Γ results from Γ, A and $\Gamma, \neg A$ by a cut. By induction hypothesis there are $\gamma < \omega^2, \alpha < \omega \cdot 2$ and $r < \omega$ such that

$$\forall \mathrm{n} \in \mathbb{N}^m \quad \mathrm{F}_\gamma[\mathrm{n}] \overset{\alpha}{\underset{r}{\vdash}} \Gamma(\underline{\mathrm{n}}), (\neg)A(\underline{\mathrm{n}}, \underline{0})$$

By choosing $r > \mathrm{rk}(A)$ we obtain the claim by an application of (cut).
5. The missing cases are covered by lemma 2.7 or easy (rules for $\vee$ and $\wedge$).

Theorem 2.1. *Let A be a prime formula such that $PA \vdash \forall x \exists y\, A(x, y)$ and $FV(A) \subseteq \{x, y\}$. Then there is a $\gamma < \varepsilon_0$ such that*

$$\forall x \in \mathbb{N}\, \exists y < F_\gamma(x) \quad \mathbb{N} \models A(\underline{x}, \underline{y})$$

Proof by embedding 2.8, iterated cut-elimination 2.5, inversion 2.3 (iii) and collapsing 2.6.

Remark: The methods of this paper yield also classifications of the provable recursive functions of the fragments $(I\Sigma_{n+1})$ of PA and of PA+TI($\prec \lceil$). The extension to KPω has recently been carried out in full detail by the authors.

Acknowledgement. We would like to thank W. Buchholz for valuable advice.

References

1. W. Ackermann. Zur Widerspruchsfreiheit der reinen Zahlentheorie. *Mathematische Annalen 117*, 1940.
2. T. Arai. A slow growing analogue to Buchholz' proof. *Annals of Pure und Applied Logic 54* (1991), 101-120.
3. T. Arai. Proof theory for theories of ordinals. Preprint, Hiroshima, 1995.
4. B. Blankertz and A. Weiermann. A uniform characterization of the provably total functions of KPM and some of its subsystems. Preprint, Münster (1995). To appear in: *Proceedings of the Logic Colloquium 1994*.
5. W. Buchholz. Three contributions to the conference on "Recent advances in Proof Theory". Oxford, April, 10-16, 1980, (mimeographed).
6. W. Buchholz. An independence result for $(\Pi^1_1 - CA) + BI$. *Annals of Pure and Applied Logic 33* (1987), 131-155.
7. W. Buchholz. A simplified version of local predicativity. *Proof Theory: A Selection of Papers from the Leeds Proof Theory Meeting 1990*, P. Aczel, H. Simmons, S.S. Wainer, editors. Cambridge University Press (1992), 115-148.
8. W. Buchholz. Notation systems for infinitary derivations. *Archive for Mathematical Logic 30* 5/6 (1991), 277-296.
9. W. Buchholz und S.S. Wainer. Provably computable functions and the fast growing hierarchy. *Logic and Combinatorics. Contemporary Mathematics 65*, American Mathematical Society (1987), 179-198.
10. W. Buchholz, A. Cichon and A. Weiermann. A uniform approach to fundamental sequences and hierarchies. *Mathematical Logic Quarterly 40* (1994), 273-286.
11. H. Friedman and M. Sheard. Elementary descent recursion and proof theory. *Annals of Pure and Applied Logic 71* (1995), 1-45.
12. K. Gödel. Über eine bisher noch nicht benützte Erweiterung des finiten Standpunkts. *Dialectica 12* (1958), 280-287.
13. P. Hájek and P. Pudlák. *Metamathematics of First-Order Arithmetic*, Springer (1993).
14. G. Kreisel. On the interpretation of non-finitist proofs II. *The Journal of Symbolic Logic 17* (1952), 43-58.
15. W. Pohlers. Proof Theory. An Introduction. *Lecture Notes in Mathematics 1407* (1989). Springer, Berlin.
16. W. Pohlers. Proof theory and ordinal analysis. *Archive for Mathematical logic 30* 5/6 (1991), 311-376.

17. W. Pohlers. A short course in ordinal analysis. *Proof Theory: A Selection of Papers from the Leeds Proof Theory Meeting 1990*, P. Aczel, H. Simmons, S.S. Wainer, editors. Cambridge University Press (1992), 27-78.
18. M. Rathjen. Proof-theoretic analysis of KPM. *Archive for Mathematical Logic 30* (1991), 377-403.
19. M. Rathjen. Proof theory of reflection. *Annals of Pure and Applied Logic* (1994), 181-224.
20. M. Rathjen. Ordinal analysis of $\Pi^1_2 - (CA)$ and related systems. Preprint, Stanford, 1995.
21. H. E. Rose. *Subrecursion: Functions and Hierarchies*. Oxford University Press 1984.
22. K. Schütte. *Proof Theory*. Springer, Berlin 1977.
23. H. Schwichtenberg. Eine Klassifikation der ε_0-rekursiven Funktionen. *Zeitschrift für Mathematische Logik und Grundlagen der Mathematik* (1971), 61-74.
24. H. Schwichtenberg. Proof theory: Some applications of cut-elimination. *Handbook of Mathematical Logic,* J. Barwise (ed.), North-Holland (1977), 867-895.
25. W.W. Tait. Infinitely long terms of transfinite type. *Formal Systems and Recursive Functions,* Crossley, Dummet (eds.), North-Holland (1965), 176-185.
26. G. Takeuti. *Proof Theory*, 2nd edition. North-Holland, Amsterdam 1987.
27. S.S. Wainer. A classification of the ordinal recursive functions. *Archiv für Mathematische Logik 13* (1970), 136-153.
28. A. Weiermann. How to characterize provably total functions by local predicativity. Preprint, Münster 1993. To appear in: *The Journal of Symbolic Logic*.
29. A. Weiermann. A strongly uniform termination proof for Gödel's T by methods from local predicativity. Preprint, Münster 1995, submitted.

Completeness has to be restricted: Gödel's interpretation of the parameter t *

Giora Hon

Department of Philosophy
University of Haifa, Mt. Carmel
Haifa 31905, Israel

> ...even such successful theories as relativity and quantum mechanics have proven to be vulnerable, for, implicit in each is a latent completeness postulate. Once one realizes that a physical theory should remain open and therefore incomplete, the role of the associated mathematics becomes a crucial issue.
>
> W. Yourgrau[1]

In 1949 Gödel presented Einstein with a new solution of the field equations of the general theory of relativity: an exact solution which allows for the bizarre possibility of time-travel.[2] This discovery of a universe with a strange time-structure which is consistent with the general theory of relativity did not however surprise Einstein, or so at least he leads the reader to believe. In his reply to Gödel's contribution, Einstein intimated that the strange feature of a possible closed time-like lines in which the distinction "earlier-later" should be abandoned, "disturbed... [him] already at the time of the building up of the general theory of relativity, without... [him] having succeeded in clarifying it." The weird result of Gödel constituted in Einstein's view "an important contribution to the general theory of relativity, especially to the analysis of the concept of time."[3]

Unlike the general theory, the special theory of relativity has been accepted by the mid-century as an undisputed part of theoretical physics. It has been commonly presented as the invariance theory of the Maxwell equations under the Lorentz group of transformations. However, Einstein seems to have conceived of the special theory as the theory of a special type of gravitational field: the uniform one. This recognition motivated him to go further and require generalization by the inclusion of gravitational phenomena.[4] In

---

* This paper is in its final form and no similar paper has been or is being submitted elsewhere.

[1] Yourgrau, 1969, p.80.

[2] Gödel, 1949b.

[3] Schilpp, 1970, p.687.

[4] Kerszberg, 1989, pp.71–72.

this move one wishes that the "causal" structure of space-time in general relativity should manifest locally the same qualities as the flat space-time of special relativity, though globally significant differences are expected to emerge. However, the local nature of general relativity which deals with gravitational phenomena did not combine easily with the global nature of the invariance principles of the electrodynamics of moving bodies. Thus, in Einstein's 1917 cosmological solution of the field equations of his general theory time is conceived of as linear, separated from the three-dimensional structure of space. The solution presents a quasi-absolute time and a preferred coordinate system. Gödel challenged precisely this conception of time — a conception of time which does not cohere satisfactorily with the results of special relativity.

Gödel underlined the "surprising insights into the nature of time" which the special theory of relativity has revealed. "The very starting point of special relativity theory consists in the discovery of a new and very astonishing property of time, namely the relativity of simultaneity, which to a large extent implies that of succession." Simultaneity loses its objective meaning. Special relativity provides, in Gödel's words, "an unequivocal proof" for a philosophical position which "den[ies] the objectivity of change and consider[s] change an illusion or an appearance due to our special mode of perception."[5] In his first paper on this subject, Gödel observed that,

> all cosmological solutions with non-vanishing density of matter ... have the common property that... they contain an "absolute" time coordinate, owing to the fact that there exists a one-parametric system of three-spaces everywhere orthogonal on the world lines of matter.[6]

He pointed out that in general relativity,

> the existence of matter,... as well as the particular kind of curvature of space-time produced by it, largely destroy the equivalence of different observers and distinguish some of them conspicuously from the rest, namely, those which follow in their motion the mean motion of matter.

He then continued to make the incisive remark that

> in all cosmological solutions of the gravitational equations... the local times of all *these* observers fit together into one world time, so that apparently it becomes possible to consider this time as the "true" one, which lapses objectively.[7]

Gödel objected strongly to this notion of world, or "true", time which comprises local times associated with the class of privileged observers.

[5] Gödel, 1949b, p.557.
[6] Gödel, 1949a. See Gödel, 1990, p.190.
[7] Gödel, 1949b, p.559 (emphasis in the original). *Cf.*, North, 1965, p.359.

To retain consistently the relativity of simultaneity, one has to show that there are solutions in the general theory, that is, universes, in which there cannot be such a "world-time", such a "true" time. In these universes the experienced "lapse of time can exist without an objective lapse of time," so that "no reason can be given why an objective lapse of time should be assumed at all." [8] Put differently, it would be wrong, according to Gödel, to believe that the postulated equivalence of inertial observers also enables us to recover equivalence with regard to a unique, absolute time function. The special theory does not allow us to force becoming into any space-time background.

In his solution of the field equations, Gödel sought to implement consistently this result of the special theory: it should be impossible to fit chosen local times into a world-time. By eliminating the Einsteinian "system of three-spaces", he was able to do away with the notion of "absolute" time and recover thereby the result of the special theory of relativity. "It is easily seen," Gödel observed, "that the non-existence of such a system of three- spaces is equivalent with a rotation of matter relative to the compass of inertia."[9]

As *a consequence* of eliminating the three-space system, Gödel obtained a universe in which matter rotates — by all accounts, an absolute *spatial* feature. Matter in the Gödel Universe rotates relative to the path that a test particle follows if it is given an initial radial velocity. One may refer to the tangent of such a path as the "compass of inertia". Thus the compass of inertia rotates relative to the matter of the Gödel Universe, or vice versa.[10] In the Gödel solution rotational mass phenomena occur as a result of the denial that cosmic time would be everywhere orthogonal to the three-dimensional space. By exploiting the geometrization of time, Gödel introduced a limiting case: the parameter t must be interpreted in a certain restrictive sense. Gödel believed he had made the implications of relativity theory, *vis-à-vis* spatialization, inescapable.[11] The existence of closed timelike curves in the Gödel solution deprives time of its unique direction and makes it behave like space, distinguished to be sure from the other three dimensions, but *only geometrically*. Hence the possibility of time-travel: "By making a round trip on a rocket ship in a sufficiently wide curve, it is possible in these worlds," Gödel speculated, "to travel into any region of the past, present, and future,

[8] Gödel, *ibid.*, p.561.

[9] Gödel, 1990, p.190.

[10] Sklar clarifies this rotation with the following analogy: the path of a particle moving in a straight line out from the center over a phonograph record spinning on a turntable will mark a spiral groove on the record indicating thereby that the record rotates. In the Gödel solution each observer could count himself as central to the spinning of the cosmic matter — an absurdity in the Machian view. Sklar, 1995, p.78. See also Gödel, 1949a; 1950; 1990, pp.190, 212. Cf., Adler *et al.*, 1965, p.377. Hawking and Ellis, 1973, pp.168–70.

[11] Yourgrau, 1991, p.11.

and back again, exactly as it is possible in other worlds to travel to distant parts of space."[12]

It is worth noting that Gödel's interest in this result went beyond the purely theoretical; witness his remarks concerning the physical meaning of the solution and the compilation of astronomical observations. Gödel made a rough estimation of the velocity the rocket ship will have to attain as well as the amount of fuel that would be needed for the journey. The point of these calculations, as Gödel remarked, was to demonstrate that the velocities which would be necessary in order to complete the voyage in a reasonable length of time are far beyond everything that can be expected ever to become a practical possibility; therefore, the result cannot be excluded *a priori* on the ground that the space-time structure of the so-called real world does not allow in principle for such a voyage, after all, the demarcation between difficulties in practice and difficulties in principle is not at all fixed in this situation. Furthermore, there appear to be two notebooks in Gödel's *Nachlass* in which he tabulated angular orientations of galaxies, perhaps in the hope of finding some observational evidence for his solution.[13]

It is important to stress that Gödel did not detect any inconsistency in the notion of cosmic time as expressed originally in Einstein's solution of the field equations. Gödel chose to proceed in a different way than the standard approach. He sought a categorically different solution which was in his view closer to the spirit of relativity.[14] Gödel derived a possible structure of the physical universe from a space-time metric compatible with general relativity, instead of starting from a requirement about this structure.[15] He thereby established a new type of solution of the general theory which indicates in turn a possible existence of a structure of these different types of solution, a structure whose understanding, Gödel believed, could illumine the origin of inertia and the nature of time.

Regarding the further development of the general theory, Gödel did not expect an extension in the sense that the theory should comprehend a broader range of facts, but rather "a mathematical analysis of the equations which would make it possible to take hold of their solutions systematically and to recognize general properties of the solutions." Gödel probably expected that such an analysis of the equations would lead to *some* kind of general theorems about the structure of their solutions, in the sense in which the conservation laws are general theorems about the structure of the solutions of the Newtonian equations: "So far we do not even know the analogue of

[12] Gödel, 1990, p.205. On time-travel and the spatial sense of time see Malament, 1984, 1987; Pfarr, 1981: the existence of closed timelike world lines in general relativity is not restricted to the peculiar case of Gödel's universe, see, p.1074. *Cf.*, Stein, 1970; Yourgrau, 1991, pp.10–15, 20, 32–38, 43–49.

[13] Gödel, 1990, pp.197–198, 205 fn.11. See, Malament, 1984; 1987, p.2429; Pfarr, 1981, pp.1089–90; Wang, 1991, p.117.

[14] See North, 1965, p.360.

[15] Kerszberg, 1989, p.374.

the fundamental integral formulas of the Newtonian theory," Gödel observed and continued confidently that these integral formulas "must unquestionably exist." He was of the opinion that a deeper physical understanding of the general theory would be obtained if we were to know these integral formulas; or, conversely, if a more precise analysis of the physical content of the theory could lead to such mathematical theorems.[16]

Gödel succeeded in disturbing Einstein's original relativistic cosmology *from within*. While claiming no new principle, he undermined the Mach principle which constituted according to Einstein's initial view one of the three principles of general relativity.[17] According to Grünbaum, "Einstein named his own organic fusion of Riemann's and Mach's ideas 'Mach Principle'." [18] The geometry, or metric relations, of a continuous manifold is determined by forces *extrinsic* to the manifold. Thus, any inertial forces experienced by accelerating bodies and systems must have their origin in an interaction with the material contents of the universe. Indeed, "in a consistent theory of relativity there can be no inertia *relatively to 'space'*, but only an inertia of masses *relatively to one another*." These are Einstein's words. [19] Inertia is associated with the reciprocal interaction of *all* the potentially observable masses in the universe. Einstein's search for a theory that has the feature of completeness comes here to the fore.

This formulation has the immediate implication that the only meaningful idea of motion is motion relative to other material objects, thus the local inertial frame should not rotate with respect to the frame defined by the distant stars; there is no meaning to an intrinsic rotation of the material content of the universe. As it stands, the Gödel solution is inconsistent with this implication of the Mach principle. Gödel's exact solution of Einstein's *modified* field equations of the general theory of relativity presents a completely homogeneous, but not isotropic, *finite* universe filled with matter of constant density (pressure-free perfect fluid) which rotates rigidly relative to the local compass of inertia: the so-called rotating mass solution. The co-moving matter of the Gödel universe undergoes an intrinsic uniform rotation. This result contradicts the Mach principle which expresses the requirement that the bulk matter of the universe determines the compass of inertia, and therefore the two cannot rotate relative to each other. To put it in an equivalent formulation, the demand of the principle that the local inertial compass and the light compass (of the fixed stars) must coincide does not hold in this solution.[20]

The solution implies the disturbing fact that the general theory of relativity may not be after all completely relatavised as Einstein hoped to achieve. Unlike the current popular view, Einstein's general theory of relativity has

[16] Quoted by Wang, 1991, p.155; *cf.*, p.156.
[17] Einstein, 1918.
[18] Grünbaum, 1974, pp.419–20.
[19] Einstein, 1917, in Lorentz et al., 1952, p.180 (emphases in the original).
[20] Adler *et al.*, 1965, p.377.

not resolved the controversy between the absolute and relative conception of space and time in favour of the latter, at least with respect to the implementation of the Mach principle.[21]

It may seem that there is no connection between Gödel's contribution to logic and mathematics and his interest in the relativity theory. To be sure, the two fields are certainly distinct and the results which Gödel obtained in these fields have different technical impacts.[22] However, these results share a fundamental quality: in both fields Gödel went directly and wholeheartedly to the questions which lie at the very centre of the problems at stake.[23] Thus in logic and mathematics as well as in physics Gödel had a similar intention: to attain a perspicacious view of the very foundations of the fields and to focus on the central, crucial questions. He sought, in a word, foundational problems in an attempt to arrive at definite results on broad conceptual issues. Hence, his interest in the notions of proof and truth and their relation to decidability, and in the notion of time and its relation to physical structure. Such an encompassing approach leads inevitably to the problem of completeness and its mirror image — the limits of knowledge, by all accounts a philosophical issue. Gödel's contributions to logic and mathematics as well as to physics are the fruitful results of a methodology which seeks to delimit the notion of completeness: in logic and in physics, if the sentence and the physical probe, were to be both actor and spectator, both assertion and referent, then the claim to completeness would founder.[24] In a strict sense, Gödel was not a philosopher, but he did see himself as contributing not only to his fields of expertise, but also to philosophy at large. He was concerned with essentially philosophical questions.[25]

Addressing the very foundation of logic and mathematics, Gödel succeeded in forcing a distinction between proof and truth. He arrived at this result by reconsidering the notion of completeness of formal systems. It is only with respect to completeness that the syntactic concepts of derivability and consistency coincide with the semantic concepts of validity and satisfiability. This is not the case when the system is shown to be incomplete. Gödel's "Incompleteness Theorem" is genuinely limitative. He demonstrated that there is an intrinsic limitation on the semantic that can be imposed on the syntactic predicates in Hilbert's formal systems.[26]

Just as the "Incompleteness Theorem" demonstrates in regard to the Hilbert programme that, in that context, mathematical truth cannot be simulated by formal proof, so in regard to Einstein's theory, the construction

[21] Grünbaum, 1974, p.422.

[22] See Gödel's letter to Seelig (1955) in which he referred to problems in general relativity as "very remote from [his]...own area of work." (Quoted by Wang, 1991, p.155.)

[23] Wang, 1991, pp.2–4, 108–9, 152–54.

[24] Schlegel, 1967, pp.197–98.

[25] Wang, 1991, p.151. *Cf.*, Pais, 1982, p.13.

[26] Yourgrau, 1991, pp.12–13.

of a formal- mathematical model of the Gödel universe demonstrates that, in this context, t cannot be given the intuitive, contentful interpretation of equating the experienced successive, unfolding time with "world-time" or "true" time — an objective temporal becoming.[27] With the derivation of the Gödel Universe as a solution to the field equations of general relativity, Gödel constructed a limiting case for the relativistic geometrization of time. That is, he produced a formal model that essentially limits the possible intuitive, contentful interpretations it can support. As Yourgrau remarks, this construction of limiting cases is "in the great philosophical-mathematical tradition of employing limit concepts to enable the mind to grasp or delimit the most distant reaches of reality."[28]

Gödel's limiting cases originated in a dialectic of the formal and the intuitive. Mathematics for Gödel is not just syntax. Mathematical realism and the expression it finds in Platonism guided Gödel's intuition in logic and mathematics. Gödel explicitly remarked that he did not see any reason why we should have less confidence in mathematical intuition, than in sense perception, which induces us to build up physical theories and to expect that future sense perceptions will agree with them, and, moreover, to believe that a question not decidable now has meaning and may be decided in the future.[29] There are, in Gödel's view, more similarities than differences between sense perceptions and the perceptions of concepts. When one realizes that there are two different sharp concepts mixed together in the original, intuitive concept, then the paradox which has arisen disappears. Similarly, one may not distinguish two neighboring stars a long distance away, but using a high-resolution telescope one can see that there are indeed two stars. This objectivist conception of mathematics marked the entire work of Gödel.[30]

The case of the vicious-circle principle is a good illustration of this point. According to Russell, who gave the principle its original formulation, if an object is defined in terms of some totality, then, if that object were in the totality, one would have a vicious circle. The principle states then that "whatever involves *all* of a collection must not be one of the collection;" or, conversely: "if, provided a certain collection had a total, it would have members definable only in terms of that total, then the said collection has no total."[31] A member of the totality is singled out with the paradoxical property that this member cannot be a member of that totality. Russell intended the principle to rule out as illegitimate and meaningless quantification over the totality of all sets and all statements about the set of all sets. [32] Gödel, on his part, formulated the principle somewhat differently: "no totality can contain members definable only in terms of this totality, or members involving or presupposing this

[27] *Ibid.*, pp.14–15.
[28] *Ibid.*, p.17.
[29] Gödel, 1990, p.170.
[30] Wang, 1974, pp.8–10, 85, 324.
[31] Wang, 1991, p.317.
[32] Chihara, 1973, pp.4, 6–7. *Cf.*, Grim, 1991, pp.28–31; Priest, 1995, p.150.

totality."[33] It therefore appears that there must exist a definition, that is, a description of construction, which does not refer to the totality to which the object at stake belongs. The construction of an object simply cannot be based on the totality of objects to which the object to be constructed belongs. But Gödel's way out of the circularity is different. It is very characteristic of his approach:

> If, however, it is a question of objects that exist *independently of our constructions*, there is nothing in the least absurd in the existence of totalities containing members which can be described (i.e., uniquely characterized) only by reference to this totality. [34]

This is very striking. Gödel's argument may be regarded, by analogy, as an application of an abstracted version of the Mach principle in mathematics: a member of a totality is referred to uniquely by the totality itself without any resort to an external element (= matter in the universe, the totality, determines the inertial frame at a certain point within the said totality). "It seems to me," Gödel continued,

> that the assumption of such [concrete mathematical] objects is quite as legitimate as the assumption of physical bodies and there is quite as much reason to believe in their existence. They are in the same sense necessary to obtain a satisfactory system of mathematics as physical bodies are necessary for a satisfactory theory of our sense perceptions.

Just as physical objects are natural and necessary for organizing our physical experience, mathematical objects are natural and necessary for organizing our mathematical experience. Thus, according to Gödel, if classes and concepts were to be conceived of as real objects, "objects that exist *independently of our constructions*," objects that have been there from the very beginning, then they could have been members of a certain totality, even infinite totality that permits infinite iterations, without falling into the trap of the vicious-circle principle. According to Gödel, "the set-theoretical paradoxes are hardly any more troublesome for mathematics than deceptions of the senses are for physics." He argued that the vicious-circle principle creates such deceptions; it is objectively not true.[35]

---

[33] Gödel, 1990, p.125; 1944, pp.123–153. See also Wang, 1991, p.317.

[34] Gödel, 1990, pp.127–28 (emphasis has been added).

[35] Chihara, 1973, p.63. Ramsey pointed out, that "we may refer to a man as the tallest in a group, thus identifying him by means of a totality of which he is himself a member without there being any vicious circle." (Ramsey, 1965, p.41.) Ramsey indeed challenged the validity of the principle itself arguing that to express a proposition indirectly by a reference to a totality "is certainly... a circuitous process, but there is clearly nothing vicious about it." (p.42.) He criticized Whitehead and Russell for using a "rather vague" principle in "a rather slopy way." (pp.76, 24.)

It transpires that a case can be made for the claim that a similar epistemology and ontology guided Gödel's intuition and methodology in the respective fields of logic, mathematics and physics. The Mach principle is a case in point. That Gödel proved its inconsistency with the general theory, does not diminish the force of this principle in inspiring and directing the search for the origin of inertia, the comprehension of time and the construction of cosmological models. After all, the Mach principle is an expression of a belief in completeness, in the totality of objects which exist "independently of our constructions," as Gödel would have it, and determine the course of physical phenomena. It is thus still a moot question whether time, to use Gödel's formulation, "depends on the particular way in which matter and its motion are arranged in the world."[36]

References

1. Adler, R., Bazin, M. and Schiffer, M., 1965, INTRODUCTION TO GENERAL RELATIVITY, McGraw-Hill, New York.
2. Barbour, J. and Pfister, H. (eds.), 1995, MACH'S PRINCIPLE: FROM NEWTON'S BUCKET TO QUANTUM GRAVITY, Einstein's Studies Vol.6, Birkhäuser, Boston, Basel, Berlin.
3. Bernstein, J. and Feinberg, G. (eds.), 1986, COSMOLOGICAL CONSTANTS: PAPERS IN MODERN COSMOLOGY, Columbia University Press, New York.
4. Chihara, C. S., 1973, ONTOLOGY AND THE VICIOUS-CIRCLE PRINCIPLE, Cornell University Press, Ithaca.
5. Einstein, A., 1917, "Kosmologische Betrachtungen zur allgemeinen Relativitätstheorie," *Sitzungsberichte der Preussischen Akad. d. Wiss., Berlin*, pp.142–52; for an English translation see Bernstein and Feinberg, 1986, pp.16–26; Lorentz *et al.*, 1952, pp.175–88.
6. Einstein, A., 1918, "Prinzipielles zur allgemeinen Relativit'atstheorie," *Annalen der Physik,* 4th ser., 55, pp.241–44.
7. Friedman, M., 1983,FOUNDATIONS OF SPACE-TIME THEORIES, Princeton University Press, Princeton, New Jersey.
8. Gödel, K., 1944, "Russell's mathematical logic," in Schilpp, P. A. (ed.), 1944, THE PHILOSOPHY OF BERNARD RUSSELL, Northwestern University Press, Chicago, pp.123–153.
9. Gödel, K., 1949a, "An example of a new type of cosmological solutions of Einstein's field equations of gravitation," *Rev. Mod. Phys.*, 21, pp.447–50.
10. Gödel, K., 1949b, "A remark about the relationship between relativity theory and idealistic philosophy," in Schilpp, 1970, pp.557–62.
11. Gödel, K., 1950, "Rotating universes in general relativity theory," in Graves, L. M., Hille, E., Smith, R. A. and Zariski, O. (eds.), PROCEEDINGS OF THE 1950 INTERNATIONAL CONGRESS OF MATHEMATICS, Providence RI, American Mathematical Society, 1952, I, pp.175– 181.
12. Gödel, K., 1962, ON FORMALLY UNDECIDABLE PROPOSITIONS OF PRINCIPIA MATHEMATICA AND RELATED SYSTEMS, Meltzer, B. (tr.) with Introduction by Braithwaite, R. B., Basic Books, New York.

[36] Gödel, 1949b, p.562. *Cf.*, Stein, 1970, pp.594–95. See also Pais, 1982, p.288.

13. Gödel, K., 1990, Collected Works, Feferman, S. *et. al.* (eds.), Vol. 2: Publications 1938–1974. Oxford University Press, Oxford.
14. Grim, P., 1991, The Incomplete Universe: Totality, Knowledge and Truth, a Bradford Book, The MIT Press, Cambridge, Mass., London.
15. Grünbaum, A., 1974 Philosophical Problems of Space and Time, Boston Studies in the Philosophy of Science, XII, Reidel, Dordrecht.
16. Hawking, S. and Ellis, G. F. R., 1973, The Large Scale Structure of Space-Time, Cambridge University Press, Cambridge.
17. Kerszberg, P., 1989, The Invented Universe: The Einstein-De Sitter controversy (1916–17) and the rise of relativistic cosmology, Clarendon Press, Oxford.
18. Lorentz, H. A., Einstein A., Minkowski, H. and Weyl, H., 1952, The Principle of Relativity, with notes by A. Sommerfeld, Dover, New York.
19. Malament, D., 1984, "'Time travel' in the Gödel-universe," *Proceedings of the Philosophy of Science Associations*, **2**, pp.91–100.
20. Malament, D., 1987, "A note about closed timelike curves in Gödel space-time," *J. Math. Phys.*, **28**, pp.2427–30.
21. North, J. D., 1965, The Measure of the Universe, Clarendon Press, Oxford.
22. Pais, A., 1982, Subtle is the Lord, Oxford University Press, Oxford.
23. Pfarr. J., 1981, "Time travel in Gödel's space," *General Relativity and Gravitation*, **13**, pp.1073–91.
24. Priest, G., 1995, Beyond the Limits of Thought, Cambridge University Press, Cambridge.
25. Ramsey, F., 1965, The Foundations of Mathematics, Routledge & Kegan Paul, London.
26. Schilpp, P. A. (ed.), 1970, Albert Einstein: Philosopher-Scientist, Open court, La Salle, Illinois.
27. Schlegel, R., 1967, Completeness in Science, Appleton-Century-Crofts, New York.
28. Sklar, L., 1995, Philosophy of Physics, Oxford University Press, Oxford.
29. Stein, H., 1970, "On the paradoxical time-structures of Gödel," *Phil. of Science*, **37**, pp.589–601.
30. Wang, H., 1974, From Mathematics to Philosophy, Routledge, London.
31. Wang, H., 1991, Reflections on Kurt Gödel, the MIT Press, a Bradford Book, Cambridge, Massachusetts (3rd printing).
32. Yourgrau, P., 1991, The Disappearance of Time: Kurt Gödel and the Idealistic Tradition in Philosophy, Cambridge University Press, Cambridge.
33. Yourgrau, W., 1969, "Gödel and physical theory," *Mind*, **78**, pp.77–90

Added for first reprinting. The author wishes to acknowledge that the four sentences referred to by footnotes 26 and 27 include several direct quotations from the cited source (Yourgrau, 1991).

A Bounded Arithmetic Theory for Constant Depth Threshold Circuits*

Jan Johannsen

IMMD 1, Universität Erlangen-Nürnberg, Germany
email: `johannsen@@informatik.uni-erlangen.de`

Summary. We define an extension $\bar{R}^0_2$ of the bounded arithmetic theory R^0_2 and show that the class of functions Σ^b_1-definable in $\bar{R}^0_2$ coincides with the computational complexity class TC^0 of functions computable by polynomial size, constant depth threshold circuits.

1. Introduction

The theories S^i_2, for $i \in \mathbb{N}$, of Bounded Arithmetic were introduced by Buss [3]. The language of these theories is the language of Peano Arithmetic extended by symbols for the functions $\lfloor \frac{1}{2}x \rfloor$, $|x| := \lceil \log_2(x+1) \rceil$ and $x\#y := 2^{|x|\cdot|y|}$. A quantifier of the form $\forall x \leq t$, $\exists x \leq t$ with x not occurring in t is called a *bounded quantifier*. Furthermore, a quantifier of the form $\forall x \leq |t|$, $\exists x \leq |t|$ is called *sharply bounded*. A formula is called (sharply) bounded if all quantifiers in it are (sharply) bounded.

The class of bounded formulae is divided into an hierarchy analogous to the arithmetical hierarchy: The class of sharply bounded formulae is denoted Σ^b_0 or Π^b_0. For $i \in \mathbb{N}$, Σ^b_{i+1} (resp. Π^b_{i+1}) is the least class containing Π^b_i (resp. Σ^b_i) and closed under conjunction, disjunction, sharply bounded quantification and bounded existential (resp. universal) quantification.

Now the theory S^i_2 is defined by a finite set $BASIC$ of quantifier-free axioms plus the scheme of *polynomial induction*

$$A(0) \wedge \forall x\, (A(\lfloor \frac{1}{2}x \rfloor) \to A(x)) \;\to\; \forall x A(x)$$

for every Σ^b_i-formula $A(x)$ (Σ^b_i-$PIND$).

For a class of formulae Γ, a number-theoretic function f is said to be Γ-definable in a theory T if there is a formula $A(\bar{x}, y) \in \Gamma$, describing the graph of f in the standard model, and a term $t(\bar{x})$, such that T proves

$$\forall \bar{x}\, \exists y \leq t(\bar{x})\, A(\bar{x}, y)$$

$$\forall \bar{x}, y_1, y_2\, A(\bar{x}, y_1) \wedge A(\bar{x}, y_2) \to y_1 = y_2$$

The main result of [3] relates the theories S^i_2 to the Polynomial Time Hierarchy PH of Computational Complexity Theory (cf. [9]):

* This paper is in its final form, and no version of it will be submitted for publication elsewhere

The class of functions that are Σ^b_{i+1}-definable in S^{i+1}_2 coincides with $FP^{\Sigma^p_i}$, the class of functions computable in polynomial time with an oracle from the ith level of the PH.

In particular, the functions Σ^b_1-definable in S^1_2 are precisely those computable in polynomial time.

The theories R^i_2 were defined in various disguises by several authors [1, 10, 5]. Their language is the same as that of S^i_2 extended by additional function symbols for subtraction $\dot{-}$ and $MSP(x,i) := \lfloor \frac{x}{2^i} \rfloor$. They are axiomatized by an extended set $BASIC$ of quantifier-free axioms plus the scheme of *polynomial length induction*

$$A(0) \wedge \forall x\, (\, A(\lfloor \frac{1}{2} x \rfloor) \rightarrow A(x)\,) \;\rightarrow\; \forall x A(|x|)$$

for every Σ^b_i-formula $A(x)$ (Σ^b_i-$LPIND$).

R^1_2 is related to the complexity class NC, the class of functions computable in polylogarithmic parallel time with a polynomial amount of hardware:

The Σ^b_1-definable functions of R^1_2 are exactly those in NC.

In [10] it was shown that R^0_2 is equivalent to S^0_2 in the extended language, which is trivially equivalent to the theory given by the $BASIC$ axioms and the scheme of *length induction*

$$A(0) \wedge \forall x\, (A(x) \rightarrow A(Sx)) \rightarrow \forall x\, A(|x|)$$

for every Σ^b_0-formula $A(x)$ (Σ^b_0-$LIND$).

TC^0 denotes the class of functions computable by uniform polynomial size, constant depth families of threshold circuits (cf. [2]). This class can be viewed as the smallest reasonable complexity class, e.g. it is the smallest class known to contain all arithmetical operations: integer multiplication is complete for it under a very weak form of reducibility.

Let B be the set of functions containing all projections, the constant 0, $s_0(x) := 2x$, $s_1(x) := 2x + 1$, $Bit(x,i)$ giving the value of the ith bit in the binary representation of x, $\#$ and multiplication. The class TC^0 was characterized in [6] as the smallest class of functions that contains the initial functions in B and is closed under composition and the operation of *concatenation recursion on notation* (CRN), where a function f is defined by CRN from g and h_0, h_1 if

$$\begin{aligned} f(\bar{x},0) &= g(\bar{x}) \\ f(\bar{x},s_0(y)) &= 2 \cdot f(\bar{x},y) + h_0(\bar{x},y) \qquad \text{for } y > 0 \\ f(\bar{x},s_1(y)) &= 2 \cdot f(\bar{x},y) + h_1(\bar{x},y) \end{aligned}$$

provided that $h_i(\bar{x},y) \leq 1$ for all $\bar{x}, y$ and $i = 0,1$. It follows from this characterization by methods from [4] that the characteristic function of any predicate defined by a Σ_0^b-formula in the language of R_2^0 is in TC^0, and that TC^0 is closed under *sharply bounded minimization*, i.e. if $g \in TC^0$, then f defined by $f(x) := \mu i \leq |x| \; g(i) = 0$ is also in TC^0.

We shall define an extension $\bar{R}_2^0$ of R_2^0 the Σ_1^b-definable functions of which are exactly the functions in TC^0. In [6], an arithmetical theory TTC^0 is presented that also characterizes TC^0. We shall compare our work to this in the final section of the paper.

2. Definition of $\bar{R}_2^0$

Before the theory $\bar{R}_2^0$ can be defined, we have to develop R_2^0 a little. To be able to talk about the bits of a number, we first define $Mod2(x) := x \dot{-} 2 \cdot \lfloor \frac{1}{2}x \rfloor$ and then $Bit(x,i) := Mod2(MSP(x,i))$. In R_2^0, a number is uniquely determined by its bits, as the extensionality axiom

$$|a| = |b| \wedge \forall i < |a| \; (Bit(a,i) = Bit(b,i)) \rightarrow a = b$$

can be proved in R_2^0 (see [7] for a proof).

We shall need the possibility to define a number by specifying its bits. So for a class of formulae Γ, let the Γ-comprehension scheme be the axiom scheme

$$\exists y < 2^{|t|} \; \forall i < |t| \; (Bit(y,i) = 1 \leftrightarrow A(i))$$

for every formula $A(i) \in \Gamma$.

Next we need the possibility of coding pairs and short sequences. The coding used is based on the one presented in [5], but we need a refined analysis to show its accessibility in R_2^0.

First let $\bar{sg}(x) := 1 \dot{-} x$, and then $[x \leq y] := \bar{sg}(x \dot{-} y)$. Obviously, $[x \leq y] = 1$ iff $x \leq y$ and $[x \leq y] = 0$ else. Further let $[x < y] := [Sx \leq y]$, and then define

$$\max(x,y) := [x \leq y] \cdot y + [y < x] \cdot x \, .$$

Let now $x \frown y := x \cdot 2^{|y|} + y$, then we define

$$\langle x,y \rangle := (2^{|\max(x,y)|} + x) \frown (2^{|\max(x,y)|} + y) \, .$$

We go on to define $DMSB(x) := x \dot{-} 2^{||\lfloor \frac{1}{2}x \rfloor||}$, $front(x) := MSP(x, \lfloor \frac{1}{2}|x| \rfloor)$ and $back(x) := x \dot{-} front(x) \cdot 2^{|front(x)|}$, and finally

$$(x)_1 := DMSB(front(x)) \text{ and } (x)_2 := DMSB(back(x)) \, .$$

Using extensionality, one can prove in R_2^0 that $(\langle x,y \rangle)_1 = x$ and $(\langle x,y \rangle)_2 = y$, hence these functions form a pairing system. The pairing function is not surjective, but its range can be described by

$$pair(x) :\leftrightarrow x > 2 \wedge Mod2(|x|) = 0 \wedge Bit(x, \lfloor \frac{1}{2}|x| \rfloor \dot{-} 1) = 1 \, .$$

Inductively we can define $(x)^{(2)}_i := (x)_i$ for $i = 1, 2$, and for $n \geq 2$ and $j \leq n$

$$\begin{aligned} \langle x_1, \ldots, x_n, x_{n+1}\rangle &:= \langle\langle x_1, \ldots, x_n\rangle, x_{n+1}\rangle \\ (x)^{(n+1)}_j &:= ((x)_1)^{(n)}_j \\ (x)^{(n+1)}_{n+1} &:= (x)_2 \end{aligned}$$

Note that all the functions defined up to now are *terms* in the language of R^0_2. Furthermore, they are all in TC^0, since the function symbols in the language represent functions in TC^0.

We define a restricted form of division for small numbers by the formula

$$z = LenDiv(x, y) :\leftrightarrow (y = 0 \wedge z = 0) \vee (y > 0 \wedge z \cdot y \leq |x| \wedge (Sz) \cdot y > |x|)\ ,$$

then in R^0_2 we can prove $\forall x, y\, \exists z \leq |x|\; z = LenDiv(x, y)$ as follows: Consider the following instance of Σ^b_0-$LIND$:

$$b \cdot 0 < S|a| \wedge \forall x\, (b \cdot x < S|a| \to b \cdot Sx < S|a|) \to \forall x\; b \cdot |x| < S|a|$$

Since $b > 0 \to \neg\forall x\; b \cdot |x| < S|a|$ is provable, and $b \cdot 0 \geq S|a|$ can be refuted, we get from the contrapositive of the above

$$b > 0 \to \exists x\, (b \cdot x \leq |a| \wedge b \cdot Sx > |a|)$$

from which the claim follows easily. The uniqueness of a z with $z = LenDiv(x, y)$ is also easily proved in R^0_2.

Now the formula $z = LenDiv(x, y)$ is Σ^b_0, and z is always bounded by $|x|$, hence we can extend the language by a function symbol for $LenDiv$ such that any sharply bounded formula in the extended language is equivalent to a Σ^b_0-formula in the original language.

Let $LenMod(x, y) := |x| \dot{-} y \cdot LenDiv(x, y)$. For readability, we write $\lfloor \frac{|x|}{y} \rfloor$ for $LenDiv(x, y)$ and $|x| \mathbf{mod} y$ for $LenMod(x, y)$. Let furthermore $LSP'(x, y) := x \dot{-} MSP(x, |y|) \cdot 2^{|y|}$; we also write $LSP(x, |y|)$ for this, where $LSP(x, i)$ is intended to be the number consisting of the rightmost i bits of x, i.e. $x \mathbf{mod} 2^i$. Now we define a coding for sequences of numbers of length less than $|a|$ by

$$\begin{aligned} Seq_a(w) &:\leftrightarrow |w| \mathbf{mod} |a| = 0 \wedge \forall i < \lfloor \tfrac{|w|}{|a|} \rfloor\; Bit(w, (i+1) \cdot |a|) = 1 \\ Len_a(w) &:= \lfloor \tfrac{|w|}{|a|} \rfloor \\ \beta_a(w, i) &:= DMSB(LSP(MSP(w, (i \dot{-} 1) \cdot |a|), |a|)) \end{aligned}$$

Note that $\beta_a(w, i)$ is a term, and $Seq_a(w)$ as well as any sharply bounded formula containing Len_a are equivalent to a Σ^b_0-formula. Finally we define

$$\begin{aligned} Seq(w) &:\leftrightarrow pair(w) \wedge Seq_{(w)_1}((w)_2) \\ Len(w) &:= Len_{(w)_1}((w)_2) \\ \beta(w, i) &:= \beta_{(w)_1}((w)_2, i) \end{aligned}$$

The remarks above concerning β_a, Seq_a and Len_a also apply to β, Seq and Len. Finally we need a term $SqBd(x, y)$ such that a sequence of length $|x|$ all of whose entries are bounded by y has a code less than $SqBd(x, y)$. For this we can set $SqBd(x, y) := 4(x\#2y)^2$.

By using sharply bounded minimization, one sees that the functions $LenDiv$ and $LenMod$, and hence also the sequence coding operations, are in TC^0.

Now for a class of formulae Γ, the Γ-replacement axiom scheme is

$$\forall x \leq |s| \, \exists y \leq t(x) \, A(x, y) \to \exists w < SqBd(2s, t(|s|)) \, [Seq(w) \wedge$$

$$\wedge \, Len(w) = |s| + 1 \wedge \forall x \leq |s| \, \beta(w, Sx) \leq t(x) \wedge A(x, \beta(w, Sx))] \ ,$$

for every formula $A(x, y) \in \Gamma$.

Finally, the theory $\bar{R}^0_2$ is defined as R^0_2 extended by the schemes of Σ^b_0-comprehension and Σ^b_0-replacement. A result in [7] shows that this extension is proper.

3. Definability of TC^0-functions

For every Σ^b_1-formula $A(\bar{a})$ we define a formula $\text{WITNESS}_A(w, \bar{a})$ (to be read as "w witnesses $A(\bar{a})$") inductively as follows: If $A(\bar{a})$ is a Σ^b_0-formula, then

$$\text{WITNESS}_A(w, \bar{a}) \quad :\equiv A(\bar{a}).$$

If $A(\bar{a}) \equiv B(\bar{a}) \circ C(\bar{a})$ for $\circ \in \{\wedge, \vee\}$, then

$$\text{WITNESS}_A(w, \bar{a}) \quad :\equiv \text{WITNESS}_B((w)_1, \bar{a}) \circ \text{WITNESS}_C((w)_2, \bar{a}).$$

If $A(\bar{a}) \equiv \exists x \leq t(\bar{a}) \, B(\bar{a}, x)$ and $A(\bar{a})$ is not a Σ^b_0-formula, then

$$\text{WITNESS}_A(w, \bar{a}) \quad :\equiv (w)_2 \leq t(\bar{a}) \wedge \text{WITNESS}_B((w)_1, \bar{a}, (w)_2).$$

If $A(\bar{a}) \equiv \forall x \leq |s(\bar{a})| \, B(\bar{a}, x)$ and $A(\bar{a})$ is not a Σ^b_0-formula, then

$$\text{WITNESS}_A(w, \bar{a}) \quad :\equiv Seq(w) \wedge Len(w) = |s(\bar{a})| + 1 \, \wedge$$
$$\wedge \, \forall x \leq |s(\bar{a})| \, \text{WITNESS}_B(\beta(w, x + 1), \bar{a}, x).$$

If $A(\bar{a}) \equiv \neg B(\bar{a})$ and $A(\bar{a})$ is not a Σ^b_0-formula, then let $A^*(\bar{a})$ be a formula logically equivalent to $A(\bar{a})$ obtained by pushing the negation side inside by de Morgan's rules, and let

$$\text{WITNESS}_A(w, \bar{a}) \quad :\equiv \text{WITNESS}_{A^*}(w, \bar{a}).$$

Clearly, $\text{WITNESS}_A(w, \bar{a})$ is equivalent Σ^b_0-formula for every Σ^b_1-formula $A(\bar{a})$.

Proposition 3.1. *For every Σ^b_1-formula $A(\bar{a})$ there is a term $t_A(\bar{a})$ such that:*

1. $\bar{R}^0_2 \vdash \text{WITNESS}_A(w,\bar{a}) \to A(\bar{a})$
2. $\bar{R}^0_2 \vdash A(\bar{a}) \to \exists w \leq t_A(\bar{a})\ \text{WITNESS}_A(w,\bar{a})$

This is proved by a straightforward induction on the complexity of the formula $A(\bar{a})$. For part (*ii*), in the case where $A(\bar{a})$ starts with a sharply bounded universal quantifier, Σ^b_0-replacement is needed.

Proposition 3.2. *The Σ^b_1-replacement axioms are provable in $\bar{R}^0_2$.*

Proof. By Prop. 3.1, every Σ^b_1-formula $A(x,y)$ is equivalent in $\bar{R}^0_2$ to a formula of the form $\exists z \leq u(x,y)\ B(x,y,z)$ for some term $u(x,y)$ and $B(x,y,z) \in \Sigma^b_0$, hence it suffices to deduce the replacement axiom for such a formula.

From the premise of the replacement axiom for this formula we can now easily conclude $\forall x \leq |s|\ \exists p \leq \langle t(x), u(x,t(x))\rangle\ B(x,(p)_1,(p)_2)$, and an application of Σ^b_0-replacement yields

$$(*)\exists v \leq SqBd(2s, \langle t(|s|), u(|s|, t(|s|))\rangle)\ \big[Seq(v) \wedge Len(v) = |s| + 1 \wedge$$

$$\wedge\, \forall x \leq |s|\ \beta(v,Sx) \leq \langle t(x), u(x,t(x))\rangle \wedge B(x,(\beta(v,Sx))_1,(\beta(v,Sx))_2)\big]\ .$$

Next we need the following

Lemma 3.1. *For every term $t(x)$ the following is provable in $\bar{R}^0_2$:*

$$\forall v\ Seq(v) \to$$

$$\exists w\ \big[Seq(w) \wedge Len(w) = Len(v) \wedge \forall i \leq Len(w)\ \beta(w,Si) = t(\beta(v,Si))\big]\ .$$

This lemma, which is easily proved by Σ^b_0-replacement, for $t(x) = (x)_1$ applied to the v from $(*)$ yields a sequence as required in the conclusion of the replacement axiom. □

Now we are ready to show

Theorem 3.1. *Every function in TC^0 is Σ^b_1-definable in $\bar{R}^0_2$.*

Proof. It is trivial that the Σ^b_1-definable functions in $\bar{R}^0_2$ comprise the initial functions in B and are closed under composition, hence it remains to prove that they are closed under CRN.

So let f be defined by CRN from g, h_0 and h_1, let g and h_i be Σ^b_1-defined by the formulae $C(\bar{x},y)$ and $B_i(\bar{x},y,z)$ resp. and the terms $s(\bar{x})$ and $t_i(\bar{x},y)$, for $i = 0,1$.

First we show the existence of the sequence of those values of the functions h_i that are needed in the computation of $f(x,y)$ by CRN, i.e. we prove in $\bar{R}^0_2$

$$\exists w \leq SqBd(2y, m(\bar{x},y))\ Seq(w) \wedge Len(w) = |y| + 1 \wedge$$

$$\wedge\, \forall i \leq |y|\ \big[\,\big(Bit(y,i) = 0 \wedge B_0(\bar{x}, MSP(y,|y| \dot{-} i), \beta(w,i+1))\big) \vee$$

$$\vee\, \big(Bit(y,i) = 1 \wedge B_1(\bar{x}, MSP(y,|y| \dot{-} i), \beta(w,i+1))\big)\big]\ ,$$

where $m(\bar{x},y) := \max(t_0(\bar{x},y), t_1(\bar{x},y))$. This follows by Σ_1^b-replacement from

$$\begin{array}{rl} \forall i < |y|\, \exists z \leq m(\bar{x},y)\, [& \big(Bit(y,i) = 0 \wedge B_0(\bar{x}, MSP(y, |y| \dot{-} i), z) \big) \vee \\ \vee & \big(Bit(y,i) = 1 \wedge B_1(\bar{x}, MSP(y, |y| \dot{-} i), z) \big) \big] , \end{array}$$

which is easily obtained from the existence conditions in the Σ_1^b-definitions of h_0 and h_1.

Now we show that for every sequence w and number a there is a number consisting of a concatenated with the least significant bits of the terms of w, i.e.

$$\begin{array}{rl} \forall a, w\ Seq(w) \rightarrow & \exists z \leq 1\#aw \,\big[\, |z| = |a| + Len(w) \wedge \\ \wedge\, \forall i < |z| & \big(i < Len(w) \wedge Bit(z,i) = Mod2(\beta(w, i+1)) \big) \\ \vee & \big(i \geq Len(w) \wedge Bit(z,i) = Bit(a, i \dot{-} Len(w)) \big) \big] \end{array}$$

which is easily deduced in $\bar{R}_2^0$ by use of Σ_0^b-comprehension. Setting $g(\bar{x})$ for a and the sequence from above for w yields the existence condition for a Σ_1^b-definition of f, with the bounding term $1\#s(\bar{x}) \cdot SqBd(2y, m(\bar{x},y))$. The uniqueness is easily proved by use of extensionality. □

4. Witnessing

The converse of Thm. 3.1 is proved by a witnessing argument as in [3]. For this, $\bar{R}_2^0$ has to be formulated in a sequent calculus with special rules for the introduction of bounded quantifiers, the $BASIC$, comprehension and replacement axioms as initial sequents and the Σ_0^b-$LIND$ rule

$$\frac{A(b), \Gamma \Longrightarrow \Delta, A(Sb)}{A(0), \Gamma \Longrightarrow \Delta, A(|t|)} \,,$$

where the free variable b must not occur in the conclusion, except possibly in the term t.

Since the formulae in the initial sequents are all Σ_1^b, we can, by a standard cut elimination argument, assume that every formula appearing in the proof of a Σ_1^b-statement is in $\Sigma_1^b \cup \Pi_1^b$. Therefore we can prove the following witnessing theorem by induction on the length of a proof:

Theorem 4.1. *Let Γ, Δ be sequences of Σ_1^b-formulae and Π, Λ sequences of Π_1^b-formulae such that*

$$\bar{R}_2^0 \vdash \Gamma, \Pi \Longrightarrow \Delta, \Lambda \ =: S,$$

let furthermore all free variables in S be among the $\bar{a}$. Let $G :\equiv \bigwedge \Gamma \wedge \bigwedge \neg\Lambda$ and $H :\equiv \bigvee \Delta \vee \bigvee \neg\Pi$. Then there is a function $f \in TC^0$ such that

$$\mathbb{N} \models \textsc{Witness}_G(w, \bar{a}) \rightarrow \textsc{Witness}_H(f(w,\bar{a}), \bar{a})$$

Proof. The induction base has four cases: A logical axiom $A \Longrightarrow A$, where A is an atomic formula, is trivially witnessed, and likewise the initial sequents stemming from the *BASIC* axioms. A function witnessing a Σ_0^b-comprehension axiom

$$\exists y < 2^{|t|}\, \forall i < |t|\ (Bit(y,i) = 1 \leftrightarrow A(i))$$

can be defined by CRN from the characteristic function of the predicate $A(i)$, which is in TC^0 since $A(i)$ is a Σ_0^b-formula.

A witness for the left hand side of a Σ_0^b-replacement axiom

$$\forall x \leq |s|\, \exists y \leq t(x)\ A(x,y) \Longrightarrow \exists w < SqBd(2s, t(|s|))\ \big[Seq(w) \wedge$$

$$\wedge\, Len(w) = |s| + 1 \wedge \forall x \leq |s|\, \beta(w, Sx) \leq t(x) \wedge A(x, \beta(w, Sx))\big]\ ,$$

is a sequence of length $|s| + 1$ whose ith term is a pair $\langle \ell_i, r_i \rangle$, where ℓ_i is a witness for $A(i-1, r_i)$. Similar to Lemma 3.1 we obtain the sequence $R := \langle r_i \rangle_{i \leq |s|+1}$. This sequence satisfies the matrix $B(w) := [\ldots]$ of the right hand side of the replacement axiom, and since $B(w)$ is equivalent to a Σ_0^b-formula, this can be witnessed by any value. Thus $\langle 0, R \rangle$ witnesses $\exists w \leq SqBd(2s, t(|s|))\ B(w)$.

In the induction step there is a case distinction corresponding to the last inference in the proof. In the cases of bounded quantifier inferences, we further have to distinguish whether the principal formula of the inference is Σ_0^b or not. Most of the cases are straightforward or easily adapted from existing witnessing proofs like the proof of the main theorem in [3].

The only more difficult cases are ($\forall\leq$: *right*) where the principal formula is not Σ_0^b, and *LIND*. W.l.o.g. we can assume that a ($\forall\leq$: *right*) inference is of the form

$$\frac{b \leq |t|, \Gamma \Longrightarrow \Delta, A(b)}{\Gamma \Longrightarrow \Delta, \forall x \leq |t|\ A(x)}$$

with Γ, Δ consisting of Σ_1^b-formulae. Then the induction hypothesis yields a function $f \in TC^0$ such that $f(w,b)$ witnesses $\bigvee \Delta \vee A(b)$ provided that w witnesses $b \leq |t| \wedge \bigwedge \Gamma$.

We need a function g such that $g(w)$ witnesses $\bigvee \Delta \vee \forall x \leq |t| A(x)$ whenever w witnesses $\bigwedge \Gamma$. Let now $w' := \langle 0, (w)_1^{(|\Gamma|)}, \ldots, (w)_{|\Gamma|}^{(|\Gamma|)} \rangle$ and let

$$g(w) := \Big\langle \big(f(w',0)\big)_1^{(|\Delta|+1)}, \ldots, \big(f(w',0)\big)_{|\Delta|}^{(|\Delta|+1)}, s(w,t) \Big\rangle$$

where $s(w,t)$ is a code for the sequence $\big\langle \big(f(w,i)\big)_{|\Delta|+1}^{(|\Delta|+1)} \big\rangle_{i \leq |t|}$. The function s can be defined by use of CRN, and thus g is in TC^0. Now it is easily verified that g has the desired witnessing property.

Finally we consider a *LIND*-inference of the form

$$\frac{A(b), \Gamma \Longrightarrow \Delta, A(Sb)}{A(0), \Gamma \Longrightarrow \Delta, A(|t|)}\ ,$$

with Γ, Δ as above. Since $A(b)$ is Σ_0^b, by induction there is $f \in TC^0$ such that for each w, b with w witnessing $A(b) \wedge \bigwedge \Gamma$, either $f(w,b)$ witnesses $\bigvee \Delta$ or $A(Sb)$ holds. Now define

$$g(w) := f(w, \mu y \leq |t| \text{ WITNESS}_{\bigvee \Delta}(f(w,y))) ,$$

then for w witnessing $A(0) \wedge \bigwedge \Gamma$, either $g(w)$ witnesses $\bigvee \Delta$ and we are done, or for every $y \leq |t|$ $f(w,y)$ does not witness $\bigvee \Delta$. Since w also witnesses $A(y) \wedge \bigwedge \Gamma$, we can conclude $A(Sy)$ from this for every such y, hence we can conclude $A(|t|)$ inductively from $A(0)$ then. Since $A(|t|)$ is Σ_0^b, it is then trivially witnessed. □

From this witnessing theorem we obtain the converse of Thm. 3.1:

Corollary 4.1. *Every function Σ_1^b-definable in $\bar{R}_2^0$ is in TC^0.*

Proof. If f is Σ_1^b-definable in $\bar{R}_2^0$, there is a Σ_1^b-formula $A(\bar{a}, b)$ and a term $t(\bar{a})$ such that $\bar{R}_2^0$ proves $\exists y \leq t(\bar{a})\ A(\bar{a}, y)$. Then by Thm. 4.1 there is a function $g \in TC^0$ such that $g(\bar{a})$ witnesses this. But then $(g(\bar{a}))_2$ satisfies $A(\bar{a}, (g(\bar{a}))_2)$ for every $\bar{a}$, and hence $f(\bar{a}) = (g(\bar{a}))_2$, and thus $f \in TC^0$. □

Together with Thm. 3.1 we get the characterization of the functions in TC^0:

Theorem 4.2. *The Σ_1^b-definable functions in $\bar{R}_2^0$ are exactly those in TC^0.*

5. Conclusion

We have characterized the class TC^0 as the Σ_1^b-definable functions in $\bar{R}_2^0$. From this characterization, we can conclude things like

> *If $\bar{R}_2^0 = R_2^1$, then $TC^0 = NC$, and $\bar{R}_2^0 = S_2^1$ implies $TC^0 = FP$.*

or, viewed from a different perspective:

> *Under the hypothesis that $TC^0 \neq FP$ (or $TC^0 \neq NC$), S_2^1 (resp. R_2^1) is not conservative over $\bar{R}_2^0$ w.r.t. $\forall\Sigma_1^b$-sentences.*

In [6], a theory TTC^0 is defined that also yields a characterization of TC^0. For the purpose of comparison, we recall the definition of TTC^0: The language is the same as that of $\bar{R}_2^0$. To state its axioms we first need a technical definition:

A formula A is called *essentially sharply bounded*, or *esb*, in a theory T, if A is in the smallest class Γ of formulae s.t.

1. every atomic formula is in Γ.
2. Γ is closed under propositional connectives and sharply bounded quantification.

3. if $A(\bar{x}, y)$ and $B(\bar{x}, y)$ are in Γ, and $\forall y, z \leq t(\bar{x})\, A(\bar{x}, y) \wedge A(\bar{x}, z) \rightarrow y = z$ and $\forall \bar{x}\, \exists y \leq t(\bar{x})\, A(\bar{x}, y)$ are provable in T, then the formulae

$$\exists y \leq t(\bar{x})\, A(\bar{x}, y) \wedge B(\bar{x}, y) \quad \text{and} \quad \forall y \leq t(\bar{x})\, A(\bar{x}, y) \rightarrow B(\bar{x}, y)$$

are in Γ.

Now the theory TTC^0 is given by the $BASIC$ axioms, *esb*-$LIND$ and the *esb*-comprehension scheme, i.e. TTC^0 is the least theory T that contains the basic axioms and has the property that whenever $A(x)$ is *esb* in T, then

$$A(0) \wedge \forall x\, (A(x) \rightarrow A(x+1)) \rightarrow \forall x\, A(|x|)$$

and

$$\exists y < 2^{|t|}\, \forall i < |t|\; (Bit(y, i) = 1 \leftrightarrow A(i))$$

are axioms of T.

The theory TTC^0 characterizes TC^0 in the following way: TC^0 coincides with the class of *esb*-definable functions in TTC^0. Compared to this characterization, the one in the present paper is, in the author's opinion, much more natural.

First, the notion of Σ_1^b-definability is a more useful one than that of *esb*-definability, since it delineates the functions in TC^0 among a probably larger class of functions (those whose graph is in NP vs. those whose graph is in TC^0). This might be easily remedied since it could be the case that the Σ_1^b-definable functions of (some extension of) TTC^0 also coincide with TC^0.

But second, the theory TTC^0 itself has a quite cumbersome definition. We think that the axiomatization of a theory should be such that the set of axioms is easily decidable. This is not the case with TTC^0: It seems that for a $\forall\Sigma_1^b$-sentence, determining whether it is an axiom of TTC^0 is as difficult as deciding its provability in TTC^0.

There is of course the possibility that TTC^0 is equivalent to $\bar{R}_2^0$, but this seems to be unlikely, or at least difficult to prove, in view of the following fact: A crucial step in the obvious proof of equivalence would be to show that every *esb*-formula is equivalent to a Σ_0^b-formula in TTC^0. Now the *esb*-formulae in TTC^0 describe exactly the predicates in TC^0. But in [8] it was shown that the class of predicates definable by Σ_0^b-formulae in (a variant of) the language of R_2^0 is a proper subclass of P. Hence a proof of equivalence as above would separate TC^0 from P, and thus solve a difficult open problem in Complexity Theory.

References

1. B. Allen. Arithmetizing uniform NC. *Annals of Pure and Applied Logic*, 53:1–50, 1991.

2. D. A. M. Barrington, N. Immermann, and H. Straubing. On uniformity within NC^1. *Journal of Computer and System Sciences*, 41:274–306, 1990.
3. S. R. Buss. *Bounded Arithmetic*. Bibliopolis, Napoli, 1986.
4. P. Clote. On polynomial size Frege proofs of certain combinatorial principles. In P. Clote and J. Krajíček, editors, *Arithmetic, Proof Theory and Computational Complexity*, volume 23 of *Oxford Logic Guides*, pages 162–184. Clarendon Press, Oxford, 1993.
5. P. Clote and G. Takeuti. Bounded arithmetic for NC, $ALogTIME$, L and NL. *Annals of Pure and Applied Logic*, 56:73–117, 1992.
6. P. Clote and G. Takeuti. First order bounded arithmetic and small boolean circuit complexity classes. In P. Clote and J. Remmel, editors, *Feasible Mathematics II*, pages 154–218. Birkhäuser, Boston, 1995.
7. J. Johannsen. A note on sharply bounded arithmetic. *Archive for Mathematical Logic*, 33:159–165, 1994.
8. S.-G. Mantzivis. Circuits in bounded arithmetic part I. *Annals of Mathematics and Artificial Intelligence*, 6:127–156, 1992.
9. L. Stockmeyer. The polynomial-time hierarchy. *Theoretical Computer Science*, 3:1–22, 1976.
10. G. Takeuti. $RSUV$ isomorphisms. In P. Clote and J. Krajíček, editors, *Arithmetic, Proof Theory and Computational Complexity*, volume 23 of *Oxford Logic Guides*, pages 364–386. Clarendon Press, Oxford, 1993.

Information content and computational complexity of recursive sets *

Lars Kristiansen **

Department of Informatics, University of Oslo, Pb 1080 Blindern, 0316 Oslo, Norway

An honest function is, roughly speaking, a unary, recursive, and strictly increasing function with a very simple graph. Thus if f is an honest function, then the growth of f reflects the computational complexity of f. The honest elementary degrees are the degree structure induced on the honest functions by the reducibility relation "being (Kalmár) elementary in". (Other subrecursive reducibility relations will also work, for example "being primitive recursive in", but not "being polynomial time computable in the *length of input*". "Being polynomial time computable in the *input*" might work, at least in some respects.) A recursive function turns out to be total iff it is elementary in some honest function. Thus, since the set of functions elementary in a particular honest function constitutes a complexity class, the structure of honest elementary degrees will provide a measure for the computational complexity of any total recursive function f. If f is not elementary in a honest function of degree **a**, it is because f is too hard to compute, i.e. it requires more resources to compute f than the honest degree **a** allows.

The structure of subrecursive honest degrees is studied, explicitly or implicitly, in Meyer and Ritchie [11], Basu [2], Machtey [8] [9] [10], Simmons [16], and Kristiansen [5] [6]. Machtey shows that the structure of elementary honest degrees is a lattice with strong density properties, for instance between any degrees $\mathbf{a}, \mathbf{b}$ such that $\mathbf{a} < \mathbf{b}$ there are two incomparable degrees. Kristiansen studies a jump operator on the structure. Among other results he shows that it is possible to invert the jump; there exist low degrees; there exist degrees which are neither high nor low; every situation compatible with $\mathbf{a}' \cup \mathbf{b}' \leq (\mathbf{a} \cup \mathbf{b})'$ is realized in the structure; every situation compatible $\mathbf{a} \leq \mathbf{b} \Rightarrow \mathbf{a}' \leq \mathbf{b}'$ is realized in the structure, e.g. we have incomparable degrees $\mathbf{a}, \mathbf{b}$ such that $\mathbf{a}' < \mathbf{b}'$ and incomparable degrees $\mathbf{a}, \mathbf{b}$ such that $\mathbf{a}' = \mathbf{b}'$ etcetera. Moreover there is a close relationship between the elementary honest degrees and the subrecursive hierarchies described in the book of Rose [15]. Let $\mathbf{0}$ be the degree of the elementary functions, let $\cdot'$ be the jump operator from [6], and let $\mathcal{E}^0, \mathcal{E}^1, \mathcal{E}^2, \ldots$ denote the classes in the Grzegorczyk hierarchy. Then the class $\mathcal{E}^3$ is exactly the functions elementary in an honest function of degree $\mathbf{0}$, the class $\mathcal{E}^4$ is exactly the functions elementary in an honest function of degree $\mathbf{0}'$, the class $\mathcal{E}^5$ corresponds to the degree $\mathbf{0}''$ and so on. By introducing an ω-jump in an obvious way, we will be able to

* This paper is in its final form, and no similar paper has been or is being submitted elsewhere.

** The author wishes to thank Wolfgang Merkle and Stan S. Wainer

climb beyond the Grzegorczyk classes and generate elementary honest degrees that correspond to higher levels in the transfinite hierarchies. Heaton and Wainer [4] study the relationship between subrecursive hierarchies and a jump operator similar to $\cdot'$.

The most popular current subrecursive degree theory is *not* a theory of honest degrees, but a theory that is concerned with reducibility between recursive sets. Due to its importance in computer science, "being polynomial time computable in" is the most popular reducibility relation between the sets. In our discussion it is convenient to use the reducibility relation "being (Kalmár) elementary in", and we use $\leq_E$ to denote this relation. Thus the degrees in the set-degree theory are the equivalence classes induced on the recursive sets by the $\leq_E$-relation. (For the time being it is not important whether we are talking about m-degrees or T-degrees.) This set-degree theory is in many respects different from the honest degree theory induced by the same reducibility relation. The proof methods are different, the degree structures are different, and it seems that a theory of set-degrees does not admit a jump operator.

In a classical paper Ladner [7] proves that neither the polynomial m- nor T-degrees of recursive sets are a lattice. (They are just upper semi-lattices.) He also shows density results and minimal pair results for the same structures. Ladner's proof methods are based on traditional recursion theoretic constructions, and his methods and results generalize to a wide variety of subrecursive reducibilities, e.g. to "being elementary in" and "being primitive recursive in". After Ladner researchers have used refinements of his methods in further studies of the structure of polynomial time degrees of recursive sets. See Ambos-Spies [1] for an overview and further references. I believe the techniques developed in the area can be transferred to all other reasonable subrecursive reducibility relations between sets, and as far as I know all the techniques involve some kind of constructions. In the study of the honest degrees it is possible to obtain a lot of results without doing any constructions at all. Instead in our proofs we exploit that there exists a bound on the growth of the functions in a honest degree. For instance all the results in Kristiansen [6] are achieved by such means. Exactly how far we can get in the study of the honest degrees without constructions, is an interesting question.

I have argued that the honest degrees are degrees of computational complexity, and I am about to argue that the set-degrees are not. Let B be a recursive non-elementary set, and let $\mathcal{B} = \{A \mid A \leq_E B \text{ and } A \text{ is a set}\}$. Then $\mathcal{B}$ is not a complexity class in the sense that every set computable within a certain amount of resources belongs to $\mathcal{B}$. No, $A \leq_E B$ iff B contains enough information to decide membership in A within elementary time. So $\mathcal{B}$ is the class of sets that is computable in elementary time if we have access to the information in B, i.e. if we do not have to compute B. It is reasonable to view the set-degrees as degrees of information content and not as degrees of computational complexity. If a set A is not subrecursively reducible to a set in the set-degree **a**, it is because A contains too much information, i.e. more information than the degree **a** permits.

The objective of this paper is to obtain some relationships between the computational complexity and the information content of sets. Let the set A be of a certain computational complexity, i.e. A is of a certain honest degree. How much information can A possibly contain, i.e. which sets are subrecursively reducible to A? This paper gives answers to such questions, and hopefully this paper contributes to bridge a gap between the two different approaches to subrecursive degree theory.

1. General preliminaries and definitions

I assume the reader is familiar with the most basic concepts of classical recursion theory. An introduction and survey can be found in the books [12] and [14]. I also assume acquaintance with subrecursion and, in particular, with the elementary functions. An introduction to this subject can be found in [13] or [15]. Here I just state some important basic facts and definitions. See [13] and [15] for proofs.

The initial elementary functions are the successor ($\mathcal{S}$), projections ($\mathcal{I}_i^n$), zero (0), addition (+), and modified subtraction ($\dot{-}$) functions. *The elementary schemes* are *composition*, i.e. $f(\mathbf{x}) = h(g_1(\mathbf{x}), \dots, g_m(\mathbf{x}))$ and bounded *sum* and *product*, i.e. $f(\mathbf{x}, y) = \sum_{i<y} g(\mathbf{x}, i)$ and $f(\mathbf{x}, y) = \prod_{i<y} g(\mathbf{x}, i)$. *The class of elementary functions* is the least class which contains the initial elementary functions and is closed under the elementary schemes. A *relation* or a *predicate* $R(\mathbf{x})$ *is elementary* when there exists an elementary function f with range $\{0, 1\}$ such that $f(\mathbf{x}) = 0$ iff $R(\mathbf{x})$ holds. That a function f has an *elementary graph* means that the relation $f(\mathbf{x}) = y$ is elementary. If we can define a function g from the function f plus the initial elementary functions by the elementary schemes, we say that *g is elementary in f*.

The definition scheme $(\mu z < x)[\dots]$ is called the *bounded μ-operator*, and $(\mu z < y)[R(\mathbf{x}, z)]$ denotes the least $z < y$ such that the relation $R(\mathbf{x}, z)$ holds. Let $(\mu z < y)[R(\mathbf{x}, z)] = 0$ if no such z exists. The elementary functions are closed under the bounded μ-operator. If f is defined by a primitive recursion over g and h and $f(\mathbf{x}, y) \leq j(\mathbf{x}, y)$, we say that f is a *limited recursion* over g, h and j. (It is convenient to think about limited recursion as a scheme with g, h and j as parameters, although the j is actually not used to generate f.) The elementary functions are closed under limited recursion, but not under primitive recursion. Moreover, the elementary relations are closed under the operations of the propositional calculus and under bounded quantification, i.e. $(\forall x < y)[R(x)]$ and $(\exists x < y)[R(x)]$.

The class of elementary functions is the closure of $\{0, \mathcal{S}, \mathcal{I}_i^n, 2^x, \max\}$ under composition and limited recursion. A proof of this characterization of the elementary functions can be found in [15] or [3]. They prove that the elementary functions equal the third Grzegorczyk class $\mathcal{E}^3$, and $\mathcal{E}^3$ is defined to be the class $\{0, \mathcal{S}, \mathcal{I}_i^n, E_2, \max\}$ closed under composition and limited recursion. It is easy to see that the class $\mathcal{E}^3$ remains the same if we use 2^x in place of E_2 in the definition. ($E_2(x) = E_1^x(2)$ where $E_1(x) = x^2 + 2$.) Thus it follows that the class of functions elementary in f is the closure of $\{0, \mathcal{S}, \mathcal{I}_i^n, 2^x, \max, f\}$ under composition and limited recursion.

All the closure properties of the elementary functions can be proved by using Gödel numbering and coding techniques. Uniform systems for coding the finite sequences of natural numbers are available inside the class of elementary functions. Let $F_f(x)$ be the code for the sequence $\langle f(0), f(1), \ldots f(x)\rangle$. Then F_f belongs to the elementary functions if f does. We are quite informal and indicate the use of coding functions with the notations $\langle \ldots \rangle$. Our coding system is monotone, i.e. $\langle x_0, \ldots, x_n\rangle < \langle x_0, \ldots, x_n, y\rangle$ for every value of y, and $\langle x_0, \ldots, x_i, \ldots, x_n\rangle < \langle x_0, \ldots, x_i + 1, \ldots, x_n\rangle$.

A relation $\sim$ between two functions holds *almost everywhere* (a.e.) iff there is a number k such that for all $x > k$ we have $g(x) \sim f(x)$. Notation: $g(x) \overset{(a.e.)}{\sim} f(x)$. The function f^k is the k'th iterate of the unary function f, i.e. $f^0(x) = x$ and $f^{k+1}(x) = ff^k(x)$.

2. Theorems on total recursive functions

Definition 2.1. *A function f is* honest *iff (i) f is unary, (ii) $f(x) \geq 2^x$, (iii) f is monotone (nondecreasing), and (iv) f has an elementary graph.*

It is clause (iv) in the definition which is essential. We require that an honest function has an elementary graph because we want no "hidden complexity" in the function. We want the growth of the function to mirror the computational complexity of the function. The structure of honest degrees would be the same if an honest function were not required to satisfy (i), (ii), and (iii), but those requirements are needed for other purposes. Meyer and Ritchie [11] give some characterizations of the honest functions.

We define *recursive function*, *recursive index*, *computation tree*, and other well-known concepts in the usual way. When e is a recursive index for the function f, we adopt the traditional abuse of notation and write $\{e\}(\mathrm{x})$ both for (i) the computation of $f(\mathrm{x})$ associated with e and for (ii) the eventual result of the computation. Let $\mathcal{U}$ be the function such that $\mathcal{U}(\langle x_1, \ldots, x_m\rangle) = x_m$, i.e. a function that gives the last coordinate of a sequence number. When y codes the empty sequence, or when y is not a sequence number at all, let $\mathcal{U}(y) = 0$. Let $\mathcal{T}_n$ be the Kleene predicate for $n = 0, 1, 2, \ldots$, i.e. the predicate $\mathcal{T}_n(e, x_1, \ldots, x_n, t)$ holds iff t is a computation tree for $\{e\}(x_1, \ldots, x_n)$. The relation $\mathcal{T}_n$ is elementary. (According to Rose [15], this was one of the main motivations for introducing the elementary functions in the first place.) The function $\mathcal{U}$ is also elementary, and for each total recursive f we have

$f(x_1, \ldots, x_n) = \{e\}(x_1, \ldots, x_n) = \mathcal{U}(\mu z[\mathcal{T}_n(e, x_1, \ldots, x_n, z)])$ when e is a recursive index for f. We will now state refined versions of the Kleene Normal Form Theorem and the Second Recursion Theorem. The proofs are close to the usual proofs of the original theorems, just a little bit of additional book-keeping is required.

Theorem 2.1 (The Normal Form Theorem). *An n-ary function g is elementary in an honest function f iff there exist a recursive index e for g and a fixed number k such that*

$$\{e\}(x_1, \ldots, x_n) = \mathcal{U}(\mu y < f^k(\max(x_1, \ldots, x_n))[\mathcal{T}_n(e, x_1, \ldots, x_n, y)]) .$$

Proof. Suppose $g(\mathbf{x}) = \{e\}(\mathbf{x}) = \mathcal{U}(\mu y \leq f^k(\max(\mathbf{x}))[\mathcal{T}_n(e, \mathbf{x}, y)])$. Then it is trivial that g is elementary in f; the Kleene predicate $\mathcal{T}_n$ is elementary, the functions $\mathcal{U}$ and max are elementary, and the elementary functions are closed under composition and the bounded μ-operator. To prove the other direction of the equivalence, assume that g is elementary in the honest function f. Then g can be generated from the functions $0, \mathcal{S}, \mathcal{I}_i^n$, max and f by composition and limited recursion. Complete the proof of the theorem by induction on such a generation of g.

Theorem 2.2 (The Recursion Theorem). *Let g be an $n+1$-ary function elementary in an honest function f. Let $\mathbf{x} = x_1, \ldots, x_n$. Then there exists a recursive index e and a fixed number k such that*

$$g(e, \mathbf{x}) = \{e\}(\mathbf{x}) = \mathcal{U}(\mu z < f^k(\max(\mathbf{x}))[\mathcal{T}_n(e, \mathbf{x}, z)]) .$$

Proof. Prove a refined version of the S_n^m-theorem. Such proof of has a structure similar to the proof of the ordinary S_n^m-theorem. Then carry out a proof that is similar to Kleenes proof of the original Second Recursion Theorem. No surprises come up, and we leave the details.

Definition 2.2. $\{e\}^f(\mathbf{x}) \stackrel{\text{def}}{=} \mathcal{U}(\mu t < f(\max(\mathbf{x}))[\mathcal{T}_n(e, \mathbf{x}, t)])$.

Under this notation the Normal Form Theorem says that a function g is elementary in an honest function f iff there exists a recursive index e (for g) and a fixed number k such that $g(\mathbf{x}) = \{e\}^{f^k}(\mathbf{x}) = \{e\}(\mathbf{x})$. The Recursion Theorem says that if g is elementary in f, then there exist numbers e, k such that $g(e, \mathbf{x}) = \{e\}^{f^k}(\mathbf{x})$. The following implication is trivial:

$$g(\max(\mathbf{x})) \geq f(\max(\mathbf{x})) \wedge \{e\}(\mathbf{x}) = \{e\}^f(\mathbf{x}) \Rightarrow \{e\}(\mathbf{x}) = \{e\}^g(\mathbf{x}) .$$

3. The honest functions and the elementary degrees

Definition 3.1. *Let $f \leq_E g$ denote that f is elementary in g, let $f <_E g$ denote that $f \leq_E g$ and $g \not\leq_E f$, and let $f \equiv_E g$ denote that $f \leq_E g$ and*

$g \leq_E f$. The equivalence classes induced by $\equiv_E$ are the elementary degrees. *We let $deg(f) \stackrel{\text{def}}{=} \{g \mid g \equiv_E f\}$ and refer to $deg(f)$ as* the degree of f. *We use $<, \leq$ for the ordering induced on the degrees by $<_E, \leq_E$. An elementary degree* $\mathbf{a}$ *is* honest *iff $\mathbf{a} = deg(f)$ for some honest f. We will use small bold-faced letters early in the Latin alphabet, i.e. $\mathbf{a}, \mathbf{b}, \mathbf{c}, \ldots$, to denote honest elementary degrees. If $\mathbf{a} \leq \mathbf{b} \leq \mathbf{c}$ then $\mathbf{a}$ is a degree* below $\mathbf{b}$, *and $\mathbf{b}$ is a degree* above $\mathbf{a}$, *and $\mathbf{b}$ is a degree* between $\mathbf{a}$ *and* $\mathbf{c}$. *Every degree in this paper is an* honest elementary degree. *If we just say* degree *or* honest degree, *we do really mean* honest elementary degree. *From now on we reserve the letters f and g to denote honest functions only.*

The next theorem is important. It enables us to avoid the usual recursion theoretic constructions when we are proving results on honest degrees and functions.

Theorem 3.1 (The Growth Theorem). *Let f and g be honest functions. Then*

$$g \leq_E f \quad \Leftrightarrow \quad g(x) < f^k(x) \text{ for some fixed } k\ .$$

Proof. The left-right direction of the equivalence follows trivially from the Normal Form Theorem. Now suppose that $g(x) < f^k(x)$. Since g is honest, the relation $g(x) = y$ is elementary. We have $g(x) = (\mu y < f^k(x))[g(x) = y]$. Hence $g \leq_E f$ since the elementary functions are closed under composition and the bounded μ-operator.

Definition 3.2. *Let f be an honest function. We define the function f' by $f'(x) \stackrel{\text{def}}{=} f^x(x)$. We call $\cdot'$ the* jump operator. *Let $\mathbf{a}$ be an honest elementary degree. Then $\mathbf{a}' \stackrel{\text{def}}{=} deg(f')$ where f is some honest function in $\mathbf{a}$. (The degree $\mathbf{a}'$ does not depend on the choice of f in $\mathbf{a}$. This follows from the Growth Theorem.) We let $\mathbf{0}$ denote the honest degree $deg(2^x)$, i.e. $\mathbf{0}$ is the class of elementary functions. Further we let*

$$[\langle d, k\rangle]^f(\mathbf{x}) \stackrel{\text{def}}{=} \{d\}^{f^k}(\mathbf{x}) \stackrel{\text{def}}{=} \mathcal{U}(\mu t < f^k(\max(\mathbf{x}))[\mathcal{T}_n(d, x_1, \mathbf{x}, t)])\ .$$

*We say that e is an f-*elementary index for *ψ whenever $\psi(\mathbf{x}) = [e]^f(\mathbf{x})$ and f is some honest function.*

We may also define a meet and a join operator on the honest elementary degrees since the structure is a lattice, but we do not need such operators in this paper. Anyway, the function $\max(f(x), g(x))$ is the l.u.b. of the honest functions f and g, and the function $\min(f(x), g(x))$ is the g.l.b. of the honest functions f and g. See [6] and [9].

The jump operator seems a bit arbitrary, but it is very natural. The next few lemmas tell us that it is indeed an analogue to the jump operator on the Turing degrees.

Lemma 3.1. *Let f be an honest function. Then $\{[e]^f\}_{e\in\omega}$ is an enumeration of the functions elementary in f.*

Proof. Let e be an arbitrary natural number. Let $\psi = [e]^f$ and let $e = \langle d, k\rangle$. Then, straightaway from the definitions, we have

$$\psi(\mathbf{x}) = \{d\}^{f^k}(\mathbf{x}) = \mathcal{U}(\mu t < f^k(\max(\mathbf{x}))[\mathcal{T}_n(d, \mathbf{x}, t)]) .$$

Thus ψ is elementary in f since $\mathcal{U}, \mathcal{T}_n$ etc. are elementary. Further, suppose ψ is elementary in f. By the Normal Form Theorem there exists a recursive index d for ψ and a fixed number k such that $\psi(\mathbf{x}) = \{d\}^{f^k}(\mathbf{x})$. Let $e = \langle d, k\rangle$. Then we have $\psi = [e]^f$.

Lemma 3.2. *Let $\mathcal{J}(f)(\langle x_1, x_2\rangle) = [x_1]^f(x_2)$. Then $f' \equiv_E \mathcal{J}(f)$ whenever f is an honest function.*

Proof. First we prove that f' is elementary in $\mathcal{J}(f)$. The k'th iterate of f, i.e. f^k, is elementary in f for all $k \in \omega$. There exists an elementary function ψ such that $f^k(x) = [\psi(k)]^f(x)$. Thus we have $f'(x) = \mathcal{J}(f)(\langle \psi(x), x\rangle)$ and thereby $f' \leq_E \mathcal{J}(f)$. Next we prove that $\mathcal{J}(f)$ is elementary in f'. Let $a\langle x_1, x_2\rangle = x_1$ and $b\langle x_1, x_2\rangle = x_2$. Then

$$\mathcal{J}(f)(x) = [ax]^f(bx) = \{aax\}^{f^{bax}}(bx) = \mathcal{U}(\mu t < f^{bax}(bx)[\mathcal{T}_1(aax, bx, t)] .$$

(The first equality holds by the definition of $\mathcal{J}(f)$, the second by the definition of $[\cdot]^f$, and the third by the definition of $\{\cdot\}^{f^k}$.) It is trivial that the function $f^{bax}(bx)$ is elementary in f'. Thus $\mathcal{J}(f) \leq_E f'$ since $\mathcal{U}, \mathcal{T}_1$ etc. all are elementary functions.

Definition 3.3. *A binary function ρ is a* universal function *for an honest degree* $\mathbf{a} = deg(f)$ *iff for all unary $\xi \leq_E f$ there exists an n such that $\xi(x) = \rho(n, x)$. Let f and g be honest functions. We write $f \ll g$ when there is a universal function ρ for the degree deg(f) such that $\rho \leq_E g$. We also write $\ll$ for the corresponding relation on the degrees.*

That the relation $\mathbf{a} \ll \mathbf{b}$ holds means that in some sense $\mathbf{b}$ lies far above $\mathbf{a}$. The situation $\mathbf{a} \ll \mathbf{b}$ implies that $\mathbf{a} < \mathbf{b}$, but there exist degrees $\mathbf{a}, \mathbf{b}$ such that $\mathbf{a} < \mathbf{b}$ and $\mathbf{a} \not\ll \mathbf{b}$. The next theorem gives a characterization of the $\ll$-relation. Meyer and Ritchie [11] prove related results.

Theorem 3.2. *Let g and f be honest functions. Then (1) $g \ll f$, (2) $(\exists m)(\forall k)[g^k(x) \overset{(a.e.)}{<} f^m(x)]$, and (3) $(\exists \psi \leq_E f)(\forall \phi \leq_E g)[\phi(x) \overset{(a.e.)}{<} \psi(x)]$ are equivalent.*

Proof. (2) $\Rightarrow$ (3): Assume (2). Then we can pick a number m such that $g^k(x) \overset{(a.e.)}{<} f^m(x)$ for all k. By the Growth Theorem we have that $f^m \leq_E f$, and that every function elementary in g is bounded by g^k for some fixed k. Thus (3) follows. (3) $\Rightarrow$ (1): Assume (3). Then there exists a function ψ elementary in f which majorizes (a.e.) the function g^m, i.e. $g^m(x) \overset{(a.e.)}{<} \psi(x)$ for every fixed m. This implies that for every m there exists an n such that

$g^m(x) < n + \psi(x)$ (*). Let ξ be any unary function elementary in g. By the Normal Form Theorem we have a recursive index d for ξ and a fixed number m such that

$$\begin{aligned}\xi(x) &= \{d\}^{g^m}(x) = \mathcal{U}((\mu t < g^m(x))[\mathcal{T}_1(d,x,t)]) \\ &\overset{(*)}{=} \mathcal{U}((\mu t < n + \psi(x))[\mathcal{T}_1(d,x,t)])\,.\end{aligned}$$

Let $\rho(\langle d,n\rangle,x) \overset{\text{def}}{=} \mathcal{U}((\mu t < n+\psi(x))[\mathcal{T}_1(d,x,t)])$. Then $\rho \leq_E f$, and for every unary function $\xi \in \deg(g)$ there exists a number k such that $\xi(x) = \rho(k,x)$. Thus (1) holds. (1) $\Rightarrow$ (2): Let $\psi(x) = (\max_{i\leq x}\max_{j\leq x}\rho(i,j)) + 1$, where ρ is a universal function for $\deg(g)$ and $\rho \leq_E f$. Now $\psi \leq_E f$, so the Growth Theorem yields a fixed m such that $\psi(x) < f^m(x)$. It is easy to verify that ψ majorizes (a.e.) every function which is elementary in g. Since $g^k \leq_E g$ for every fixed k, we have $g^k(x) \overset{(a.e.)}{<} \psi(x) < f^m(x)$ for every fixed k. Thus (2) holds.

4. Main results

Definition 4.1. *A* set *is a unary function with range* $\{0,1\}$. *We will use the first few capital letters in the Latin alphabet to denote sets, and we will write* $x \in A$ *instead of* $A(x) = 0$ *etcetera. We denote the sets in the* lower cone *of the honest degree* $\mathbf{a}$ *by* $\leq_{\mathcal{E}}(\mathbf{a})_*$, *i.e.*

$$\leq_{\mathcal{E}}(\mathbf{a})_* \overset{\text{def}}{=} \{A \mid A \text{ is a set and } A \leq_E f \text{ for some } f \text{ such that } deg(f) = \mathbf{a}.\}$$

A set A *is (elementarily)* m-reducible *to a set* B *iff there exists an elementary function* ψ *such that* $x \in A \Leftrightarrow \psi(x) \in B$. *A set* B *is* $\leq_{\mathcal{E}}(\mathbf{a})_*$-hard *iff every set in* $\leq_{\mathcal{E}}(\mathbf{a})_*$ *is* m*-reducible to* B.

Fix an honest function f *such that* $deg(f) = \mathbf{a}$. *A set* B *is* effectively $\leq_{\mathcal{E}}(\mathbf{a})_*$-hard *iff there exists an elementary function* ψ *such that for every* f*-elementary index* e *for a set* $A \in \leq_{\mathcal{E}}(\mathbf{a})_*$ *we have* $x \in A \Leftrightarrow \psi(e,x) \in B$.

Theorem 4.1. *The cone* $\leq_{\mathcal{E}}(\mathbf{b})_*$ *contains an effectively* $\leq_{\mathcal{E}}(\mathbf{a})_*$*-hard set iff* $\mathbf{a}' \leq \mathbf{b}$.

Proof. Let $B \overset{\text{def}}{=} \{\langle e,x\rangle \mid [e]^g(x) = 0\}$ where g is an honest function. We have $B \leq_E g'$ by Lemma 3.2. Let $A \leq_E g$, and let e be a g-relative index for A. Then $x \in A \Leftrightarrow \langle e,x\rangle \in B$. Thus, whenever $\mathbf{b}$ lies above $\mathbf{a}'$, the cone $\leq_{\mathcal{E}}(\mathbf{b})_*$ contains an effectively $\leq_{\mathcal{E}}(\mathbf{a})_*$-hard set.

Now assume that $\mathbf{a}' \not\leq \mathbf{b}$ and that $B \in \leq_{\mathcal{E}}(\mathbf{b})_*$ is an effectively $\leq_{\mathcal{E}}(\mathbf{a})_*$-hard set. We shall derive a contradiction from these assumptions. Let $\mathbf{a} = \deg(g)$ and $\mathbf{b} = \deg(f)$. Further let $A_i(\langle e,x\rangle) \overset{\text{def}}{=} 1 \dot{-} \{e\}^{g^i}(x)$ if $\{e\}$ is a unary function, and let $A_i(\langle e,x\rangle) \overset{\text{def}}{=} 1$ if $\{e\}$ is not unary. Then $A_i \leq_E g$ for every $i \in \omega$. Thus there exists an elementary function ξ such that for all fixed i

we have $x \in A_i \Leftrightarrow \xi(e,x) \in B$ whenever e is a g-elementary index for A_i. Let $\psi(i)$ give a g-elementary index for A_i. Note that ψ is elementary. Let $\phi(y,x) \stackrel{\text{def}}{=} B(\xi(\psi(\langle y,x\rangle),\langle y,x\rangle))$. Then we have $\phi(y,x) = A_{\langle y,x\rangle}(\langle y,x\rangle)$ (i). We also have $\phi \leq_E f$ because $B \leq_E f$. Thus the Recursion Theorem yields a recursive index e and a fixed k such that $\phi(e,x) = \{e\}^{f^k}(x) = \{e\}(x)$ (ii). Since $\deg(g') = \mathbf{a}' \not\leq \mathbf{b} = \deg(f)$, the Growth Theorem says that for every fixed k there exist infinitely many x such that $g^x(x) = g'(x) \geq f^k(x)$ (iii). When we put (i), (ii) and (iii) together we get

$$\begin{aligned}
\phi(e,x) &\stackrel{\text{(ii)}}{=} \{e\}^{f^k}(x) & \\
&= \{e\}^{g^x}(x) & \text{(iii) for some large } x \\
&= \{e\}^{g^{\langle e,x\rangle}}(x) & \langle\cdot,\cdot\rangle \text{ is monotone} \\
&\neq 1 \dot{-} \{e\}^{g^{\langle e,x\rangle}}(x) & \\
&= A_{\langle e,x\rangle}(\langle e,x\rangle) & \text{def. of } A_i \\
&= \phi(e,x)\,. & \text{(i)}
\end{aligned}$$

So there exist e, x such that $\phi(e,x) \neq \phi(e,x)$. Contradiction!

Lemma 4.1. *Assume that f, g are honest functions and that there exist infinitely many x such that $g^i(x) > f^{k+1}(x)$. Then, for every number m there exist infinitely many x such that $g^i(\langle m,x\rangle) > f^k(\langle m,x\rangle)$.*

Proof. The pairing function $\langle\cdot,\cdot\rangle$ is polynomial and monotone in both arguments. Let m be any fixed number. Then we have $\langle m,x\rangle \stackrel{\text{(a.e.)}}{<} f(x)$ since f is an honest function. (We have $f(x) \geq 2^x$ for any honest f.) Therefore it is possible to pick an arbitrarily large x such that

$$f^k(\langle m,x\rangle) \leq f^{k+1}(x) < g^i(x) \leq g^i(\langle m,x\rangle)\,.$$

Theorem 4.2. *Assume $\mathbf{0} \ll \mathbf{b}$. The cone $\leq_E(\mathbf{b})_*$ contains a $\leq_E(\mathbf{a})_*$-hard set iff $\mathbf{a} \ll \mathbf{b}$.*

Proof. Let g and f be honest functions such that $\deg(g) = \mathbf{a} \ll \mathbf{b} = \deg(f)$. By Theorem 3.2 there exists a universal function ρ for the degree $\mathbf{a}$ such that $\rho \leq_E f$. Let $B \stackrel{\text{def}}{=} \{\langle y,x\rangle \mid \rho(y,x) = 0\}$. Then $B \leq_E f$ and thus $B \in \leq_E(\mathbf{b})_*$. Let A be any set such that $A \leq_E g$. We show that A is m-reducible to B. Since ρ is an $\mathbf{a}$-universal function there exists n such that $A(x) = \rho(n,x)$. Fix such an n and let $\xi(x) = \langle n,x\rangle$. Then ξ is elementary and $x \in A \Leftrightarrow \xi(x) \in B$. Thus A is m-reducible to B. That was the proof of the if-direction.

In order to prove the only-if-direction assume that $B \in \leq_E(\mathbf{b})_*$ is $\leq_E(\mathbf{a})_*$-hard, that $\mathbf{a} \not\ll \mathbf{b}$, and that $\mathbf{0} \ll \mathbf{b}$. We will derive a contradiction from these assumptions. Let g and f be honest functions such that $\deg(g) = \mathbf{a}$ and $\deg(f) = \mathbf{b}$. By Theorem 3.2 we have

$$(\forall k)(\exists i)\,[\, g^i(x) > f^k(x) \text{ for infinitely many } x\,] \qquad \text{(I)}$$

since $\mathbf{a} \not\ll \mathbf{b}$. Let

$$A_{i,k}(\langle y, x\rangle) \stackrel{\text{def}}{=} \begin{cases} 1 \dot{-} \{y\}^{f^k}(\langle y, x\rangle) & \text{if } \{y\} \text{ is unary} \\ & \text{and } f^k(\langle y, x\rangle) < g^i(\langle y, x\rangle) \\ 0 & \text{otherwise} \end{cases}$$

For every fixed i and k the set $A_{i,k}$ is elementary in g. (Here is an argument to support the this claim: The graph of f is elementary since f is honest. By induction on k it is easy to show that the graph of f^k is elementary. Thus the function $\xi(x) \stackrel{\text{def}}{=} (\mu z < g^i(x))[f^k(x) = z]$ is elementary in g because the relation $f^k(x) = z$ is elementary. Moreover, if $\xi(x) \neq 0$ then $f^k(x) < g^i(x)$, and if $\xi(x) = 0$ then $f^k(x) \not< g^i(x)$. Therefore it is possible to decide elementarily in g whether $f^k(x) < g^i(x)$ for fixed i and k. Now it is easy to see that it is also possible to decide elementarily in g whether $x \in A_{i,k}$. So $A_{i,k} \leq_E g$.) The set B is elementary in f. The Normal Form Theorem says that there exists a recursive index e_0 for B and a fixed number j such that $B(x) = \{e_0\}^{f^j}(x)$. By assumption B is $\leq_E(\mathbf{a})_*$-hard. Thus, for all fixed i and k there exists an elementary ψ such that $x \in A_{i,k} \Leftrightarrow \psi(x) \in B$. We have also assumed $\mathbf{0} \ll \mathbf{b} = \deg(f)$. This assumption implies that for every elementary function ψ there exists a recursive index e_1 such that $\psi(x) \stackrel{(a.e.)}{=} \{e_1\}^f(x)$. We have $A_{i,k}(x) = B(\psi(x))$ for some elementary ψ. Thus for all fixed i, k there exists a recursive index e_1 such that $A_{i,k}(x) \stackrel{(a.e.)}{=} \{e_0\}^{f^j}(\{e_1\}^f(x))$, and $\{e_0\}^{f^j}(\{e_1\}^f(x)) = \{e\}^{f^m}(x)$ for some recursive index e and fixed number m. So it is possible to pick a fixed number m such that for every i, k there exists a recursive index e such that

$$A_{i,k}(x) \stackrel{(a.e.)}{=} \{e\}^{f^m}(x). \tag{II}$$

Note that e depends on i and k, but m does not.

Now fix m such that (II) holds. Fix k such that $k > m$. Fix i such that $g^i(x) > f^{k+1}(x)$ for infinitely many x. Such an i exists by (I). Then fix a d such that $A_{i,k}(x) \stackrel{(a.e.)}{=} \{d\}^{f^m}(x)$. Such a d exists by (II). Now we have $A_{i,k}(x) \stackrel{(a.e.)}{=} \{d\}^{f^m}(x) = \{d\}^{f^k}(x)$. Thus, when we substitute $\langle d, x\rangle$ for x in the last formula, we get

$$A_{i,k}(\langle d, x\rangle) = \{d\}^{f^k}(\langle d, x\rangle) \text{ for every sufficiently large } x. \tag{III}$$

We have chosen i such that $g^i(x) > f^{k+1}(x)$ holds for infinitely many x. Hence, by Lemma 4.1, there exists an arbitrarily large number n such that $g^i(\langle d, n\rangle) > f^k(\langle d, n\rangle)$. Fix a sufficiently large such n. Now we have $A_{i,k}(\langle d, n\rangle) = 1 \dot{-} \{d\}^{f^k}(\langle d, n\rangle)$ by the definition of $A_{i,k}$, but we also have $A_{i,k}(\langle d, n\rangle) = \{d\}^{f^k}(\langle d, n\rangle)$ by (III). So we have numbers i, k, d, n such that $A_{i,k}(\langle d, n\rangle) \neq A_{i,k}(\langle d, n\rangle)$, i.e. we have a contradiction.

Let $\mathbf{a}$ be any honest degree such that $\mathbf{0} \ll \mathbf{a}$. The structure of the sets in the lower cone of $\mathbf{a}$ is obviously an ideal (with respect to the partial ordering $\leq_E$). By the previous theorem we may infer that such a structure is never a principal ideal since $\mathbf{b} \not\ll \mathbf{b}$ for all honest degrees $\mathbf{b}$.

We know that there exists a whole $\ll$-dense set of honest degrees between $\mathbf{a}$ and $\mathbf{a}'$. (See Meyer and Ritchie [11], Simmons [16] and Kristiansen [5].) Unfortunately it is not known whether the elementary honest degrees are $\ll$-dense. (This is stated as an open problem in Meyer and Ritchie [11].) Anyway, let us suppose that the answer to this open question is positive. Suppose also that $\mathbf{0} \ll \mathbf{a}$. Then there isn't any least degree $\mathbf{b}$ such that $\leq_E(\mathbf{b})_*$ contains a $\leq_E(\mathbf{a})_*$-hard set. This is a consequence of the previous theorem. In contrast Theorem 4.1 says that there is a least degree $\mathbf{b}$ such that $\leq_E(\mathbf{b})_*$ contains an *effectively* $\leq_E(\mathbf{a})_*$-hard set, namely $\mathbf{b} = \mathbf{a}'$. No degree $\mathbf{b}$ strictly below or incomparable to $\mathbf{a}'$ is such that $\leq_E(\mathbf{b})_*$ contains an effectively $\mathbf{a}$-hard set. (It is a pity that we need the condition $\mathbf{0} \ll \mathbf{b}$ in the previous theorem. Is it possible that $\leq_E(\mathbf{a})_*$ possesses a $\leq_E(\mathbf{0})_*$-hard set when $\mathbf{a}$ is some honest degree very close to $\mathbf{0}$?)

Let $\mathbf{b}$ be an arbitrary set which lies strictly above $\mathbf{a}$, i.e. $\mathbf{a} < \mathbf{b}$. The previous theorem says that $\leq_E(\mathbf{b})_*$ does not necessarily possess a $\leq_E(\mathbf{a})_*$-hard set. (We may have $\mathbf{a} < \mathbf{b}$, but not $\mathbf{a} \ll \mathbf{b}$.) This leads us to ask if it is necessarily the case that $\leq_E(\mathbf{b})_*$ contains any set at all which isn't also contained in $\leq_E(\mathbf{a})_*$; or if $\mathbf{a}$ and $\mathbf{b}$ are incomparable, is it necessarily the case that $\leq_E(\mathbf{a})_*$ and $\leq_E(\mathbf{b})_*$ are incompatible? The next theorem answers these questions.

Theorem 4.3. *We have the equivalence* $\leq_E(\mathbf{a})_* \subseteq \leq_E(\mathbf{b})_* \;\Leftrightarrow\; \mathbf{a} \leq \mathbf{b}$.

Proof. If $\mathbf{a} \leq \mathbf{b}$, then $\leq_E(\mathbf{a})_* \subseteq \leq_E(\mathbf{b})_*$ follows trivially. Assume $\mathbf{a} \not\leq \mathbf{b}$ and that $\mathbf{a} = \deg(f)$ and $\mathbf{b} = \deg(g)$. Let $A(\langle y,x\rangle) \stackrel{\text{def}}{=} 1 \dot{-} \{y\}^f(x)$ whenever y is an index for a unary function. (Let $A(\langle y,x\rangle) \stackrel{\text{def}}{=} 1$ if $\{y\}$ is not a unary function.) It is obvious that A is elementary in f, i.e. $A \in \leq_E(\mathbf{a})_*$. We will now derive a contradiction from the assumption that $A \in \leq_E(\mathbf{b})_*$. So assume $A \leq_E g$. Then the Recursion Theorem proclaims the existence of e and k such that $A(\langle e,x\rangle) = \{e\}^{g^k}(x)$ (*). But since $f \not\leq_E g$ we may use the Growth Theorem and pick an m such that $f(m) \geq g^k(m)$ (**). Now the following contradiction emerges:

$$A(\langle e,m\rangle) \stackrel{(*)}{=} \{e\}^{g^k}(m) \stackrel{(**)}{=} \{e\}^f(m) \neq 1 \dot{-} \{e\}^f(m) = A(\langle e,m\rangle)\,.$$

References

1. K. Ambos-Spies. *On the structure of polynomial time degrees of recursive sets.* (Habilitationsschrift) Forschungsbericht Nr. 206/1985. Universität Dortmund, Dortmund, Germany, 1985. (P.O. Box 500500, D-4600 Dortmund 50)

2. S. K. Basu. On the structure of subrecursive degrees. *Journal of Computer and System Sciences* 4 (1970), 452-464.
3. A. Grzegorczyk. Some classes of recursive functions. *Rozprawy Matematyczne*, No. IV, Warszawa, 1953.
4. A. J. Heaton and S. S. Wainer. Axioms for subrecursion theories. In: *Computability, enumerability, unsolvability*, (eds. Cooper, Slaman, Wainer) 123-138. LMS *Lecture Note Series* 224, Cambridge University Press, 1996.
5. L. Kristiansen. On some classes of subrecursive functions. *Norsk Informatikkonferanse '94*. ISBN 82-519-1428-0, 33-52.
6. L. Kristiansen. A jump operator on honest subrecursive degrees. Submitted.
7. R. E. Ladner. On the structure of polynomial time reducibility. *Journal of the Association for Computing Machinery* 22 (1975), 155-171.
8. M. Machtey. Augmented loop languages and classes of computable functions. *Journal of Computer and System Sciences* 6 (1972), 603-624.
9. M. Machtey. The honest subrecursive classes are a lattice. *Information and Control* 24 (1974), 247-263.
10. M. Machtey. On the density of honest subrecursive classes. *Journal of Computer and System Sciences* 10 (1975), 183-199.
11. A. R. Meyer and D. M. Ritchie. A classification of the recursive functions. *Zeitschr. f. math. Logik und Grundlagen d. Math. Bd.* 18 (1972), 71-82.
12. P. Oddifreddi. *Classical recursion theory*. North-Holland, 1989.
13. R. Péter. *Rekursive Funktionen*. Verlag der Ungarischen Akademie der Wissenschaften, Budapest, 1957. [English translation: Academic Press, New York, 1967]
14. H. Rogers. *Theory of recursive functions and effective computability*. McGraw Hill, 1967.
15. H. E. Rose. *Subrecursion. Functions and hierarchies*. Clarendon Press, Oxford, 1984.
16. H. Simmons. *A density property of the primitive recursive degrees*. Technical Report Series UMCS-93-1-1 Department of Computer Science, University of Manchester, 1993.

Kurt Gödel and the Consistency of $R^{\#\#}$*

Robert K. Meyer

Automated Reasoning Project, RSISE
Australian National University
Canberra, ACT 0200, Australia

Summary. This paper continues my investigations of arithmetics formulated relevantly. (See references [1], [10], [12], and [9].) It is proved again that relevant **Peano** arithmetic $\mathbf{R}^{\#}$ and relevant **true** arithmetic $\mathbf{R}^{\#\#}$ (with the ω-rule) are **demonstrably consistent** by simple finitary arguments. E.g., it requires little more than truth-tables to show that $0 = 1$ is a non-theorem. This **removes** much of the sting from Gödel's **second** theorem. Regard for relevance **bounds** the harm that even potential contradictions can do. But Gödel still collects **his dues**, since proving **negation-consistency** remains annoyingly (and ineluctably) **non-constructive**. To the extent that $\sim$ in *Formalese* is unlike **not** in *English*, as it seems to be, Gödel's theorems are **dirty tricks**.

1.

Being mainly self-educated in beginning logic, I was alarmed to read in [2] that Gödel had shown that elementary number theory is either inconsistent or incomplete. "What," thought I to myself, "could this possibly **mean**?" Could it be in **doubt** that $2+2 = 4$? Might one **multiply** 27 and 37 and get 998? What is **going on** here?

More mature reflection convinces one that what is going on is a logical **dirty trick**. Speaking at his most persuasive, Gödel in [3] conned a certain sentence G into saying of itself that it was unprovable.[1] We all know where the story goes from there, at least intuitively. If G is *false*, then it is provable after all, which engenders contradiction. So G had better be true. And, as this reasoning can be carried out in any sufficiently strong, consistent and effective system **S**, the moral **is** (or is **alleged** to be) that

(I) **S** is incomplete, containing an unprovable truth (Gödel's **first** theorem), and

(II) **S** lacks the means to formalize a proof of **its own** consistency (Gödel's **second** theorem).

I shall throw no stones at (I) here. But (II) is another matter. The idea behind it is said to be (in, e. g., [3] and [4]) that we can **formalize** the proof of (I). This leaves us with the following **S**-theorem:[2]

* This paper is in its final form and no similar paper has been or is being submitted elsewhere.

[1] G is 17 gen r, says [3].

[2] We follow [3] in using "Wid" (for "widerspruchsfrei") for "consistent".

(A) $\vdash_{\mathbf{S}}$ Wid (**S**) $\supset G$.

In view of the rule $\supset$E of *modus ponens*, if we could prove in **S** that **S** is consistent then we could also prove in **S** its non-theorem G. As this is impossible (unless things have gone **very badly wrong**), we cannot prove Wid (**S**) either.

There is no doubt that [3] is an amazing and an incredible achievement in logic, and that it has been seen as such since its publication over 60 years ago. Still, there is something about the result that does not seem to ring true.[3] What I wish to offer here are **philosophical corrections** of the logic that induces the false chime. The first crux lies in the little sign '$\sim$'. This sign is supposed to mean 'not'. The real import of Gödel's arguments may be summed up succinctly thus: '$\sim$' **never** means what it is supposed to mean, **within** a particular sufficiently strong system intended seriously to formalize mathematics.

To be sure, we **can** claim to give a semantic interpretation of an effectively presented arithmetic, and tell the world that on this interpretation '$\sim$' means 'not'. The world will then ask how it comes about that on some occasions on which A is false, we cannot prove $\sim A$ in the system. We shall perhaps reply that relative to the interpretation the system is (negation-)incomplete. But how much more accurate it would be to reply instead that, because of certain formal anomalies in the technical engineering, we just **cannot** so fix things that '$\sim$' works within the system the way that 'not' is to be taken as working in English.[4]

We make this point clear, on Gödelian grounds, with respect to (standard classical Peano arithmetic) $\mathbf{P}^{\#}$.[5] We suppose some standard coding (Gödel numbering) that assigns to each formula A a unique natural number I. We assume that this coding is an effective mapping from formulas to natural numbers, and we shall henceforth use A^I for the formula with Gödel number I.[6] (The scheme of [4] will do for the purpose of furnishing such a (Gödel) numbering.) Let **N** be the set of all natural numbers; and, taking our formal system **S** abstractly, we **identify** natural numbers with the corresponding **numerals** of the system.

---

[3] So much so, I was informed by van Fraassen when he was editor of the **Journal of Philosophical Logic**, that a chief source of manuscripts submitted to that journal were "fix-ups" for Gödel's theorems.

[4] In view of the Liar Paradox and associated anomalies, we might more candidly have to admit that we don't know how 'not' really works in English, either.

[5] Take $\mathbf{P}^{\#}$ to be the system **S** of [4, p. 102f]; equivalently, let it be the system **PA** of [7].

[6] Even more elegant is the course of simply **identifying** the **formula** A^I with the **number** I. One might read [3] as suggesting this very course. For as it doesn't really matter what the **formal objects** constituting a denumerable formal system are, they **might as well be** the **natural numbers**. This is **Gödel numbering** with a vengeance!

Then, as is well-known, there is an open formula Px,[7] with sole free variable x, which serves as a **provability predicate** for $\mathbf{P}^{\#}$ in the following sense:

(1) For all natural numbers I, PI is a theorem of $\mathbf{P}^{\#}$ iff A^I is a theorem of $\mathbf{P}^{\#}$.

Taking truth in the standard model $\mathbf{N}$ in the usual Tarskian sense, we have also

(2) For all natural numbers I, PI is true iff A^I is a theorem of $\mathbf{P}^{\#}$.

So on our semantic understanding of $\sim$ we have immediately from (2),

(3) For all natural numbers I, $\sim PI$ is true iff A^I is **not** a theorem of $\mathbf{P}^{\#}$.

But on Gödelian grounds we do **not** have

(4) For all natural numbers I, $\sim PI$ is a theorem of $\mathbf{P}^{\#}$ iff A^I is **not** a theorem of $\mathbf{P}^{\#}$.

A counterexample to (4) is our old friend "I am unprovable," alias 17 gen r.

Viewed extrinsically, as in (3), we may perhaps think of '$\sim$' as a formal counterpart of 'not'. Viewed systematically, and given (1), what clearer demonstration could we ask than (4) of the proposition that '$\sim$' doesn't work formally in $\mathbf{P}^{\#}$ the way that 'not' works intuitively in English?[8] And we now fix the considerations and notation set out above for the rest of the paper (including the standard Gödel numbering, which we need not specify further). These considerations are central to an examination of the character and import of Gödel's second theorem. It is of little interest that we cannot **prove** the consistency of $\mathbf{P}^{\#}$ within itself, unless $\mathbf{P}^{\#}$ has the **vocabulary** to say that it is consistent (and moreover that what it says in this vocabulary is in fact unprovable).

Let us reflect. **First**, nobody ever expected $\mathbf{P}^{\#}$ to be muttering introspectively about itself at all. $\mathbf{P}^{\#}$ was constructed to say that $5+3=8$, that every number > 1 has a prime divisor, and the like. It was **not** constructed to say that nobody ever loved it before Hilbert, that it often wishes it were complete, or (for present purposes) that it is consistent.

Of course, after Gödel we are **all** prepared to believe that $\mathbf{P}^{\#}$ does introspect, in code. And this has made its psychoanalysis (or whatever the equivalent process is in formal systems) a regular element in training logicians. As in all psychoanalysis, there is a certain indistinctness in the method. When, e. g., $\mathbf{P}^{\#}$ seems to be saying, "Every natural number is the sum of 4 squares," we may suppose that it is bragging, "Show me a sentential tautology I can't prove!" Or maybe it is complaining, "I can only demonstrate

[7] In [3] we have "Bew" (for "beweisbar") instead of "P" (for "provable").

[8] Recall that (4) **must** fail, on pain of **bankruptcy** otherwise for the standard arithmetical mythology!

Fermat's Last Theorem for regular primes." And this should lead to a little humility on our part. Stripped of the code, $\mathbf{P}^{\#}$ is still **saying** "Every natural number is the sum of 4 squares." The rest we **read into** what it says; and, since $\mathbf{P}^{\#}$ is incomplete, we err even in simple arithmetic if we try to interpret what it says **categorically**.

Second, we must exercise unusual care in finding a technical form for Gödel's second theorem.[9] We must formalize the statement "I am consistent," uttered by $\mathbf{P}^{\#}$ about itself. There are a couple of ways in which our previous worries about negation will come to the fore. For all of our classically equivalent ways (1)-(6) in section 3 below of saying that formal arithmetic is consistent will assert that something-or-other is **not** provable. If we are worried about our capacity to **express** 'not' in $\mathbf{P}^{\#}$ then we shall have **most distressing worries** about how to say that arithmetic is consistent, even in code.

Third, one begins to wonder what Gödel's second theorem adds to his first theorem. No one expects us to be able to **prove** what we cannot **say**. And it then seems otiose to claim that any effective, finitary proof of the consistency of formal arithmetic would yield a proof in $\mathbf{P}^{\#}$ of a formula that, so far as $\mathbf{P}^{\#}$ is concerned, is only a dubious candidate for the role of being **the statement** in the vocabulary of $\mathbf{P}^{\#}$ which expresses its consistency. We don't linger over these issues. We even take a rather orthodox stand with respect to them. But we do note that they are exacerbated when we ask "What particular form, even in English, should the statement that $\mathbf{P}^{\#}$ (or any formal arithmetic) is consistent take?" We devote the next two sections to some of these problems.

2.

We ask ourselves first why we (or Hilbert, or Gödel) should **care** whether arithmetic[10] is consistent. We answer immediately that, mainly, we do **not** care. Put optimistically, we are so strongly convinced that arithmetic is consistent that demonstrating its consistency is just a game—the game of seeing how little, or how much, is required for a formal consistency proof. After all, it was the reliability of mathematical **analysis** that truly worried Hilbert and others. And since the ultimate effect of the great 19th and early 20th century programs was to substitute insecurity in reasonings about sets for insecurity in reasonings about infinitesimals and series, the neck-wringing that Gödel administered to these programs in 1931 leaves us having registered no gain on the main point.[11]

[9] Feferman said so in [5], and Gödel appended a footnote to the same effect to the translation of [3].

[10] or at least any part of it of which **serious** use is going to be made

[11] Indeed, they have led to a certain abandonment of the main point, all hands being needed (as Reid aptly puts it in [6]) to defend the homeland of arithmetic.

But while we are playing the game, what we are presumably concerned to show is that our intuitive arithmetic is **reliable**, by establishing that our carefully chosen formal counterparts of that arithmetic are reliable. So long as we accept Gödel's first theorem, part of that task remains beyond us. For according to that theorem **no** effectively presented formal system in the ordinary first-order vocabulary will serve as a fully acceptable formal counterpart of intuitive arithmetic.

So any effective formal arithmetic is at best a partial arithmetic. We can **improve** this situation in a couple of ways. First, it is clear that being **partial** is not to be confused with being **unreliable**. It is sad to have to relate that this is the point at which the usual appeals to Gödel's second theorem tend to descend into perversity. For it is claimed that no bag of mathematical tricks can be demonstrated to be reliable, except on appeal to some trick that isn't in the bag. So we are confronted with a picture (on the usual story) on which the reliability of any mathematical system (save such as are inadequate for whole number arithmetic) can only be demonstrated in some system **less reliable**, *prima facie*, than the system from which we began.

This picture, if accurate, severs mathematical logic from its chief foundational purpose—namely, making possible a rigorous reconstruction of intuitive mathematics. One gets the impression rather that even the reconstruction of simple arithmetic is dubious enough, and that every step on becomes even more dubious. And it is accordingly no wonder that many mathematical logicians have gone off to live in a world of their own—a world, frankly, that has little relevance to mathematics, even less to the philosophy of mathematics, and almost none to general philosophy. For the depressing picture is that **more** than intuitive mathematics must be assumed in order to reconstruct intuitive mathematics. Chauvinist mathematicians (e. g., Poincaré), who always bridled at the suggestion that their discipline was just pure logic, may find cause to rejoice in this picture. But logicians must weep, for it denigrates exact thought for the sake of the old mumbo-jumbo.

We began to talk about consistency, but we have slipped in this section into talk about reliability. Consistency is a formal property of formal systems (though, depending on the author, it may be any one of several properties, not necessarily related). Reliability is an intuitive property, measuring a formal system against the purposes for which it was designed. And let it be clear from the outset that it is **reliability** that is most desired. We wish that our formal systems shall be adequate to their purposes. And whatever formal property we decide to identify with the honorific 'consistent', it is of interest only insofar as possession of this property is a guide to (and hopefully a guarantee of) the system's reliability.

3.

Now let us pick up a few stones. Already in [4] was Feferman's [5] cited, which does present a plausible candidate for Wid (**S**) which is a theorem of suitable **S**. What are we to conclude from that? Only that the proof of (A) of section 1 will then break down, unless (God forbid) **S** is already inconsistent. But, in choosing our stones, let us **forget** (temporarily) about trying to find a way to say **in S** that **S** is consistent. What, **in English**, is a reasonable way of saying this? Here are some.

(1) There exists a formula A such that A is a non-theorem of **S**
(2) $0 = 1$ is a non-theorem of **S**
(3) All **numerically incorrect** equations[12] A are non-theorems of **S**
(4) All **algebraically incorrect** polynomial equations[13] A are non-theorems of **S**
(5) $\sim(0 = 0)$ is a non-theorem of **S**
(6) For no formula A are both A and $\sim A$ theorems of **S**

These are intended as **increasingly stringent** criteria for consistency.[14] But except for the option (7) of the last footnote, these criteria all come classically to the **same thing**. For intuitively (6) implies (5), and so forth until we get to the fact that (2) implies (1). But (1) implies (6), completing the circle, on account of the implicational paradox $A \& \sim A \supset B$. So classical logic **blurs** what are intuitively clear distinctions.

Relevant and other substructural logics **L** exist that **excise** some of the evils of classical logic. Perhaps if we choose one of these **L** in which to formulate our formal arithmetic **S**, the distressing equivalence of **all** of (1)-(6) will disappear. And we need go no further than the Church-Anderson-Belnap system **R** of [8] and [9] to reach this goal.[15] Here for example is a list of postulates to be added to the first-order relevant logic $\mathbf{R}^{\forall\exists x}$ of [9][16] to formulate the first-order relevant Peano arithmetic $\mathbf{R}^{\#}$.

R#1 $x = y \rightarrow x' = y'$
R#2 $x = y \rightarrow (x = z \rightarrow y = z)$
R#3 $x + 0 = x$

[12] A formula $t = u$ without free variables is correct if so by primary school arithmetic, else it is **incorrect**. $27 \times 37 = 999$ is **correct**. $2 + 2 = 5$ is **incorrect**.

[13] A polynomial equation t=u is **correct** if so by high school algebra, else it is **incorrect**. Thus, e. g., $(x + y)(x + y) = xx + xy + yy$ is incorrect. Correct is $(x + y)(x + y) = xx + 2xy + yy$.

[14] Another criterion begs for admission here—namely, (7) no **arithmetic falsehood** is a theorem of **S**. But (7) jumps from the Deep End into the Standard Numerical Mythology. We do **not** jump with it—**yet**!

[15] As Mortensen and I point out in [10], the stronger Dunn-McCall system **RM** will also do.

[16] Before [9], $\mathbf{R}^{\forall\exists x}$ was called **RQ**. See [10]-[12] for the vocabulary of $\mathbf{R}^{\#}$ and its notational conventions.

R#4 $x + y' = (x + y)'$
R#5 $x \times 0 = 0$
R#6 $x \times y' = (x \times y) + x$
R#7 $x' = y' \to x = y$
R#8 $\sim(x' = 0)$
R#9 $A0 \mathbin{\&} \forall x(Ax \to Ax') \to \forall x A x$[17]

As rules we take

$\to$E From $A \to B$ and A infer B
$\forall$I From A infer $\forall x A$

Unfortunately for $\mathbf{R}^{\#}$, it is even less satisfactory than $\mathbf{P}^{\#}$ as a vehicle for formal arithmetic.[18] But there is a repair $\mathbf{R}^{\#\#}$, which adds the following ω-rule:

$\forall 012\ldots$ Infer $\forall x Ax$ from **all** of $A0, A1, \ldots, An, \ldots$, for every numeral n.

We borrow from previous results to show that $\mathbf{R}^{\#\#}$ (and *a fortiori* $\mathbf{R}^{\#}$) is consistent.

4.

Where n is any natural number, we let F_κ be the integers $\{0, 1, \ldots, n-1\}$ modulo n. Note that $+$ and $\times$ are naturally defined on F_κ. And we let **S3** be the **3-valued** matrix on $\{+1, 0, -1\}$ defined as follows:[19]

$\to$	-1	0	$+1$
-1	$+1$	$+1$	$+1$
$*0$	-1	0	$+1$
$*+1$	-1	-1	$+1$

The $*$'ed elements 0 and $+1$ are *designated values*, and we set $m \leq n$ in **S3** if $m \to n$ is designated. (Note that $\leq$ is just the usual order on $-1, 0, +1$.) Letting $\to$ be defined by the table above, define the other connectives on **S3** by

$$\sim m = -m,$$
$$m \mathbin{\&} n = \min(m, n), \text{ and}$$
$$m \vee n = \max(m, n).$$

Where D is a domain of individuals, we can also interpret the quantifiers in **S3** by setting

[17] This **induction** postulate may be stated as a rule **RMI** for $\mathbf{R}^{\#}$. [13] rightly prefers **RMI** in weaker logics.

[18] Ackermann's rule γ ($\supset$E) fails for $\mathbf{R}^{\#}$. See Friedman and Meyer's [11].

[19] The 3-valued Sugihara matrix **S3** appears early and often in [8] and [9].

$$\forall x Ax = \min(Ad : d \in D)$$
$$\exists x Ax = \max(Ad : d \in D),$$

where for each $d \in D$ we have Ad as the "truth-value" in **S3** that results from evaluating Ax *at* d. We can now **complete a standard interpretation** I of each of the $\mathcal{F}_\kappa$ in **S3** by specifying the value of I on atomic sentences. In the simplest case, where $n = 2$ and $\mathcal{F}_\kappa = \{0, 1\}$, we set

(=2t) $I(0 = 0) = I(1 = 1) = 0$, and
(=2f) $I(0 = 1) = I(1 = 0) = -1$.

It is readily observed that all theorems of $\mathbf{R}^{\#\#}$ take non-negative values on I.[20] This immediately shows that $\mathbf{R}^{\#\#}$ (and *a fortiori* $\mathbf{R}^{\#}$) is consistent in senses (1) and (2) of section 3 , by a completely elementary argument. So much for the demise of the Hilbert program purportedly brought on by Gödel!

What about the notions (3)-(6) of **III**? (3) is in principle no more difficult than (2); for any number equation $t = u$ will reduce to one $m = n$, where m and n are numerals; assuming without loss of generality that $m \leq n$, we can further reduce an **incorrect** number equation to $n - m = 0$, where $n - m$ is a **positive** integer. We generalize (=2t) and (=2f) above to define standard I by

(=nt) $I(k = k) = 0$ for all $k \in \mathcal{F}_\kappa$
(=nf) $I(j = k) = -1$, for all distinct $j, k \in \mathcal{F}_\kappa$

And it is now clear that all theorems of $\mathbf{R}^{\#\#}$ take a designated value on each standard I in **S3** while for each incorrect number equation it is trivial to find a $\mathcal{F}_\kappa$ that refutes it. That's consistency in sense (3). And we can extend this proof to get consistency in sense (4), observing that every **incorrect** polynomial equation has a **substitution instance** in natural numbers which is **numerically incorrect**. So $\mathbf{R}^{\#\#}$ is **polynomially consistent** as well.

We have, however, taken these methods about as far as we can. In particular, we **cannot** get from the $\mathcal{F}_\kappa$ a simple proof that $\mathbf{R}^{\#\#}$ is negation-consistent or even that $\sim(0 = 0)$ is unprovable, which are (6) and (5) respectively of **III**. Consider the latter. As our standard interpretation I will assign 0 to each $k = k$, it will similarly assign $-0 = 0$ to $\sim(k = k)$. As 0 is itself a designated value in **S3**, this is no way to show that negated identities are unprovable. Moreover since $A \,\&\, {\sim}A \rightarrow {\sim}(0 = 0)$ is a theorem scheme of $\mathbf{R}^{\#\#}$ (indeed, already of $\mathbf{R}^{\#}$), we cannot use the $\mathcal{F}_\kappa$ to establish negation-consistency either.

Still, if one thing doesn't work, we can try another. Defining material implication $\supset$ as usual by

(D$\supset$) $A \supset B =_{\text{df}} {\sim}A \vee B$,

[20] The nice thing here is that we have actually made the ω-rule $\forall 012\ldots$ finitary. For, mod 2, the quantifiers are just ranging over 0 and 1!

we already have provable in $\mathbf{R}^{\#}$

(8) $\sim(0 = 0) \supset 0 = 1$,

in view of the theoremhood of $0 = 0$. Of course (8) is also provable in $\mathbf{R}^{\#\#}$. But $\mathbf{R}^{\#\#}$, unlike $\mathbf{R}^{\#}$, admits $\supset$E, as [12] demonstrates. So, as 0=1 has a finitary refutation *mod* 2, we can rest content that $\sim(0 = 0)$ is also a non-theorem; similarly, since if any contradiction were a theorem then so also would $\sim(0 = 0)$ be provable, we can assert that $\mathbf{R}^{\#\#}$ is negation-consistent.

So we can. But the pleasant finitary character of our appeal to **S3** through the $F_{\aleph}$ is gone and lost forever[21] when we go through the $\supset$E proof of [12]. Granted, on Gödelian grounds a charge will be made **somewhere** for a proof of negation-consistency. And the charge in this case is the **non-constructive** character of the argument of [12].

5.

What, I have been asked, is the relation between $\mathbf{R}^{\#}$ and $\mathbf{R}^{\#\#}$ and more familiar systems like classical $\mathbf{P}^{\#}$? When I first published a few of my results about relevant arithmetic in the abstract [1], it was my hope that $\mathbf{R}^{\#}$ would exactly contain $\mathbf{P}^{\#}$, in the sense that every theorem A of $\mathbf{P}^{\#}$ would be a theorem of $\mathbf{R}^{\#}$, on direct truth-functional translation. This was not to be, on account of Friedman's contribution to [11]. But there are nonetheless several exact translations on which $\mathbf{P}^{\#}$ is a truth-functional subsystem of $\mathbf{R}^{\#}$. Thus all theorems of $\mathbf{P}^{\#}$ are on these translations theorems of $\mathbf{R}^{\#}$, whence all classical metatheory goes through.

$\mathbf{R}^{\#\#}$ is quite another kettle of fish. It stands to $\mathbf{R}^{\#}$ as classical **true arithmetic** $\mathbf{P}^{\#\#}$ stands to $\mathbf{P}^{\#}$. It is proved in [12] (though already noted in [1]) that $\mathbf{P}^{\#\#}$ **is contained** in $\mathbf{R}^{\#\#}$ on direct truth-functional translation. This disposes immediately of many questions that one **might have had** about $\mathbf{R}^{\#\#}$. Its theorems are not recursively enumerable; nor is it recursively axiomatizable. (Why? Because the **truths** of $\mathbf{P}^{\#\#}$ aren't r.e.)

A referee has called attention to an apparent **incoherence** between section 1 (in which we said that Gödelian troubles arose because '$\sim$' does **not** mean 'not') and section 3 (in which we complained that what is most wanted in formal systems for arithmetic and other stuff is **reliability**). This is a good point, though I wish here to reiterate both claims. Naturally we want our systems to be reliable; and I say that relevant systems are (demonstrably) **more** reliable than the standard brands. For we can rest reasonably content with what **has** been shown here. We have sidestepped the layers and layers of Gödelian uncertainty by showing that some sorts of mistakes **just can't happen**, if we formulate our theories relevantly. This has been a point **all too easily overlooked** in the debates for and against the adequacy of

[21] Just like "my darling Clementine"

truth-functional insights. To have **faith** is a wonderful thing (perhaps). But in logic, **reason** is supposed to rule. And not merely actual contradiction but also potential contradiction can undermine a system that is formulated truth-functionally. Regard for relevance **bounds** the harm that even **potential** contradictions can do.

But Gödel bites the truth-functionalist even more deeply. What are we to make of a statement

$$\sim(721 \text{ is provable})$$

when this is the very statement #721? As I said at the outset, this is a **dirty trick**. And dirty tricks ought not to be confused with profound metaphysical insights about 'not'. We rest our case!

Acknowledgement. Thanks are due to numerous colleagues for discussions of the topics of this paper over the years. Among them are the referees, Dunn, Belnap, McRobbie, Martin, Restall, Slaney, Mares, Friedman, Paris, Thistlewaite, Mortensen and many others. Surendonk's help in preparing the text was another godsend.

References

1. R. Meyer. Relevant arithmetic. (abstract), *Bulletin of the section of logic 5* (1976), 133–137.
2. W. Quine. *Methods of Logic.* Holt, New York, 1950.
3. K. Gödel. Über formal unentscheidbare Sätze der Principia mathematica und verwandter Systeme. *Monatshefte für Mathematik und Physik 38* (1931), pp. 173-98, reprinted with English tr. in S. Feferman, J.W. Dawson, S.C. Kleene, G.H. Moore, R.M. Solovay and J. van Heijenoort (eds.), *Kurt Gödel, Collected Works (vol. I)*, Oxford, 1986, pp. 144-95.
4. E. Mendelson. *Introduction to mathematical logic.* van Nostrand, Princeton, 1964.
5. S. Feferman. Arithmetization of metamathematics in a general setting. *Fundamenta mathematicae 49* (1960), pp. 35-92.
6. C. Reid. *Hilbert.* Springer, N. Y., 1970.
7. G. Boolos. *The logic of provability.* Cambridge, 1993.
8. A.R. Anderson and N.D. Belnap, Jr. *Entailment (vol. I).* Princeton, 1975.
9. A.R. Anderson, N.D. Belnap, Jr. and J.M. Dunn. *Entailment (vol. II).* Princeton, 1992.
10. R. Meyer and C. Mortensen. Inconsistent models for relevant arithmetics. *The journal of symbolic logic 49* (1984), 917-929.
11. H. Friedman and R. Meyer. Whither relevant arithmetic? *The journal of symbolic logic 57* (1992), 824-31.
12. R. Meyer. ⊃E is admissible in "true" relevant arithmetic. forthcoming, *Journal of philosophical logic.*
13. G. Restall. *Logics without contraction.* PhD thesis, U. of Queensland 1994.

Best possible answer is computable for fuzzy SLD-resolution *

Leonard Paulík **

Mathematical Institute, Slovak Academy of Sciences,
Jesenná 5, SK-041 54 Košice, Slovakia
e-mail: lepaulik@kosice.upjs.sk,
tad/fax (+42 95)6 228 291

Summary. This is a direct continuation of a joint work with P. Vojtáš in which we proved soundness and completeness of fuzzy SLD-resolution for arbitrary many-valued logic with two continuous conjunctions. Using this result here we prove that the maximal value of grade for a fuzzy answer is attained during fuzzy SLD-resolution in logic with only one continuous conjunction. Based on this result we prove better characterization of the least fuzzy Herbrand model of fuzzy definite logic program, which allows us to give a refinement of the completeness part.

1. Introduction

In [12] authors (P. Vojtáš and L. Paulík) consider theoretical (mathematical) model of extended logic programming in many valued logic with arbitrary triple of connectives $(\mathrm{seq}, \mathrm{et}_1, \mathrm{et}_2)$, where et_1 evaluates modus ponens containing the implication seq, and et_2 is the conjunction from bodies of clauses. Declarative semantics is based on generalization of P. Hájek's RPL and RQL. Let us make several remarks concerning these logics in more general way.

It is worth mentioning here, that well-known non-classical logics was proposed by Łukasiewicz and Gödel and that main interest was payed on 1-tautologies in these logics. J. Pavelka made further development of Łukasiewicz logic on propositional level ([10]). He considered not only single formula φ but also a numerical value r (grade) connected with the formula and proposed deduction calculus for maintaining graded formulas $(\varphi; r)$ meaning truth value of φ is at least r. He introduced notion of graded proof and norma $|\varphi|_T$ means the supremum of values for graded proof of φ from a theory T. Norma $\|\varphi\|_T$ means the infimum of values which formula φ gets in models of T. J. Pavelka proved completeness theorem of the form $|\varphi|_T = \|\varphi\|_T$. Further development of Pavelka's ideas was done by V. Novák on predicate level [9]. Substantial simplification of these logics was achieved in P. Hájek's RPL and RQL, see [3],[5].

Recently P. Hájek and D. Švejda in [7] proved strong completeness for finitely axiomatized theory in Łukasiewicz logic and as a consequence also

* This paper is in final form and no similar paper has been or is being submitted elsewhere.

** This work was supported by the grant 2/1224/95 of the Slovak Grant Agency for Science.

completeness for finite Pavelka's Rational logic (if T is a finite fuzzy theory and r is a rational number such that $\|\varphi\|_T = r$, then $T \vdash (\varphi; r)$). This means that the supremum in the definition of $|\varphi|_T$ is attained in deduction process.

Let us note that completeness theorem ($|\varphi|_T = \|\varphi\|_T$) does not generally hold for Gödel and product logic (see e.g. [4]).

Complete axiomatization for 1-tautologies in product logic was proposed in [6].

In [12] we introduced a procedural semantics for SLD-resolution with two continuous conjunctions (as we mentioned above) and prove soundness and completeness theorem (in the sense ($|\varphi|_P = \|\varphi\|_P$). But in [12] is not proved that the best possible answer is really attained during computation. We only proved that we can obtain answers with values which are arbitrary close to the best one (see Theorem 11). Now we are able to prove that during computation with one continuous conjunction ($\mathrm{et}^{\bullet} = \mathrm{et}_1^{\bullet} = \mathrm{et}_2^{\bullet}$) the best possible answer is really attained (see Theorem 14 and Theorem 17).

Moreover, we can state some more general remark. If the variant of fuzzy logic uses deduction rules based on t-norms, in the sense that the value of a consequent of a deduction step is t-norm value from values of antecedents, (it is not possible to increase value during deductions), then as a corollary of our Lemma 13 we have: If a fuzzy theory $T : Fml \to [0,1]$ is such that range of T, $T(Fml)$ is finite (even when theory T itself is infinite), then

$$|\varphi|_T = \max\{r; T \vdash (\varphi; r)\}$$

for every formula $\varphi \in Fml$. (We can write *max* instead of *sup* for $|\varphi|_T$.)

For information and references concerning motivations and applications of fuzzy logic programming see e.g. [12].

2. Declarative and procedural semantics

Let us recall several definitions and theorems from [11], [12], which we will need in the next sections.

Let $\mathcal{L}$ be a first order language containing variables, function symbols, predicate symbols, constants, quantifiers and connectives $\neg$, seq and et (intended meaning is that seq is an implication — the leftarrow writing version is qes and et is a conjunction). Connectives usually preserve rationality, i.e. if r, q are rational, then the value $\#^{\bullet}(r, q)$ is rational. The syntactical level is not touched by many valuedness of semantics.

We base our declarative semantics only on fuzzy Herbrand interpretations (skipping here arbitrary interpretations). Herbrand universe $U_{\mathcal{L}}$ consists of all ground terms, having function symbols we are going to interpret them crisp, Herbrand base $B_{\mathcal{L}}$ consists of all ground atoms. Note, that this step is not touched by fuzziness. An n-ary predicate symbol should be interpreted as a fuzzy subset of $U_{\mathcal{L}}^n$, i.e. as a mapping from $U_{\mathcal{L}}^n$ into the unit interval

$[0,1]$. Gluing together all fuzzy predicates we interpret all of them at once by a mapping $f : B_{\mathcal{L}} \to [0,1]$.

For a connective $\#$ the corresponding truth value function will be denoted by $\#^{\cdot}$ (i.e. a dot over the very connective). For arbitrary $x, y \in [0,1]$ put $\neg^{\cdot} x = 1 - x$ and connective $\mathrm{et}^{\cdot} : [0,1]^2 \to [0,1]$ is arbitrary t-norm, (i.e. commutative, associative, monotone in both coordinates, and with 1 as a neutral element). Let us make a notational agreement: for a conjunction et which is binary we often harm the arity, using associativness denoting multiple composition. The implication $\mathrm{seq}^{\cdot} : [0,1]^2 \to [0,1]$ is coupled with $\mathrm{et}^{\cdot}$ in such a way that modus ponens

$$\frac{(B,x),(B \,\mathrm{seq}\, A, y)}{(A, \mathrm{et}^{\cdot}(x,y))}$$

is a sound rule (see [2] and [3, 5]). This means that whenever $f(\varphi) \geq x$ and $f(\varphi \text{ seq } \psi) \geq y$, then $f(\psi) \geq \mathrm{et}^{\cdot}(x,y)$; denote this by MP(seq, et). Recall that for seq there is the largest $\mathrm{et}^{\cdot}_{\mathrm{seq}}(x,y) = \inf\{z; \mathrm{seq}^{\cdot}(x,z) \geq y\}$ for which is modus ponens sound and $\mathrm{seq}^{\cdot}(x,y) = \sup\{z : \mathrm{et}^{\cdot}_{\mathrm{seq}}(x,z) \leq y\}$ holds ([2]). So our assumption that $\mathrm{et}^{\cdot}$ evaluates modus ponens with $\mathrm{seq}^{\cdot}$ in a sound way means that $\mathrm{et}^{\cdot} \leq \mathrm{et}^{\cdot}_{\mathrm{seq}}$ holds.

Let $f : B_{\mathcal{L}} \to [0,1]$ be a fuzzy Herbrand interpretation. The truth value for ground atoms $A \in B_{\mathcal{L}}$ is defined to be $f(A)$. For arbitrary formula φ and an evaluation of variables $e : Var \to U_{\mathcal{L}}$ the truth value $f(\varphi)[e]$ is calculated using following rules along the complexity of formulas:

$$\begin{aligned} f(p(t_1,\ldots,t_n))[e] &= f(p(t_1[e],\ldots,t_n[e])) \\ f(\neg\varphi)[e] &= \neg^{\cdot}(f(\varphi)[e]) \\ f(\varphi \,\mathrm{seq}\, \psi)[e] &= \mathrm{seq}^{\cdot}(f(\varphi)[e], f(\psi)[e]) \\ f(\varphi \,\mathrm{et}\, \psi)[e] &= \mathrm{et}^{\cdot}(f(\varphi)[e], f(\psi)[e]) \\ f((\forall x)\varphi)[e] &= \inf\{f(\varphi)[e'] : e' =_x e\} \end{aligned}$$

where $e' =_x e$ means that e' can differ from e only at x.

Finally let truth value of a formula φ under fuzzy Herbrand interpretation f be same as that of its generalization and does not depend on evaluation:

$$f(\varphi) = f(\forall\varphi) = \inf\{f(\varphi)[e] : e \text{ arbitrary}\}.$$

Definition 2.1. (See [3, 5].) *A* fuzzy theory *is a partial mapping T assigning formulas a rational number . Partiality of the mapping T we understand as of being defined constantly zero outside of the domain* $\mathrm{dom}(T)$. *A fuzzy Herbrand interpretation f is a* model *of a fuzzy theory T if for all formulas $\varphi \in \mathrm{dom}(T)$ we have $f(\varphi) \geq T(\varphi)$.*

A (seq, et)-definite program clause is a formula $\forall((B_1 \,\mathrm{et} \cdots \mathrm{et}\, B_n) \text{ seq } A)$, where $A, B_1, \ldots, B_n$ are atoms. We often write it in the leftarrow form as A qes $B_1, \ldots, B_n$, where qes is the leftarrow writing of seq, commas in the antecedent denote conjunction et. Similarly we define (seq, et)-facts and goals. The empty clause is denoted by $\Box$.

Let the symbol $\approx$ denote the following equivalence on the set of all formulas: $\varphi \approx \psi$ if φ is a variant of ψ.

Definition 2.2. *A fuzzy theory P is called a* fuzzy (seq, et)-definite program, *if*

1. $\mathrm{dom}(P)$ *is a set of* (seq, et)*-definite program clauses or facts,*
2. $\mathrm{dom}(P)/_{\approx}$ *is finite*
3. *for* $\varphi \approx \psi$ *and* $\varphi \in \mathrm{dom}(P)$ *we have* $\psi \in \mathrm{dom}(P)$ *and* $P(\varphi) = P(\psi) > 0$.

Let us recall several notions and facts concerning procedural semantics ([11], [12]). Following P. Hájek ([3],[5]) we define a graded formula being a pair $(\varphi; r)$, where φ is a formula and $r \in [0, 1]$ is a rational number. Especially, $(A \text{ qes}; r)$, $(A \text{ qes } B_1, \ldots, B_n; r)$, and $(\text{qes } B_1, \ldots, B_n; r)$ are a graded fact, a graded clause, and a graded goal, respectively.

Definition 2.3. *Let* $G = (\text{qes } A_1, \ldots, A_m, \ldots, A_k; r)$ *and* $C = (A \text{ qes } B_1, \ldots, B_l;$
$q)$ *be a graded goal and a graded clause, respectively. Then a graded goal G' is f-derived from G and C using mgu θ if the following conditions hold:*

1. A_m *is an atom, called the selected atom in* G
2. θ *is a mgu of* A_m *and* A
3. $G' = (\text{qes}(A_1, \ldots, A_{m-1}, B_1, \ldots, B_l, A_{m+1}, \ldots, A_k)\theta; r \,\text{et}^{\bullet}\, q)$.

Definition 2.4. *Let P be a fuzzy* (seq, et)*-definite program and let H be a* (seq, et)*-definite goal. A pair $(\theta; r)$ consisting of a substitution θ and a rational number r is a* graded computed answer *(GCA) for P and H if there is a sequence $G_0, \ldots, G_n$ of graded goals, a sequence $D_1, \ldots, D_n$ of suitable variants of clauses from the domain of P and a sequence $\theta_1, \ldots, \theta_n$ of mgu's such that*

1. $G_0 = (H; 1)$
2. G_{i+1} *is f-derived from G_i and* $(D_{i+1}; P(D_{i+1}))$
3. $\theta = \theta_1 \circ \cdots \circ \theta_n$ *restricted to variables of H*
4. $G_n = (\Box; r)$

($G_0, \ldots, G_n$ is called a graded SLD-refutation*).*

Definition 2.5. *A pair $(x; \theta)$ consisting of a real number r and a substitution θ is a* fuzzy Herbrand correct answer *for a fuzzy* (seq, et)*-definite program P and* et*-goal* $H = \neg\exists(A_1 \text{et} \cdots \text{et} A_n)$ *if for all fuzzy Herbrand interpretations* $f : B_{\mathcal{L}} \to [0, 1]$ *which is a model of P we have* $f(\forall((A_1 \,\text{et} \cdots \text{et}\, A_n)\theta)) \geq x$.

Observation. Let us observe the connection between procedural semantics defined above and that of [12]. In [12] we used two different conjunctions et_1 and et_2, so we could not use commutativity and associativity law in full scope. We had to store values for atoms in the bodies of clauses (connected by et_2) until we have all of them and only then we used et_1 for evaluation of modus ponens. Now we have only one conjunction et, used for evaluation in the body of clause as well as for evaluation of modus ponens. Hence, from now on we can use several theorems, (which were fully proved in [12]) also for procedural semantics used here, as it is just the special case for $\text{et}_1^{\bullet} = \text{et}_2^{\bullet}$.

To finish this part (most of material is a slight modification of that in [11], [12]), let us state the Soundness Theorem (cf. [11],[12]) in the form :

Theorem 2.1 (Soundness for fuzzy (seq, et)-SLD-resolution). *Assume* MP(seq, et). *Let P be a fuzzy* (seq, et)*-definite program and H an* et*-goal. Let $(r;\theta)$ be a graded computed answer for P and H. Then $(r;\theta)$ is a fuzzy Herbrand correct answer.*

3. Approximate completeness of fuzzy SLD-resolution

In our proof we follow [11],[12] with a fuzzy analogy of classical crisp fixpoint approach of [1],[8]. Let us recall some notations (see [8],[1],[12]). Let $(L, \leq, \bot, \top)$ be a partial order with smallest $(\bot)$ and largest $(\top)$ element. L is a complete lattice if for all $X \subseteq L$ the least upper bound $\text{lub}(X)$ and greatest lower bound $\text{glb}(X)$ exists. A set $X \subseteq L$ is directed if every finite subset of X has an upper bound in X. A mapping $T : L \to L$ is monotone, if $x \leq y$ implies $T(x) \leq T(y)$ and moreover it is continuous if $T(\text{lub}(X)) = \text{lub}(T(X))$ holds for every directed subset X of L. Note that for monotone mappings $T(\text{lub}(X)) \geq \text{lub}(T(X))$ holds for all X. We say $a \in L$ is the least fixpoint of T if

$$a = \text{lfp}(T) = \text{glb}\{x : T(x) = x\} = \text{glb}\{x : T(x) \leq x\}.$$

There is another characterization of $\text{lfp}(T)$. Denote by transfinite induction

$$\begin{aligned} T{\uparrow}0 &= \bot \\ T{\uparrow}\alpha &= T(T{\uparrow}(\alpha-1)) \text{ for } \alpha \text{ successor} \\ T{\uparrow}\alpha &= \text{lub}\{T{\uparrow}\beta; \beta < \alpha\} \text{ for } \alpha \text{ limit} \end{aligned}$$

Then for a complete lattice L and a continuous mapping $T : L \to L$ holds true that $\text{lfp}(T) = T{\uparrow}\omega$.

Denote

$$\mathcal{F}_P = \{f : f \text{ is a mapping from } B_P \text{ into } [0,1]\} = [0,1]^{B_P}.$$

Let functions 0_{B_P} be constantly zero and 1_{B_P} constantly one on B_P and for $f, g \in \mathcal{F}_P$ let $f \leq g$ holds if for all $A \in B_P$ is $f(A) \leq g(A)$. Then $(\mathcal{F}_P, \leq, 0_{B_P}, 1_{B_P})$ is a complete lattice where for $X \subseteq [0,1]^{B_P}$ $\text{lub}(X)(A) = \sup\{f(A); A \in X\}$ and $\text{glb}(X)(A) = \inf\{f(A) : A \in X\}$. Moreover X is directed if for $f_1, \ldots, f_n \in X$ is $\max\{f_1, \ldots, f_n\} \in X$.

Definition 3.1. (Definition 7 in [12]) *Let P be a* (seq, et)*-definite program. The mapping $T_P : \mathcal{F}_P \to \mathcal{F}_P$ defined bellow we call the* (seq, et, P)*-operator (if the context is doubtless simply called operator).*

For $f \in \mathcal{F}_P$, let $T_P(f)$ is a mapping from B_P into $[0,1]$ defined $T_P(f)(A) = \sup\{r :$ there is $(A$ qes $A_1, \ldots, A_n)$ a ground instance of $C \in$ dom(P) and $r = P(C)\,\mathrm{et}^{\cdot} f(A_1)\,\mathrm{et}^{\cdot} \cdots \mathrm{et}^{\cdot} f(A_n)\}$.

Note, that for a fact (A qes.) $\in$ dom(P) the list $A_1, \ldots, A_n$ is empty and we understand $r = 1\,\mathrm{et}^{\cdot}\, P(C) = P(C)$ in the previous definition.

Let us also observe that range of $f \in \mathcal{F}_P$ can have infinitely many values and there can be infinitely many values for different ground instances of $A_1, \ldots, A_n$. So we cannot write *max* instead of *sup* in the previous definition.

Theorem 3.1 (Fixpoint character of the least fuzzy Herbrand model). (Theorem 10 in [12]) *Assume* $\mathrm{et}^{\cdot} = \mathrm{et}^{\cdot}_{\mathrm{seq}}$ *is continuous. Let P be a fuzzy* (seq, et)*-definite program and T_P is the corresponding* (seq, et, P) *operator. Then*

$$\mathrm{lfp}(T_P) = T_P\!\uparrow\!\omega = M_P$$

where M_P is the $([0,1]^{B_P}, \leq)$ least fuzzy Herbrand model of P.

Definition 3.2. (Definition 11 in [12]) *Define success-fuzzy Herbrand interpretation of P as $f_{s(P)} : B_P \to [0,1]$ by*

$$f_{s(P)}(A) = \sup\{r : (r, \mathrm{id}) \text{ is GCA for } P \text{ and } A\}.$$

Theorem 3.2. (Theorem 12 in [12]) *Assume* $\mathrm{et}^{\cdot} = \mathrm{et}^{\cdot}_{\mathrm{seq}}$ *is continuous. Let P be a fuzzy* (seq, et)*-definite program. Then*

$$f_{s(P)} = M_P,$$

i.e. the success-fuzzy Herbrand interpretation of P is equal to least fuzzy Herbrand model of P.

In [12] we proved completeness theorem in the form that during SLD-resolution we can obtain answers which values are arbitrary close to the best one.

Theorem 3.3. (Theorem 13 in [12]) *Assume* $\mathrm{et}^{\cdot} = \mathrm{et}^{\cdot}_{\mathrm{seq}}$ *is continuous. Let P be a fuzzy* (seq, et)*-definite program and G an* et*-definite goal. For every $(x;\theta)$ a fuzzy Herbrand correct answer for P and G and for every $\epsilon > 0$ there exists a* (seq, et)*-graded computed answer $(q;\sigma)$ for P and G such that $x - \epsilon \leq q$ and $\theta = \sigma\gamma$ for some γ.*

4. Attaining supremum value

Before we formulate additional statement to the Fixpoint characterization of the least fuzzy Herbrand model, let us observe that elements in $T_P(f)(A)$ are values of expressions like

$$q_1 \,\mathrm{et}^{\bullet}\, q_2 \,\mathrm{et}^{\bullet} \ldots \mathrm{et}^{\bullet}\, q_k,$$

where q_i are some of confidence factors of definite program clauses. Assume in general case, that $\mathrm{et}^{\bullet} = \mathbf{t}$, where $\mathbf{t}$ is an arbitrary t-norm. Observe that $x \,\mathbf{t}\, y \leq x \,\mathbf{t}\, 1 = x$ and $x \,\mathbf{t}\, y \leq 1 \,\mathbf{t}\, y = y$ for all $x, y \in [0,1]$.

Definition 4.1. *Let* $\mathbf{t}$ *be an arbitrary t-norm and* $U \subset [0,1]$. *Denote by* $V(U)$ *a set of all values of all terms (expressions)*

$$q_1 \,\mathbf{t}\, q_2 \,\mathbf{t} \cdots \mathbf{t}\, q_k,$$

for $q_1, q_2, \ldots, q_k \in U$ *and* $k = 1, 2, \ldots$

Our key lemma is the following

Lemma 4.1. *If a set* $U \subset [0,1]$ *is finite then the set* $V(U)$ *does not contain any infinite strictly increasing sequence.*

Proof. Let n be the number of elements of a set U. The proof goes by induction along n, the number of values used ($n = |U|$).

I. Let $n = 1$, i.e. $U = \{q_1\}$. In this case it is possible to form only terms like

$$q_1, \quad q_1 \,\mathbf{t}\, q_1, \quad q_1 \,\mathbf{t}\, q_1 \,\mathbf{t}\, q_1, \ldots$$

Because of monotonicity of $\mathbf{t}$ it is obvious that a longer expression cannot have bigger value than a shorter one. Hence, it is impossible to form any infinite strictly increasing sequence.

II. Let the statement of the lemma holds for all $n \leq k$. We have to prove it for $n = k+1$. Let $U = \{q_1, \ldots, q_k, q_{k+1}\}$. Thanks to commutativity and associativity of t-norm $\mathbf{t}$ we can divide the expression

$$q_{i_1} \,\mathbf{t} \cdots \mathbf{t}\, q_{i_l}, \qquad q_{i_1}, \ldots q_{i_l} \in U,\ l \in N,$$

into two parts. The first one (initial) contains only elements from the set $\{q_1, \ldots,$
$q_k\}$ and the second one (terminal) contains only several occurrences of the element q_{k+1}. Of course, every part can be empty, but not both of them in the same time.

Assume by the contrary, that the set $V(U)$ contains infinite strictly increasing sequence of elements. This sequence contains subsequence, for which the lengths of expressions are also increasing. In this subsequence we can select another one, in which at least one of the two parts must be built from subexpressions of increasing length. Let us consider the following cases:

1. There is a maximal length of the first parts, (only the second parts have increasing lengths of subexpressions). Then there is at least one instance of the initial part, for which there are infinitely many continuations in the second parts. We know that if $x \mathbf{t} y < x \mathbf{t} z$ then $y < z$. (For, if $y \geq z$ then from the definition of t-norm we have $x \mathbf{t} y \geq x \mathbf{t} z$.) Hence the second parts in these continuations form strictly increasing infinite sequence. This is a contradiction.

2. There is a maximal length of the second parts, (only the first parts have increasing lengths of subexpressions). Similarly, as in case 1, we can find infinite strictly increasing sequence of values of subexpressions in the first parts, what is again a contradiction.

3. It remains the case, when the lengths of both parts are increasing. If the first parts contain infinite strictly increasing subsequence we have an contradiction to induction hypothesis immediately. If the first parts contain infinite constant subsequence, then the second parts form an infinite strictly increasing sequence (similarly as in case 1 and 2). Again, we get a contradiction. Finally, assume that the first parts contain infinite strictly decreasing subsequence

$$b_1 > b_2 > \ldots > b_i > \ldots$$

In connection with values $c_1, c_2, \ldots, c_i, \ldots$ of the second parts we have by our assumption that

$$b_1 \mathbf{t} c_1 < b_2 \mathbf{t} c_2 < \cdots < b_i \mathbf{t} c_i < \ldots$$

If we assume that $c_i \geq c_{i+1}$ for some i then $b_i \mathbf{t} c_i \geq b_{i+1} \mathbf{t} c_i \geq b_{i+1} \mathbf{t} c_{i+1}$. We get contradiction, so $c_i < c_{i+1}$ for all $i = 1, 2, \ldots$ We have found infinite strictly increasing sequence in the second parts, what is a contradiction to induction hypothesis. Hence the lemma is proved. □

On the base of previous Lemma 13 we can state

Theorem 4.1. *For every $A \in B_P$ there is a number n_0 such that for every $n \geq n_0$*

$$(T_P{\uparrow}n_0)(A) = (T_P{\uparrow}n)(A) = (T_P{\uparrow}\omega)(A),$$

i.e. every element of $T_P \uparrow \omega$ attains his maximal value.

Proof. Let $A \in B_P$. Every element of $T_P \uparrow 0$ has value 0, hence $(T_P \uparrow 0)(A) = 0$. If the program P contains a fact Cqes., where $P(C) = r$, and A is a ground instance of C, then after first step of iteration we have $(T_P \uparrow 1)(A) = r$; otherwise $(T_P \uparrow 1)(A) = 0$. From the definition of T_P follows that every element of $T_P \uparrow \omega$ is a value of an expression $q_1 \,\mathrm{et}^{\bullet} \cdots \mathrm{et}^{\bullet}\, q_k$, where $q_1, \ldots, q_k$ are in the range of P. Recall that $\mathrm{dom}(P)/_{\approx}$ is finite, so the range of P is finite as well. Hence from our Lemma 13 we have that there is no infinite strictly increasing sequence of elements of $T_P \uparrow \omega$. It means, that after a finite number n_0 of steps of iteration, the value $(T_P \uparrow n_0)(A)$ attains the maximal value, which cannot increase during further iterations. (Of course, this value can remain 0.) □

Let us recall two lemmas from [8]:

Lemma 4.2 (Mgu Lemma). (Lemma 8.1 in [8], p. 47) *Let P be a definite program and G a definite goal. Suppose that $P \cup \{G\}$ has an unrestricted SLD-refutation, i.e. the unifiers used need not be most general. Then $P \cup \{G\}$ has an SLD-refutation of the same length such that, if $\theta_1, \ldots, \theta_n$ are the unifiers from the unrestricted SLD-refutation and $\theta'_1, \ldots, \theta'_n$ are mgu's from the SLD-refutation, then there exists a substitution γ such that $\theta_1 \ldots \theta_n = \theta'_1 \ldots \theta'_n\gamma$.*

Lemma 4.3 (Lifting Lemma). (Lemma 8.2 in [8], p. 47) *Let P be a definite program, G a definite goal and θ a substitution. Suppose there exists an SLD-refutation of $P \cup \{G\theta\}$. Then there exists an SLD-refutation of $P \cup \{G\}$ of the same length such that, if $\theta_1, \ldots, \theta_n$ are the mgu's from the SLD-refutation of $P \cup \{G\theta\}$ and $\theta'_1, \ldots, \theta'_n$ are the mgu's from the SLD-refutation of $P \cup \{G\}$, then there exists a substitution γ such that $\theta_1 \ldots \theta_n = \theta'_1 \ldots \theta'_n\gamma$.*

Now we are in the position to give promised refinement of Completeness Theorem (we can drop all things concerning ϵ).

Theorem 4.2 (Tight completeness of fuzzy (seq, et)-SLD-resolution). *Assume $\mathrm{et}^{\bullet} = \mathrm{et}^{\bullet}_{\mathrm{seq}}$ is continuous. Let P be a fuzzy (seq, et)-definite program and G an et-definite goal. For every $(x;\theta)$ a fuzzy Herbrand correct answer for P and G there exists a (seq, et)-combined graded computed answer $(q;\sigma)$ for P and G such that $x \le q$ and $\theta = \sigma\gamma$ for some γ.*

Proof. (Simplification of that from [12].) Let us observe that Theorem 10 implies completeness result for ground atoms: Assume $(x;\theta)$ is a correct answer for P and a goal consisting of ground atom A then $x \le M_P(A\theta) = M_P(A) = f_{s(P)}(A)$. From the Theorem 14 follows that $f_{s(P)}(A)$ is the maximum of computed answers (maximum is attained), we are done. (Note that wlog we can assume that our goal consists of one atom.)

Now let A be an atom (not necessarily ground) and $(x;\theta)$ a fuzzy Herbrand correct answer for P and A. Let $\{X_1, \ldots, X_n\}$ be variables of $A\theta$ and let $\{a_1, \ldots, a_n\}$ be constants distinct from everything appearing in P and A. Denote $\delta = \{X_1/a_1, \ldots, X_n/a_n\}$. Then $A\theta\delta$ is ground and $x \le M_P(\forall(A\theta)) \le M_P(A\theta\delta)$. So by the result for the ground atom $A\theta\delta$ there is a GCA $(q;\mathrm{id})$ witnessed by a derivation $G_0, \ldots, G_l$ for P and $A\theta\delta$. Replacing a_i's in G_i by X_i's we get a successful derivation G'_i for P and $A\theta$ giving the same computed answer $(q;\mathrm{id})$. Now arguing in the same way as in Lemma 15 (Mgu Lemma) and Lemma 16 (Lifting Lemma) of Lloyd ([8],p. 47) we can find a sequence of mgu's $\theta'_1 \ldots \theta'_n = \sigma$ witnessing that the derivation G''_i obtained from G'_i only changing substitutions is a successful derivation, $\theta = \sigma\gamma$ and the computed answer $(q;\sigma)$ gives the same numerical value, because clauses and facts of P used along the derivation G''_i are the same as in G'_i (and same as in G_i). □

Acknowledgement. The author thanks to P. Vojtáš for valuable discussions and many advices during preparation of this article.

References

1. K. R. Apt. Logic programming. In: J. van Leeuwen (Ed.) *Handbook of Theoretical Computer Science. Vol. B, Formal methods and semantics*, Elsevier, 1990, pp. 493–574.
2. S. Gottwald. *Mehrwertige Logik*. Akademie Verlag, Berlin, 1988.
3. P. Hájek. Fuzzy logic and arithmetical hierarchy I. *Fuzzy Sets and Systems 73* (1995) 359–363.
4. P. Hájek. Fuzzy logic from the logical point of view. In: M. Bartošek, J. Staudek and J. Wiedermann, (Eds.) *SOFSEM'95: Theory and Practice of Informatics, Lecture Notes in Comp. Sci., 1012*, Springer, Berlin, 1995, pp. 31–49.
5. P. Hájek. Fuzzy logic and arithmetical hierarchy II. To appear in *Studia Logica*, 1997.
6. P. Hájek, L. Godo and F. Esteva. A complete many-valued logic with product-conjunction. Preprint, 1996.
7. P. Hájek and D. Švejda. A strong theorem for finitely axiomatized fuzzy theories. Submitted to *Proceedings of FSTA'96*, Tatra Mt. Math. Publ.
8. J. W. Lloyd. *Foundation of Logic Programming*. Springer Verlag, Berlin, 1987.
9. V. Novák. On the syntactico-semantical completeness of first-order fuzzy logic I, II. *Kybernetika 26* (1990) 47–26, 134–152.
10. J. Pavelka. On fuzzy logic I, II, III. *Zeitschr. f. Math. Logik und Grundl. der Math. 25* (1979) 45–52, 119–134, 447–464.
11. P. Vojtáš and L. Paulík. Logic programming in RPL and RQL. In: M. Bartošek, J. Staudek and J. Wiedermann, (Eds.) *SOFSEM'95: Theory and Practice of Informatics, Lecture Notes in Comp. Sci., 1012*, Springer, Berlin, 1995, pp. 487–492.
12. P. Vojtáš and L. Paulík. Soundness and completeness of non-classical extended SLD-resolution. In: R. Dyckhoff, H. Herre, and P. Schroeder-Heister, (Eds.) *Extensions of Logic Programming, 5th International Workshop, ELP'96, Lecture Notes in Artificial Intelligence 1050*, Springer, Berlin, 1996, pp. 289–301.

The finite stages of inductive definitions *

Robert F. Stärk**

Department of Computer and Information Science, University of Pennsylvania, Philadelphia, PA 19104

Summary. In general, the least fixed point of a positive elementary inductive definition over the Herbrand universe is Π_1^1 and has no computational meaning. The finite stages, however, are computable, since validity of equality formulas in the Herbrand universe is decidable. We set up a formal system BID for the finite stages of positive elementary inductive definitions over the Herbrand universe and show that the provably total functions of the system are exactly that of Peano arithmetic. The formal system BID contains the so-called inductive extension of a logic program as a special case. This first-order theory can be used to prove termination and correctness properties of pure Prolog programs, since notions like negation-as-failure and left-termination can be turned into positive inductive definitions.

1. Why inductive definitions over the Herbrand universe?

In traditional logic programming, the semantics of a program is always given by the least fixed point of a monotonic operator over the Herbrand universe. The first example is the well-known van Emden-Kowalski operator for definite Horn clause programs in [25]. This operator is defined by a purely existential formula and is therefore continuous. The least fixed point of the operator is recursively enumerable. Moreover, the finite stages of the inductive definition are exactly what is computed by SLD-resolution.

In [11], Fitting has generalized the van Emden-Kowalski operator using three-valued logic to programs which may also contain negation in the bodies of the clauses. Although Fitting's operator is still monotonic it is no longer continuous. It follows from Blair [2] and Kunen [14] that the least fixed point of this operator can be Π_1^1-complete and that the closure ordinal can be ω_1^{CK} even for definite Horn clause programs.

The finite stages of Fitting's operator, however, are decidable and correspond to what is computed by SLDNF-resolution. This has been shown by Kunen in [15] for allowed logic programs and by the author in [21] for mode-correct programs. The class of allowed programs is considered as too restrictive in general. The class of mode-correct programs, however, contains most programs of practical interest, since a programmer has always modes in mind when he writes a program. Moreover, every allowed program is also mode-correct.

---

* This paper is in its final form and no similar paper has been or is being submitted elsewhere.

** Research supported by the Swiss National Science Foundation. This article has been written at the Department of Mathematics, Stanford University.

Even though the use of three-valued logic can be eliminated in Fitting's operator (cf. eg. [13]), the completeness results of [15] and [21] cannot be applied to existing implementations of logic programming like, for example, Prolog. The reason is that these systems use special search-strategies which depend on the order of clauses in the program and on the order of literals in the bodies of clauses. Apt and Pedreschi, however, have observed in [1] that only the order of literals in the bodies is important. They say that most programs used in practice terminate independently of the order of clauses in the program — at least for the intended inputs.

Based on this observation we assign in [22] to a predicate R of a logic program three positive elementary inductive definitions for new relations R^{s}, R^{f} and R^{t}. The relation R^{s} corresponds to Fitting's truth-value true; R^{f} corresponds to the truth-value false; R^{t} expresses left-termination in the sense of Apt and Pedreschi. The corresponding formal system is called the *inductive extension* of a logic programs and can be used for proving termination and correctness properties of Prolog programs (see also [24]).

A natural question is then: What is the proof-theoretic strength of the inductive extension? — We answer this problem below and show that the provably total functions of the inductive extension are exactly those of Peano Arithmetic. As a byproduct we obtain a proof-theoretic proof of a key lemma used in the completeness proofs in [22] and [23]. Our proof-theoretical analysis of the inductive extension in terms of provably total functions is similar to Pohlers' treatment of ID_1 in [7] and [18].

The plan of this article is as follows. After recalling some well-known facts on Clark's equality theory CET in Sect. 2, we present a framework for inductive definitions over the Herbrand universe in Sect. 3 and set up a formal system for such in Sect. 4. Using the number-theoretic ordinal functions of Sect. 5 we embed the formal system into an infinitary sequent calculus in Sect. 6. Partial cut-elimination allows then to consider the positive/negative fragment of the calculus only and to perform an asymmetric interpretation that yields the main results of the article. Sect. 7 finally provides some hints how the general framework for inductive definitions relates to the so-called inductive extension of a logic program and to notions like negation as failure and left-termination.

2. Preliminaries

Let $\mathfrak{L}$ be a set of function symbols. Constants are considered as 0-ary function symbols. We assume that $\mathfrak{L}$ contains at least one constant symbol. Function symbols are denoted by f and g. The terms built up with symbols from $\mathfrak{L}$ are denoted by small letters a, b, s, t with and without subscripts. We write $s = t$ for a formal equations and use $s \equiv t$ to express that s and t are syntactically equal. Clark's equality theory $\mathrm{CET}_{\mathfrak{L}}$ for the language $\mathfrak{L}$ comprises the following axioms (cf. [10]):

1. $x = x$
2. $x = y \to y = x$
3. $x = y \wedge y = z \to x = z$
4. $x_1 = y_1 \wedge \ldots \wedge x_m = y_m \to f(x_1, \ldots, x_m) = f(y_1, \ldots, y_m)$
5. $f(x_1, \ldots, x_m) = f(y_1, \ldots, y_m) \to x_i = y_i$
6. $f(x_1, \ldots, x_m) \neq g(y_1, \ldots, y_n)$ [if $f \not\equiv g$]
7. $x \neq t$ [if $x \in \mathrm{FV}(t)$ and $x \not\equiv t$]

Axiom (5) is called *decomposition axiom*, (6) is called *function clash axiom* and (7) is called *occurs check axiom*. Note that (7) is an axiom scheme. It can not be replaced by finitely many axioms.

Example 2.1. Let $\mathfrak{L} = \{0, s\}$, where 0 is a constant and s is a unary function symbol. Then the decomposition, function clash and occurs check axioms for $\mathfrak{L}$ are:

(a) $s(x) = s(y) \to x = y$,
(b) $s(x) \neq 0$,
(c) $s^n(x) \neq x$ for $n > 0$.

For $\mathfrak{L} = \{nil, cons\}$ we have in $\mathrm{CET}_{\mathfrak{L}}$ among others the following axioms:

(a) $cons(x, y) = cons(u, v) \to x = u$, $cons(x, y) = cons(u, v) \to y = v$,
(b) $cons(x, y) \neq nil$,
(c) $cons(x, y) \neq x$, $cons(x, y) \neq y$, $cons(cons(x, y), z) \neq x$, etc.

The theory $\mathrm{CET}_{\mathfrak{L}}$ is strongly related to unification. In fact, an equivalent axiomatization of $\mathrm{CET}_{\mathfrak{L}}$ consists of the following two schemata for arbitrary terms a, b, s_i, t_i:

1′. $s_1 = t_1 \wedge \ldots \wedge s_n = t_n \to a = b$,
   if $\sigma = \mathrm{mgu}\{s_1 = t_1, \ldots, s_n = t_n\}$ and $a\sigma \equiv b\sigma$,
2′. $\neg(s_1 = t_1 \wedge \ldots \wedge s_n = t_n)$,
   if $\{s_1 = t_1, \ldots, s_n = t_n\}$ is not unifiable.

That (1′) and (2′) follow from (1)–(7) is shown in [10]. Conversely, it is easy to see that any of the axioms (1)–(7) is an instance of (1′) or (2′). Thus the two axiomatizations are equivalent.

We denote by $U_{\mathfrak{L}}$ be the set of all closed terms built up with symbols from $\mathfrak{L}$. The set $U_{\mathfrak{L}}$ is called the *Herbrand universe* of $\mathfrak{L}$. By $\mathfrak{U}_{\mathfrak{L}}$ we denote the algebraic structure with domain $U_{\mathfrak{L}}$ and the free interpretation of the function symbols, i.e. $\mathfrak{U}_{\mathfrak{L}}(f)(t_1, \ldots, t_m) = f(t_1, \ldots, t_m)$ for all function symbols $f \in \mathfrak{L}$ and all terms $t_1, \ldots, t_m \in U_{\mathfrak{L}}$. Equality is interpreted as identity. The structure $\mathfrak{U}_{\mathfrak{L}}$ is a model of $\mathrm{CET}_{\mathfrak{L}}$ and, moreover, every model of $\mathrm{CET}_{\mathfrak{L}}$ contains an isomorphic copy of $\mathfrak{U}_{\mathfrak{L}}$. Sometimes, $\mathfrak{U}_{\mathfrak{L}}$ is called the *standard model* of $\mathrm{CET}_{\mathfrak{L}}$.

Term models of $\mathrm{CET}_{\mathfrak{L}}$ which are non-standard can be obtained by the following construction of [3]. Let $Term_{\mathfrak{L}}$ be the set of all terms of $\mathfrak{L}$ (with variables). Let Ψ be a directed set of substitutions. This means that for all substitutions $\sigma, \tau \in \Psi$ there exists a substitution $\theta \in \Psi$ such that $\sigma \leq \theta$ and $\tau \leq \theta$, where $\sigma \leq \theta$ is defined as $\exists \sigma' (\sigma \circ \sigma' = \theta)$. Define on $Term_{\mathfrak{L}}$ the congruence relation $\sim_\Psi$ by

$$s \sim_\Psi t \;:\Longleftrightarrow\; \exists \sigma \in \Psi\, (s\sigma \equiv t\sigma).$$

Then the term structure $(Term_{\mathfrak{L}}, \sim_\Psi)$ is a model of $\mathrm{CET}_{\mathfrak{L}}$, if equality is interpreted by the relation $\sim_\Psi$. Moreover, every term model of $\mathrm{CET}_{\mathfrak{L}}$ can be obtained in this way.

Models of $\mathrm{CET}_{\mathfrak{L}}$ are also called *locally free algebras*. In [16], Mal'cev gives an algorithm that transforms an arbitrary first-order formula containing the equality symbol only into a normal form over locally free algebras. A slightly more general algorithm with a different normal form is investigated by Shepherdson in [20]. From both normal form algorithms the following proposition can be derived.

Proposition 2.1 (Mal'cev [16], Shepherdson [20]).

1. *Two models of* $\mathrm{CET}_{\mathfrak{L}}$ *are elementary equivalent if, and only if, they have the same number of indecomposable elements.*
2. $\mathrm{CET}_{\mathfrak{L}}$ *is decidable.*
3. $\mathrm{CET}_{\mathfrak{L}}$ *is complete provided that* $\mathfrak{L}$ *is infinite.*
4. *The first-order theory of* $\mathfrak{U}_{\mathfrak{L}}$ *is decidable.*

In this proposition, an element is called indecomposable if it is not in the range of any function. For example, the Herbrand structure $\mathfrak{U}_{\mathfrak{L}}$ has no indecomposable elements. Same number of indecomposable elements means that either both models have the same finite number of indecomposable elements or both models have infinitely many.

If $\mathfrak{L}$ is finite, then $\mathrm{CET}_{\mathfrak{L}}$ is not always complete. Take $\mathfrak{L} = \{c\}$. Then the formula $\forall x\,(x = c)$ is true in some models but false in other models. Though the assumption of an infinite language $\mathfrak{L}$ is technically very useful because it makes $\mathrm{CET}_{\mathfrak{L}}$ a complete theory, it can be an absurd assumption in practice. Therefore we do not make any assumption on whether $\mathfrak{L}$ is finite or infinite in the following.

3. Inductive definitions over the Herbrand universe

We consider simultaneous positive elementary inductive definitions over the Herbrand universe. Our interest in such definitions comes from the declarative analysis of *negation-as-failure* and *left-termination* in logic programming (cf. Sect. 7). Both notions can be turned into positive inductive definitions of the kind we investigate in the following.

Let $R_1, \ldots, R_k$ be k relation symbols each of a given arity. An *operator form* is a formula in the language $\mathfrak{L}(R_1, \ldots, R_k, =)$ which is positive in

$R_1, \ldots, R_k$. More formally, we define Pos to be the set of formulas of the following form:

$$\top \mid \bot \mid s = t \mid s \neq t \mid R_j(\mathbf{t}) \mid A \wedge B \mid A \vee B \mid \forall x\, A \mid \exists x\, A.$$

An operator form $\mathcal{A}[\mathbf{R}, \mathbf{x}]$ is then an element of Pos. By writing $\mathcal{A}[\mathbf{R}, \mathbf{x}]$ we indicate that all the relation symbols of the formula different from the equality symbol are among the list $\mathbf{R} = R_1, \ldots, R_k$ and that all the free variables are among the list $\mathbf{x}$.

Let $\mathcal{A}_j[\mathbf{R}, \mathbf{x}_j]$ be operator forms such that the length of $\mathbf{x}_j$ is equal to the arity of R_j for $j = 1, \ldots, k$. Associated to these operator forms are monotonic operators

$$\Gamma_j : \mathcal{P}(U_{\mathfrak{L}}^{n_1}) \times \ldots \times \mathcal{P}(U_{\mathfrak{L}}^{n_k}) \to \mathcal{P}(U_{\mathfrak{L}}^{n_j}), \quad \text{for } j = 1, \ldots, k,$$

where n_j is the arity of R_j. The operator Γ_j is defined by

$$\Gamma_j(X_1, \ldots, X_k) := \{\, \langle \mathbf{y} \rangle \in U_{\mathfrak{L}}^{n_j} : \mathfrak{U}_{\mathfrak{L}} \models \mathcal{A}[X_1, \ldots, X_k, \mathbf{y}] \,\}.$$

The stages I_j^α of the simultaneous inductive definitions are defined as usual for $j = 1, \ldots, k$ (cf. [17]):

$$I_j^\alpha := \Gamma_j(I_1^{<\alpha}, \ldots, I_k^{<\alpha}), \quad I_j^{<\alpha} := \bigcup_{\beta<\alpha} I_j^\beta, \quad I_j^\infty := \bigcup_{\alpha \in On} I_j^\alpha.$$

It is clear that the relations I_j^∞ are the least fixed points of the operators Γ_j. By $\mathfrak{I}^\alpha$ we denote the structure $\langle U_{\mathfrak{L}}, I_1^\alpha, \ldots, I_k^\alpha \rangle$ and by $\mathfrak{I}^\infty$ the structure $\langle U_{\mathfrak{L}}, I_1^\infty, \ldots, I_k^\infty \rangle$. In the following we will mainly be interested in the finite stages $\mathfrak{I}^n$ for $n < \omega$, since these structures are decidable. This can be seen as follows. By induction on the natural number n we define equality formulas $R_j^{(n)}[\mathbf{x}_j]$ which characterize the nth stage I_j^n in the inductive generation of R_j. For $j = 1, \ldots, k$ and $n \in \mathbb{N}$ we define

1. $R_j^{(0)}[\mathbf{x}_j] :\equiv \bot$,
2. $R_j^{(n+1)}[\mathbf{x}_j] :\equiv \mathcal{A}_j[R_1^{(n)}, \ldots, R_k^{(n)}, \mathbf{x}_j]$.

As usual this notion means that in $\mathcal{A}_j[R_1, \ldots, R_k, \mathbf{x}_j]$ all subformulas of the form $R_i(\mathbf{t})$ are replaced by the formula $R_i^{(n)}[\mathbf{t}]$ for $i = 1, \ldots, k$. Note, that the size of the formulas $R_j^{(n)}[\mathbf{x}_j]$ grows exponentially in n. For arbitrary formulas $A[R_1, \ldots, R_k]$ we define

$$A[R_1, \ldots, R_k]^{(n)} :\equiv A[R_1^{(n)}, \ldots, R_k^{(n)}].$$

The formulas $A^{(n)}$ can also be constructed in a more explicit way. We define by main induction on n and side induction on the size of a formula A equality formulas $\mathrm{E}_n\, A$. Since we later use it only for positive formulas, we restrict the definition to this class of formulas.

$$\begin{array}{llll} \mathrm{E}_0\, R_j(\mathbf{t}) & :\equiv \bot, & \mathrm{E}_n(A \wedge B) & :\equiv \mathrm{E}_n\, A \wedge \mathrm{E}_n\, B, \\ \mathrm{E}_{(n+1)}\, R_j(\mathbf{t}) & :\equiv \mathrm{E}_n\, \mathcal{A}_j[\mathbf{R}, \mathbf{t}], & \mathrm{E}_n(A \vee B) & :\equiv \mathrm{E}_n\, A \vee \mathrm{E}_n\, B, \\ \mathrm{E}_n(s = t) & :\equiv (s = t), & \mathrm{E}_n\, \forall x\, A & :\equiv \forall x\, \mathrm{E}_n\, A, \\ \mathrm{E}_n(s \neq t) & :\equiv (s \neq t), & \mathrm{E}_n\, \exists x\, A & :\equiv \exists x\, \mathrm{E}_n\, A. \end{array}$$

Since $A^{(n)}$ is the same as $\mathrm{E}_n\, A$, we are free to use both notions in the following interchangeably. $A^{(n)}$ is used for expressing that A is true at the nth stage of the inductive definition and $\mathrm{E}_n\, A$ is used if we want to point out that it is an equality formula.

Lemma 3.1. $\mathfrak{U}_{\mathcal{L}} \models A^{(n)} \iff \mathfrak{I}^n \models A.$

From this lemma it follows immediately that the finite stages I_j^n are decidable. To test whether $\langle \mathbf{a} \rangle$ belongs to I_j^n one just has to compute the equality formula $R_j^{(n)}[\mathbf{a}]$ and then to apply the decision procedure of Proposition 2.1 for $\mathfrak{U}_{\mathcal{L}}$.

Lemma 3.2. $\mathrm{E}_n\, A$ *implies* $\mathrm{E}_{n+1}\, A$ *for all* $n \in \mathbb{N}$.

The proofs of both lemmas are standard.

4. The formal system BID

Given a list of relation symbols $\mathbf{R} = R_1, \ldots, R_k$ and operator forms $\mathcal{A}_j[\mathbf{R}, \mathbf{x}_j]$ for $j = 1, \ldots, k$ we define a first-order theory $\mathrm{BID}(R_1, \mathcal{A}_1; \ldots; R_k, \mathcal{A}_k)$. The word BID stands for *basic inductive definitions*. Sometimes we also write $\mathrm{BID}(R_1, \ldots, R_k)$ or just BID for short. The theory BID comprises the axioms of $\mathrm{CET}_{\mathcal{L}}$ and the following two principles for arbitrary formulas $B_1(\mathbf{x}_1), \ldots, B_k(\mathbf{x}_k)$ of the language $\mathcal{L}(R_1, \ldots, R_k, =)$:

$$\forall \mathbf{x}_j\, (\mathcal{A}_j[\mathbf{R}, \mathbf{x}_j] \to R_j(\mathbf{x}_j)) \quad \text{for } j = 1, \ldots, k, \tag{CLS}$$

$$\bigwedge_{j=1}^{k} \forall \mathbf{x}_j\, (\mathcal{A}_j[\mathbf{B}, \mathbf{x}_j] \to B_j(\mathbf{x}_j)) \to \bigwedge_{j=1}^{k} \forall \mathbf{x}_j\, (R_j(\mathbf{x}_j) \to B_j(\mathbf{x}_j)). \tag{MIN}$$

We write $\mathrm{BID} \vdash A$ if the formula A is derivable from $\mathrm{CET}_{\mathcal{L}}$, CLS and MIN in classical predicate logic. The principle CLS expresses that the operator forms are closed under the relations R_j and the principle MIN expresses that the relations R_j are minimal with respect to this property. It is well-known that from CLS and MIN one can derive that the relations R_j are fixed points of the operator forms. Choosing $B_j(\mathbf{x}_j) :\equiv \mathcal{A}_j[\mathbf{R}, \mathbf{x}_j]$ one derives from CLS using the monotonicity property of positive formulas

$$\bigwedge_{j=1}^{k} \forall \mathbf{x}_j\, (\mathcal{A}_j[\mathbf{B}, \mathbf{x}_j] \to \mathcal{A}_j[\mathbf{R}, \mathbf{x}_j]).$$

By an application of MIN one obtains

$$\forall \mathbf{x}_j\,(R_j(\mathbf{x}_j) \to \mathcal{A}_j[\mathbf{R},\mathbf{x}_j]) \quad \text{for } j = 1,\ldots,k. \tag{FIX}$$

Not only these fixed point principles are derivable but also the following equality axioms for relations. These axioms are usually considered as part of predicate logic:

$$x_1 = y_1 \wedge \ldots \wedge x_{n_j} = y_{n_j} \wedge R_j(x_1,\ldots,x_{n_j}) \to R_j(y_1,\ldots,y_{n_j}). \tag{EQ}$$

To see that these formulas are derivable in BID, define for $j = 1,\ldots,k$ the formulas

$$B_j[x_1,\ldots,x_{n_j}] :\equiv \forall y_1,\ldots,y_{n_j}\,(x_1 = y_1 \wedge \ldots \wedge x_{n_j} = y_{n_j} \to R_j(y_1,\ldots,y_{n_j})).$$

Then it is easy to verify that for any positive formula $A[\mathbf{R},x_1,\ldots,x_n]$ we have

$$A[\mathbf{B},x_1,\ldots,x_n] \to \forall y_1,\ldots,y_n\,(x_1 = y_1 \wedge \ldots \wedge x_n = y_n \to A[\mathbf{R},y_1,\ldots,y_n]).$$

Using CLS we obtain

$$\bigwedge_{j=1}^{k} \forall \mathbf{x}_j\,(\mathcal{A}_j[\mathbf{B},\mathbf{x}_j] \to B_j(\mathbf{x}_j)).$$

Now we can apply MIN and obtain the equality axioms (EQ).

The question is now, what is the theory BID? How does it compare to other systems? What is the proof theoretic strength of BID? What are the provably total functions of BID? What is the definition of provably total function in BID at all?

Note that unlike other formal theories for inductive definitions like ID_1 or $\widehat{\mathrm{ID}}_1$ the theory BID does not automatically include Peano Arithmetic. Therefore, in a first step, we show how one can embed PA into a theory BID(*nat*, *add*, *mul*, *leq*) for suitable predicates *nat*, *add*, *mul* and *leq*. We assume that the language $\mathfrak{L}$ contains at least the constant 0 and the successor function s. Instead of writing an operator form explicitly as

$$\mathcal{A}[nat,x] :\equiv (x = 0) \vee \exists y\,(x = s(y) \wedge nat(y))$$

we use Prolog-like notation and assume that it is implicitly given by the following clauses:

$$nat(0).$$
$$nat(s(x)) \leftarrow nat(x).$$

The operator forms for *add*, *mul* and *leq* are given by the following clauses:

$$\begin{aligned}
&add(0, y, y).\\
&add(s(x), y, s(z)) \leftarrow add(x, y, z).\\
&mul(0, y, 0).\\
&mul(s(x), y, z) \leftarrow mul(x, y, u) \land add(u, y, z).\\
&leq(0, y).\\
&leq(s(x), s(y)) \leftarrow leq(x, y).
\end{aligned}$$

It is easy to see that the theory BID(nat, add, mul, leq) proves the induction axiom for natural numbers as a special case of MIN:

$$A(0) \land \forall x\,(nat(x) \land A(x) \to A(s(x))) \to \forall x\,(nat(x) \to A(x)). \tag{4.1}$$

Using (4.1) one can then derive that the relations *add* and *mul* are the graphs of functions, eg.

$$\forall x, y\,(nat(x) \land nat(y) \to \exists! z\,(nat(z) \land add(x, y, z)).$$

Therefore one can embed PA into BID(nat, add, mul, leq) by interpreting universal quantifiers $\forall x\, A$ by $\forall x\,(nat(x) \to A)$ and existential quantifiers $\exists x\, A$ by $\exists x\,(nat(x) \land A)$.

But can we prove more in BID than in PA? — Is it, for example, possible to extend BID(nat, add, mul, leq) by a truth predicate for the structure of natural numbers? Can we inductively define a relation $tr(a, e)$ expressing that the formula a is true in $\mathbf{N}$ under the assignment e? The truth definition could then be used to prove the consistency of PA.

We show that this is not possible, at least not in the obvious way, since a truth definition typically has the following property:

$$tr(\mathit{forall}(x, a), e) \leftrightarrow \forall y\,(nat(y) \to tr(a, cons(sub(x, y), e))). \tag{4.2}$$

The relation *nat* occurs negatively on the right-hand side of (4.2). So we cannot obtain (4.2) directly from an inductive definition.

In an attempt to save the truth definition we define the complement of natural numbers by

$$notnat(x) \leftarrow x \neq 0 \land \forall y\,(x = s(y) \to notnat(y))$$

and write (4.2) as

$$tr(\mathit{forall}(x, a), e) \leftrightarrow \forall y\,(notnat(y) \lor tr(a, cons(sub(x, y), e))). \tag{4.3}$$

This equivalence can be obtained from an inductive definition, since the right-hand side is positive. But there is no way to obtain (4.2) from (4.3), since although BID proves

$$\forall x\, \neg(nat(x) \land notnat(x)), \tag{UNIQ}$$

it does not prove

$$\forall x\,(nat(x) \lor notnat(x)). \tag{TOT}$$

And for the applications we have in mind it is essential that TOT is not provable in BID (see Sect. 7).

One of the important features of BID is that if it proves a positive formula A then there exists a natural number $n \in \mathbb{N}$ such that $\mathrm{CET} \vdash \mathrm{E}_n A$ (see Theorem 6.2). Thus if TOT would be provable in CET then there would exist an $n \in \mathbb{N}$ such that TOT would already be true at the finite stage $\mathfrak{J}^n$. This, however, is not the case.

What about the provably total functions of BID(*nat*, *add*, *mul*, *leq*)? What about the provably total functions of BID in general? — First we have to define what we mean by provably total functions of BID. We assume that the language $\mathfrak{L}$ contains the constant 0 and the successor function s. Moreover, we assume that we have a unary predicate *nat* defining the natural numbers and a distinguished binary relation symbol R. We indicate this by writing $\mathrm{BID}(nat, R, \ldots)$. For $n \in \mathbb{N}$ we denote by $\overline{n}$ the term

$$\underbrace{s(\cdots(s}_{n \text{ times}}(0)\cdots).$$

Definition 4.1. A function $f: \mathbb{N} \to \mathbb{N}$ is called *provably total* in $\mathrm{BID}(nat, R, \ldots)$, if

1. $\mathrm{BID} \vdash R(\overline{m}, \overline{n})$ for all $m, n \in \mathbb{N}$ such that $f(m) = n$,
2. $\mathrm{BID} \vdash \forall x\,(nat(x) \to \exists! y\,(nat(y) \land R(x, y)))$.

This definition can be justified as follows. Let $f: \mathbb{N} \to \mathbb{N}$ be provably total in $\mathrm{BID}(nat, R, \ldots)$. Then Theorem 6.3 below says that there exists an ordinal $\alpha < \varepsilon_0$ and a function $F: \mathbb{N} \to \mathbb{N}$ which is α-recursive such that

$$\mathrm{CET} \vdash \forall x\,(\mathrm{E}_m\, nat(x) \to \exists y\,(\mathrm{E}_{F(m)}\, nat(y) \land \mathrm{E}_{F(m)}\, R(x, y))) \quad \text{for all } m \in \mathbb{N}.$$

Let $m \in \mathbb{N}$. Then $\mathrm{CET} \vdash \mathrm{E}_m\, nat(\overline{m})$ and thus

$$\mathrm{CET} \vdash \exists y\,(\mathrm{E}_{F(m)}\, nat(y) \land \mathrm{E}_{F(m)}\, R(\overline{m}, y)).$$

Since $\mathfrak{U}_{\mathfrak{L}}$ is a model of CET, there exists a closed term t such that

$$\mathfrak{U}_{\mathfrak{L}} \models \mathrm{E}_{F(m)}\, nat(t) \quad \text{and} \quad \mathfrak{U}_{\mathfrak{L}} \models \mathrm{E}_{F(m)}\, R(\overline{m}, t).$$

This implies that $nat(t)$ is true at stage $\mathfrak{J}^{F(m)}$ and therefore there exists an $n \leq F(m)$ such that $t = \overline{n}$. Thus we have

$$\mathfrak{U}_{\mathfrak{L}} \models \mathrm{E}_{F(m)}\, R(\overline{m}, \overline{n}) \quad \text{and} \quad \mathfrak{J}^{F(m)} \models R(\overline{m}, \overline{n}).$$

Since $\mathfrak{J}^{\infty} \models R(\overline{m}, \overline{n})$ and $\mathfrak{J}^{\infty}$ is a model of BID it follows that $f(m) = n$. Thus we have $f(m) \leq F(m)$ for all $m \in \mathbb{N}$. Moreover, given $m \in \mathbb{N}$ we can compute the value $f(m)$ in the following way. It is the least $n \leq F(m)$ such that

$$\mathfrak{U}_{\mathfrak{L}} \models \mathrm{E}_{F(m)}\, R(\overline{m}, \overline{n}).$$

Since the truth of equality formulas in the Herbrand universe is primitive recursively decidable (cf. Proposition 2.1), if follows that the function f is α-recursive. Thus the provably total functions of BID($nat, R, \ldots$) are exactly those computable functions which are provably total in Peano Arithmetic. What remains to show is the above mentioned Theorem 6.3.

5. Some remarks on ordinals less than ε_0

The reader familiar with the relations $<_n$ of Weiermann [26] can skip this section. The purpose of this section is to introduce number-theoretic functions $n \mapsto \theta_\alpha(n)$ which will later be used in an asymmetric interpretation.

Let $\varepsilon_0 := \min\{\xi : \omega^\xi = \xi\}$. Ordinals less than ε_0 are denoted by α, β, γ. Finite ordinals are denoted by i, j, m, n. The natural sum of α and β is denoted by $\alpha \# \beta$. The norm of an ordinal $\alpha < \varepsilon_0$ is the number of occurrences of ω in its Cantor normal form. The norm $\mathrm{N}: \varepsilon_0 \to \mathbb{N}$ is defined by

1. $\mathrm{N}(0) := 0$,
2. $\mathrm{N}(\omega^{\alpha_1} + \ldots + \omega^{\alpha_n}) := n + \mathrm{N}(\alpha_1) + \ldots + \mathrm{N}(\alpha_n)$, if $\alpha_1 \geq \ldots \geq \alpha_n$.

We write $\mathrm{N}\alpha$ for $\mathrm{N}(\alpha)$. The norm N has the following properties:

1. $\mathrm{N}\omega = 2$,
2. $\mathrm{N}\omega^\alpha = 1 + \mathrm{N}\alpha$,
3. $\mathrm{N}(\alpha \# \beta) = \mathrm{N}\alpha + \mathrm{N}\beta$.
4. $N(\alpha + n) = \mathrm{N}\alpha + n$ and $\mathrm{N}n = n$.

For each $n \in \mathbb{N}$ there are only finitely many ordinals $\alpha < \varepsilon_0$ with $\mathrm{N}\alpha \leq n$.

In the definition of the relations $<_n$ Weiermann uses (in a more general context) a recursive function $\Phi: \mathbb{N} \to \mathbb{N}$ which is a variant of the Ackermann function. For our purposes, however, it is sufficient to require the following:

$$\Phi(n) + \Phi(n) + 2 \leq \Phi(n+1) \quad \text{for all } n \in \mathbb{N}. \qquad (*)$$

For example, take $\Phi(n) := 3^{n+1}$. If Φ satisfies $(*)$, then it also has the following properties:

1. $\Phi(n) < \Phi(n+1)$ and $n \leq \Phi(n)$,
2. $m + \Phi(n) \leq \Phi(m+n)$,
3. $m \cdot n \leq \Phi(m+n)$.

Definition 5.1 (Weiermann [26]).

1. $\alpha <_n^1 \beta :\Longleftrightarrow \alpha < \beta$ and $\mathrm{N}\alpha \leq \Phi(\mathrm{N}\beta + n)$.
2. Let $<_n$ be the transitive closure of $<_n^1$.

On the interval $[\alpha, \alpha + \omega)$ the relations $<_n$ agree with $<$. If $i < j$ then $\alpha + i <_n \alpha + j$. Note, that $\alpha <_m \beta$ implies $\alpha <_n \beta$ for all n with $m \leq n$. Moreover, the usual operations on ordinals are monotonic with respect to the relations $<_n$.

Lemma 5.1 (**Weiermann [26]**).

1. $\alpha <_n \beta \implies \omega^\alpha <_n \omega^\beta$,
2. $\alpha <_n \beta \implies \alpha \# \gamma <_n \beta \# \gamma$,
3. $\alpha <_n \gamma$ *and* $\beta <_n \gamma \implies \omega^\alpha \# \omega^\beta <_n \omega^\gamma$,
4. $\alpha <_n \beta \implies \alpha + n + 1 <_0 \beta + n + 1$,
5. $m \cdot n <_n \omega + m$.

Proof. We prove assertions (1)–(5) first for the one step relations $<^1_n$. The versions with $<_n$ follow then immediately, since $<_n$ is the transitive closure of $<^1_n$. For (1), assume $\alpha <^1_n \beta$. By definition, we have $\mathrm{N}\,\alpha \leq \Phi(\mathrm{N}\,\beta + n)$. Thus,

$$\mathrm{N}\,\omega^\alpha = 1 + \mathrm{N}\,\alpha \leq 1 + \Phi(\mathrm{N}\,\beta + n) \leq \Phi(\mathrm{N}\,\beta + n + 1) = \Phi(\mathrm{N}\,\omega^\beta + n).$$

For (2), assume $\alpha <^1_n \beta$. Then we have

$$\mathrm{N}(\alpha\#\gamma) = \mathrm{N}\,\alpha + \mathrm{N}\,\gamma \leq \Phi(\mathrm{N}\,\beta + n) + \mathrm{N}\,\gamma \leq \Phi(\mathrm{N}\,\beta + n + \mathrm{N}\,\gamma) = \Phi(\mathrm{N}(\beta\#\gamma) + n).$$

For (3), assume $\alpha <^1_n \gamma$ and $\beta <^1_n \gamma$. Then we have

$$\mathrm{N}(\omega^\alpha \# \omega^\beta) \leq \Phi(\mathrm{N}\,\gamma + n) + \Phi(\mathrm{N}\,\gamma + n) + 2 \leq \Phi(\mathrm{N}\,\gamma + n + 1) = \Phi(\mathrm{N}\,\omega^\gamma + n).$$

For (4), assume $\alpha <^1_n \beta$. By definition, we have $\mathrm{N}\,\alpha \leq \Phi(\mathrm{N}\,\beta + n)$. Thus,

$$\mathrm{N}(\alpha + n + 1) \leq \Phi(\mathrm{N}\,\beta + n) + n + 1 \leq \Phi(\mathrm{N}\,\beta + n + 1) = \Phi(\mathrm{N}(\beta + n + 1)).$$

In (5), we have $\mathrm{N}(m \cdot n) = m \cdot n \leq \Phi(m+n) \leq \Phi(2+m+n) = \Phi(\mathrm{N}(\omega+m)+n)$.

□

Since the set $\{\beta : \beta <^1_n \alpha\}$ is finite for each $\alpha < \varepsilon_0$, one can define the following functions $\theta_\alpha : \mathbf{N} \to \mathbf{N}$ for $\alpha < \varepsilon_0$.

Definition 5.2. $\theta_\alpha(n) := \max\big(\{n+1\} \cup \{\, \theta_\beta(\theta_\gamma(n)) \mid \beta <^1_n \alpha,\ \gamma <^1_n \alpha \,\}\big)$.

Similar definitions can be found in [4] and [6].

Lemma 5.2. *The functions θ_α have the following properties:*

1. $n < \theta_\alpha(n)$,
2. $\theta_\alpha(n) < \theta_\alpha(n+1)$,
3. $\alpha <_m \beta \implies \theta_\alpha(m+n) < \theta_\beta(m+n)$,
4. $\alpha <_0 \gamma$ *and* $\beta <_0 \gamma \implies \theta_\alpha(\theta_\beta(n)) \leq \theta_\gamma(n)$.

Proof. (1) is trivial. (2) is proved by induction on α. If $\theta_\alpha(n) = n + 1$ then the inequality is obvious. Otherwise there exist $\beta <^1_n \alpha$ and $\gamma <^1_n \alpha$ such that $\theta_\alpha(n) = \theta_\beta(\theta_\gamma(n))$. By the induction hypothesis, we obtain $\theta_\gamma(n) < \theta_\gamma(n+1)$ and $\theta_\beta(\theta_\gamma(n)) < \theta_\beta(\theta_\gamma(n+1))$. Thus $\theta_\alpha(n) < \theta_\alpha(n+1)$, since $\beta <^1_{n+1} \alpha$ and $\gamma <^1_{n+1} \alpha$.

Assertion (3) is first shown for the one step relation $<^1_m$. Assume that $\alpha <^1_m \beta$. By (1), $\theta_\alpha(m+n) < \theta_\alpha(\theta_\alpha(m+n))$ and thus $\theta_\alpha(m+n) < \theta_\beta(m+n)$, since $\alpha <^1_{m+n} \beta$. The generatlization to $<_m$ now follows by transitivity.

In (4), assume that $\alpha <_0 \gamma$ and $\beta <_0 \gamma$. Then there exist α' and β' such that $\alpha \leq_0 \alpha' <^1_0 \gamma$ and $\beta \leq_0 \beta' <^1_0 \gamma$. Since $\alpha' <^1_n \gamma$ and $\beta' <^1_n \gamma$, it follows by (2), (3) and the definition of θ_γ that $\theta_\alpha(\theta_\beta(n)) \leq \theta_{\alpha'}(\theta_{\beta'}(n)) \leq \theta_\gamma(n)$. □

6. The infinitary system BID$_\infty$

The idea is to replace the induction scheme MIN of BID by the following infinitary rules for $j = 1, \ldots, k$. Every rule has countably many hypotheses:

$$\frac{R_j^{(n)}[\mathbf{t}] \to A \quad \text{for all } n \in \mathbb{N}}{R_j(\mathbf{t}) \to A} \qquad (\omega)$$

Similar rules are used by Cantini in [9] in a different context. Note that the nth premise grows exponentially in the size of n. The formula A is arbitrary.

How can we prove the induction scheme MIN with this rule? — Assume that $B_j(\mathbf{x}_j)$ are given formulas, $\mathbf{B} = B_1, \ldots, B_k$ and that we have

$$\forall \mathbf{x}_j \left(\mathcal{A}_j[\mathbf{B}, \mathbf{x}_j] \to B_j(\mathbf{x}_j)\right) \quad \text{for } j = 1, \ldots, k. \qquad (6.1)$$

We have to show

$$\forall \mathbf{x}_j \left(R_j(\mathbf{x}_j) \to B_j(\mathbf{x}_j)\right) \quad \text{for } j = 1, \ldots, k. \qquad (6.2)$$

In order that we can apply (ω) for proving (6.2) we show, by induction on n, that

$$\forall \mathbf{x}_j \left(R_j^{(n)}[\mathbf{x}_j] \to B_j(\mathbf{x}_j)\right) \quad \text{for } j = 1, \ldots, k \qquad (6.3)$$

is derivable from (6.1). For $n = 0$ this is trivial, since $R_j^{(0)}[\mathbf{x}_j]$ is the constant $\perp$. Under assumption of (6.3), since $R_j^{(n+1)}[\mathbf{x}_j]$ is the formula $\mathcal{A}_j[R_1^{(n)}, \ldots, R_k^{(n)}, \mathbf{x}_j]$ and positive formulas are monotonic, we obtain

$$R_j^{(n+1)}[\mathbf{x}_j] \to \mathcal{A}_j[\mathbf{B}, \mathbf{x}_j] \quad \text{for } j = 1, \ldots, k.$$

Now we can apply (6.1) and obtain

$$\forall \mathbf{x}_j \left(R_j^{(n+1)}[\mathbf{x}_j] \to B_j(\mathbf{x}_j)\right) \quad \text{for } j = 1, \ldots, k.$$

This is (6.3) for $n + 1$. Hence MIN is proved using rule (ω). We will treat this informal argument below in more detail taking care of the exact length of the derivations.

The infinitary system BID$_\infty$ is formulated in a Tait calculus. Sequents Γ and Δ are finite sets of formulas. As usual Γ, Δ stands for $\Gamma \cup \Delta$ and Γ, A for $\Gamma \cup \{A\}$. Negation is defined by DeMorgan's laws. This requires that we have also complementary relation symbols $\neq, \overline{R}_1, \ldots, \overline{R}_k$. Implication $A \to B$ is defined as $\neg A \vee B$. We define Neg to be the set of formulas of the following form:

$$\top \mid \bot \mid s = t \mid s \neq t \mid \overline{R}_j(\mathbf{t}) \mid A \wedge B \mid A \vee B \mid \forall x\, A \mid \exists x\, A.$$

The length of a formula A is the number of $\wedge$, $\vee$, $\forall$ and $\exists$ occurring in A. It is denoted by $|A|$. The rank of a formula is defined in such a way that that the rank of purely positive or purely negative formulas is zero. The rank is used to measure cut formulas.

1. $\mathrm{rk}(A) := 0$, if $A \in \mathrm{Pos} \cup \mathrm{Neg}$,
2. $\mathrm{rk}(A * B) := \max(\mathrm{rk}(A), \mathrm{rk}(B)) + 1$, if $(A * B) \notin \mathrm{Pos} \cup \mathrm{Neg}$ and $* \in \{\wedge, \vee\}$,
3. $\mathrm{rk}(Qx\, A) := \mathrm{rk}(A) + 1$, if $A \notin \mathrm{Pos} \cup \mathrm{Neg}$ and $Q \in \{\forall, \exists\}$.

The system BID_∞ is given by the derivation relation $\vdash^{\alpha}_{r} \Gamma$. This relation means that the sequent Γ is derivable with length α and cut rank r. The assignment of ordinals to proofs, however, is with respect to the relations $<_n$ and not with respect to the usual ordering relation $<$ for ordinal numbers. This assignment is due to Weiermann [26] and based on the principle of *local predicativity* of Pohlers [19] and Buchholz [5]. The axiom sequents of the infinitary system are:

1. $\top$
2. $s_1 \neq t_1, \ldots, s_n \neq t_n, a = b$
   if $\sigma = \mathrm{mgu}\{s_1 = t_1, \ldots, s_n = t_n\}$ and $a\sigma \equiv b\sigma$,
3. $s_1 \neq t_1, \ldots, s_n \neq t_n$
   if $\{s_1 = t_1, \ldots, s_n = t_n\}$ is not unifiable,
4. $\overline{R}_j(\mathbf{t}), R_j(\mathbf{t})$

The relation $\mathrm{BID}_\infty \vdash^{\alpha}_{r} \Gamma$ (or $\vdash^{\alpha}_{r} \Gamma$ for short) is defined by induction on the ordinal $\alpha < \varepsilon_0$.

Definition 6.1. $\vdash^{\alpha}_{r} \Gamma$, if

(A) $\Delta \subseteq \Gamma$ and Δ is an axiom sequent; or
($\wedge$) $(A \wedge B) \in \Gamma$ and $\vdash^{\alpha_1}_{r} \Gamma, A$ and $\vdash^{\alpha_2}_{r} \Gamma, B$ and $\alpha_1 <_0 \alpha$ and $\alpha_2 <_0 \alpha$; or
($\vee$) $(A \vee B) \in \Gamma$ and $\left(\vdash^{\alpha_1}_{r} \Gamma, A \text{ and } \alpha_1 <_0 \alpha\right)$ or $\left(\vdash^{\alpha_2}_{r} \Gamma, B \text{ and } \alpha_2 <_0 \alpha\right)$; or
($\forall$) $\forall x\, A(x) \in \Gamma$ and $\vdash^{\beta}_{r} \Gamma, A(u)$ and $\beta <_0 \alpha$ and $u \notin \mathrm{FV}(\Gamma)$; or
($\exists$) $\exists x\, A(x) \in \Gamma$ and $\vdash^{\beta}_{r} \Gamma, A(t)$ and $\beta <_0 \alpha$; or
(R) $R_j(\mathbf{t}) \in \Gamma$ and $\vdash^{\beta}_{r} \Gamma, A_j[\mathbf{R}, \mathbf{t}]$ and $\beta <_0 \alpha$; or
($\overline{R}$) $\overline{R}_j(\mathbf{t}) \in \Gamma$ and $\forall n \in \mathbf{N}\left(\vdash^{\alpha_n}_{r} \Gamma, \neg R_j^{(n)}[\mathbf{t}] \text{ and } \alpha_n <_n \alpha\right)$; or
(C) $\vdash^{\alpha_1}_{r} \Gamma, A$ and $\vdash^{\alpha_2}_{r} \Gamma, \neg A$ and $\alpha_1 <_0 \alpha$ and $\alpha_2 <_0 \alpha$ and $\mathrm{rk}(A) < r$.

Note, that the ordinal of a premise must be less in the sense of $<_0$ than the the ordinal of the conclusion except in rule ($\overline{R}$). There, the ordinal of the nth premise must be less in the sense of $<_n$.

Lemma 6.1. *1. Substitution: If* $\vdash^{\alpha}_{r} \Gamma(u)$ *then* $\vdash^{\alpha}_{r} \Gamma(t)$.

2. *Weakening: If* $\vdash^{\alpha}_{r} \Gamma$ *and* $\alpha \leq_0 \alpha'$ *and* $r \leq r'$ *then* $\vdash^{\alpha'}_{r'} \Gamma, \Delta$.
3. *Inversion of* $\wedge$*: If* $\vdash^{\alpha}_{r} \Gamma, A \wedge B$ *then* $\vdash^{\alpha}_{r} \Gamma, A$ *and* $\vdash^{\alpha}_{r} \Gamma, B$.
4. *Inversion of* $\forall$*: If* $\vdash^{\alpha}_{r} \Gamma, \forall x\, A(x)$ *then* $\vdash^{\alpha}_{r} \Gamma, A(t)$ *for all terms* t.
5. *Inversion of* $\vee$*: If* $\vdash^{\alpha}_{r} \Gamma, A \vee B$ *then* $\vdash^{\alpha}_{r} \Gamma, A, B$.

Proof. By induction on α. □

Lemma 6.2. *(Reduction) If* $\vdash^{\alpha}_{r} \Gamma, A$ *and* $\vdash^{\beta}_{r} \Delta, \neg A$ *and* $1 \leq \mathrm{rk}(A) \leq r$ *then* $\vdash^{\alpha \# \beta}_{r} \Gamma, \Delta$.

Proof. Since $\alpha \# \beta = \beta \# \alpha$ we can assume that A is either a disjunction or an existentially quantified formula. A cannot be atomic, since atomic formulas are in $\mathrm{Pos} \cup \mathrm{Neg}$ and therefore have rank 0. The formula $\neg A$ is then a conjunction or a universally quantified formula and we can apply inversion to it. The lemma is proved by induction on α. □

Lemma 6.3. *(Partial cut-elimination) If* $\vdash^{\alpha}_{r+2} \Gamma$ *then* $\vdash^{\omega^{\alpha}}_{r+1} \Gamma$.

Proof. By induction on α. □

For a sequent $\Gamma \subseteq \mathrm{Pos} \cup \mathrm{Neg}$ which consists of positive or negative formulas only and $m, n \in \mathrm{N}$ we define $\Gamma[m,n]$ to be the equality formula

$$\bigwedge_{A \in \Gamma \cap \mathrm{Neg}} \mathrm{E}_m \neg A \rightarrow \bigvee_{A \in \Gamma \cap \mathrm{Pos}} \mathrm{E}_n\, A.$$

Similar asymmetric interpretations are used in [8] and [12]. Note, that the interpretation has the following monotonicity property. If $m' \leq m$ and $n \leq n'$ then $\Gamma[m,n]$ implies $\Gamma[m',n']$. Thus $\Gamma[m,n]$ is monotonically decreasing in its first argument and monotonically increasing in its second argument.

Theorem 6.1. *If* $\vdash^{\alpha}_{1} \Gamma$ *and* $\Gamma \subseteq \mathrm{Pos} \cup \mathrm{Neg}$ *then* $\mathrm{CET} \vdash \Gamma[m, \theta_\alpha(m)]$ *for all* $m \in \mathrm{N}$.

Proof. Case (A). If $\{\overline{R}_j(\mathbf{t}), R_j(\mathbf{t})\} \subseteq \Gamma$ then $\Gamma[m, \theta_\alpha(m)]$ is provable in CET, since m is less than $\theta_\alpha(m)$ and hence $\mathrm{E}_m\, R_j(\mathbf{t})$ implies $\mathrm{E}_{\theta_\alpha(m)}\, R_j(\mathbf{t})$. If $\Delta \subseteq \Gamma$ and Δ is an axiom sequent corresponding to an axiom of CET then $\Delta[m, \theta_\alpha(m)]$ is equivalent to $\bigvee \Delta$ and thus $\Gamma[m, \theta_\alpha(m)]$ is provable in CET.

Case ($\wedge$). Assume that $(A \wedge B) \in \Gamma$, $\alpha_1 <_0 \alpha$, $\alpha_2 <_0 \alpha$,

$$\mathrm{CET} \vdash \Gamma, A[m, \theta_{\alpha_1}(m)] \quad \text{and} \quad \mathrm{CET} \vdash \Gamma, B[m, \theta_{\alpha_2}(m)] \quad \text{for all } m \in \mathrm{N}.$$

Since $\theta_{\alpha_i}(m) \leq \theta_\alpha(m)$ for $i = 1, 2$ we also have

$$\mathrm{CET} \vdash \Gamma, A[m, \theta_\alpha(m)] \quad \text{and} \quad \mathrm{CET} \vdash \Gamma, B[m, \theta_\alpha(m)].$$

If $(A \wedge B) \in \mathrm{Pos}$ then $A \in \mathrm{Pos}$, $B \in \mathrm{Pos}$ and $\mathrm{E}_{\theta_\alpha(m)}(A \wedge B)$ is the formula $\mathrm{E}_{\theta_\alpha(m)}\, A \wedge \mathrm{E}_{\theta_\alpha(m)}\, B$. If $(A \wedge B) \in \mathrm{Neg}$ then $A \in \mathrm{Neg}$, $B \in \mathrm{Neg}$ and

$E_m(\neg(A \wedge B))$ is $E_m(\neg A) \vee E_m(\neg B)$. In both cases we obtain that $\Gamma[m, \theta_\alpha(m)]$ is provable in CET for all $m \in \mathbb{N}$.

Case ($\vee$). This case goes similar to the previous one.

Case ($\forall$). In this case one needs that $\mathrm{FV}(\Gamma[m,n]) \subseteq \mathrm{FV}(\Gamma)$.

Case ($\exists$). In this case one needs that $(E_m\, A)_u[t]$ is the same as $E_m(A_u[t])$.

Case (R). Assume that $R_j(\mathbf{t}) \in \Gamma$, $\beta <_0 \alpha$ and

$$\mathrm{CET} \vdash (\Gamma, \mathcal{A}_j[\mathbf{R}, \mathbf{t}])[m, \theta_\beta(m)] \quad \text{for all } m \in \mathbb{N}.$$

The formula $\mathcal{A}_j[\mathbf{R}, \mathbf{t}]$ is positive, and since $E_{\theta_\beta(m)}\, \mathcal{A}_j[\mathbf{R}, \mathbf{t}]$ is the same as $E_{\theta_\beta(m)+1}\, R_j(\mathbf{t})$ and $\theta_\beta(m) + 1 \leq \theta_\alpha(m)$, we obtain that $\Gamma[m, \theta_\alpha(m)]$ is provable in CET.

Case ($\overline{R}$). Assume that $\overline{R}_j(\mathbf{t}) \in \Gamma$, $\alpha_n <_n \alpha$ for all $n \in \mathbb{N}$ and

$$\mathrm{CET} \vdash \Gamma, \neg E_n\, R_j(\mathbf{t})[m, \theta_{\alpha_n}(m)] \quad \text{for all } m, n \in \mathbb{N}.$$

In the special case $m = n$ we have

$$\mathrm{CET} \vdash \Gamma, \neg E_m\, R_j(\mathbf{t})[m, \theta_{\alpha_m}(m)].$$

Since $\theta_{\alpha_m}(m) \leq \theta_\alpha(m)$, we obtain that

$$\mathrm{CET} \vdash \Gamma, \neg E_m\, R_j(\mathbf{t})[m, \theta_\alpha(m)]$$

and thus $\mathrm{CET} \vdash \Gamma[m, \theta_\alpha(m)]$.

Case (C). Assume that $\mathrm{rk}(A) < 1$, $\alpha_1 <_0 \alpha$, $\alpha_2 <_0 \alpha$ and

$$\mathrm{CET} \vdash \Gamma, A[m, \theta_{\alpha_1}(m)] \quad \text{and} \quad \mathrm{CET} \vdash \Gamma, \neg A[m, \theta_{\alpha_2}(m)] \quad \text{for all } m \in \mathbb{N}.$$

Since $\mathrm{rk}(A) = 0$, we obtain that $A \in \mathrm{Pos} \cup \mathrm{Neg}$. Without loss of generality we can assume that $A \in \mathrm{Pos}$. Let $m \in \mathbb{N}$. Then we have

$$\mathrm{CET} \vdash \Gamma, A[m, \theta_{\alpha_1}(m)] \quad \text{and} \quad \mathrm{CET} \vdash \Gamma, \neg A[\theta_{\alpha_1}(m), \theta_{\alpha_2}(\theta_{\alpha_1}(m))].$$

From this we obtain $\mathrm{CET} \vdash \Gamma[m, \theta_{\alpha_1}(m)] \vee \Gamma[\theta_{\alpha_1}(m), \theta_{\alpha_2}(\theta_{\alpha_1}(m))]$. Since $m \leq \theta_{\alpha_1}(m)$, $\theta_{\alpha_1}(m) \leq \theta_\alpha(m)$ and $\theta_{\alpha_2}(\theta_{\alpha_1}(m)) \leq \theta_\alpha(m)$, by the monotonicity property of the asymmetric interpretation we obtain that $\Gamma[m, \theta_\alpha(m)]$ is provable in CET. □

In the context of finitary systems Jäger proves in [12] a similar theorem with $\Gamma[m, m+2^\alpha]$. Cantini uses in [8] an asymmetric interpretation $\Gamma[m, F(\mathcal{D}, m)]$ where F is an effective function and $\mathcal{D}$ is the code of an infinitary derivation. Also Pohlers Collapsing Lemma 43 in [19] has to be mentioned here.

Corollary 6.1. *If* $\vdash^{\alpha}_{1} A$ *and* $A \in \mathrm{Pos}$ *then* $\mathrm{CET} \vdash E_{\theta_\alpha(0)}\, A$.

Corollary 6.2. *If* $\vdash^{\alpha}_{1} \forall x\, (R(x) \to \exists y\, S(x, y))$ *then*

$$\mathrm{CET} \vdash \forall x \left(E_m\, R(x) \to \exists y\; E_{\theta_\alpha(m)}\, S(x, y)\right)$$

for all $m \in \mathbb{N}$.

Proof. Assume that $\vdash^{\alpha}_{1} \forall x\,(R(x) \rightarrow \exists y\, S(x,y))$. By inversion of $\forall$ and $\vee$, we obtain that $\vdash^{\alpha}_{1} \overline{R}(u), \exists y\, S(u,y)$. Since $\overline{R}(u) \in$ Neg and $\exists y\, S(u,y) \in$ Pos we can apply the asymmetric interpretation and obtain that the formula

$$\mathrm{E}_m\, R(u) \rightarrow \mathrm{E}_{\theta_\alpha(m)}\, \exists y\, S(u,y)$$

is provable in CET. □

What remains to show is that BID can be embedded into the infinitary system BID_∞. In a first step we show that every instance of MIN is cut-free provable in BID_∞ with length $\omega + m$ for some $m \in \mathbb{N}$ (cf. Generalized Induction Theorem 28.10 in [18]).

Lemma 6.4. *If A is an instance of MIN then there exists an $m \in \mathbb{N}$ such that $\vdash^{\omega+m}_{0} A$.*

Proof. Let $B_j(\mathbf{x}_j)$ be formulas for $j = 1, \ldots, k$ and $\mathbf{B} = B_1, \ldots, B_k$. Let $cl(\mathbf{B})$ be the formula

$$\bigwedge_{j=1}^{k} \forall \mathbf{x}_j\, (\mathcal{A}_j[\mathbf{B}, \mathbf{x}_j] \rightarrow B_j(\mathbf{x}_j)).$$

We claim that there exists a constant $c \in \mathbb{N}$ such that for all $n \in \mathbb{N}$ and all terms $\mathbf{t}$

$$\vdash^{c\cdot n}_{0} \neg cl(\mathbf{B}), \neg R^{(n)}_j[\mathbf{t}], B_j(\mathbf{t}) \quad \text{for } j = 1, \ldots, k. \tag{6.4}$$

Let $c := 2c_1 + 2c_2 + c_3 + k + 1$, where

$$c_1 := \max_{1 \le j \le k} |\mathcal{A}_j|, \quad c_2 := \max_{1 \le j \le k} |B_j|, \quad c_3 := \max_{1 \le j \le k} arity(R_j).$$

We show (6.4) by induction on n. If $n = 0$ then (6.4) holds, since $\neg R^{(0)}_j[\mathbf{t}]$ is the constant $\top$. In the induction step we obtain from (6.4), since $\mathcal{A}_j[\mathbf{R}, \mathbf{t}]$ is positive in $\mathbf{R}$,

$$\vdash^{c\cdot n+2c_1}_{0} \neg cl(\mathbf{B}), \neg \mathcal{A}_j[R^{(n)}_1, \ldots, R^{(n)}_k, \mathbf{t}], \mathcal{A}_j[B_1, \ldots, B_k, \mathbf{t}].$$

Since $\vdash^{2c_2}_{0} \neg B_j(\mathbf{t}), B_j(\mathbf{t})$, we obtain

$$\vdash^{c\cdot n+2c_1+2c_2+1}_{0} \neg cl(\mathbf{B}), \neg R^{(n+1)}_j[\mathbf{t}], B_j(\mathbf{t}), \mathcal{A}_j[\mathbf{B}, \mathbf{t}] \wedge \neg B_j(\mathbf{t}).$$

After several applications of ($\exists$) we obtain

$$\vdash^{c\cdot n+2c_1+2c_2+c_3+1}_{0} \neg cl(\mathbf{B}), \neg R^{(n+1)}_j[\mathbf{t}], B_j(\mathbf{t}), \exists \mathbf{x}_j\, (\mathcal{A}_j[\mathbf{B}, \mathbf{x}_j] \wedge \neg B_j(\mathbf{x}_j))$$

and finally after several applications of ($\vee$)

$$\vdash^{c\cdot n+2c_1+2c_2+c_3+k+1}_{0} \neg cl(\mathbf{B}), \neg R^{(n+1)}_j[\mathbf{t}], B_j(\mathbf{t}).$$

Thus we have (6.4) for $n + 1$. Since $c \cdot n <_n \omega + c$, we can apply ($\overline{R}$) and obtain

$$\vdash^{\omega+c}_{0} \neg cl(\mathbf{B}), \neg R_j(\mathbf{u}), B_j(\mathbf{u}).$$

Applications of ($\vee$), ($\forall$) and ($\wedge$) give the desired result. □

In a second step we show that if a formula is provable in BID then it is provable in the infinitary system BID_∞ with length $\omega + m$ and cut-rank r for appropriate $m, r \in \mathbb{N}$.

Lemma 6.5. *If* $\mathrm{BID} \vdash A$ *then there exist* $m, r \in \mathbb{N}$ *such that* $\mathrm{BID}_\infty \overset{\omega+m}{\underset{r}{\vdash}} A$.

Proof. By induction on the length of the proof. Remember that on the interval $[\omega, \omega+\omega)$ the relation $<$ is the same as $<_0$. □

Putting everything together we obtain the following two main theorems.

Theorem 6.2. *If* $\mathrm{BID} \vdash A$ *and* A *is positive, then there exists an* $\alpha < \varepsilon_0$ *such that* $\mathrm{CET} \vdash \mathrm{E}_{\theta_\alpha(0)}\, A$ *and hence* $\mathfrak{J}^{\theta_\alpha(0)} \models \forall(A)$.

This theorem is used in the completeness proof of LDNF-resolution (cf. [22, 23]). The bound $\theta_\alpha(0)$ is not that much important. What is important is that if a positive formula is provable in BID then it is already true at a finite stage. However, the bounds are important for the proof-theoretic strength of BID namely for characterizing the provable total functions of BID.

Theorem 6.3. *If* $\mathrm{BID} \vdash \forall x\,(nat(x) \to \exists y\,(nat(y) \wedge R(x,y)))$ *then there exists an* $\alpha < \varepsilon_0$ *such that*

$$\mathrm{CET} \vdash \forall x\,(\mathrm{E}_m\, nat(x) \to \exists y\,(\mathrm{E}_{\theta_\alpha(m)}\, nat(y) \wedge \mathrm{E}_{\theta_\alpha(m)}\, R(x,y)))$$

for all $m \in \mathbb{N}$.

The second theorem can be expressed in a more general form as well. If A and B are positive formulas such that $A \to B$ is provable in BID then there exists an $\alpha < \varepsilon_0$ such that $\mathrm{E}_m\, A \to \mathrm{E}_{\theta_\alpha(m)}\, B$ is provable in CET for all $m \in \mathbb{N}$. Since $\mathfrak{U}_\mathcal{L}$ is a model of CET, it follows that if A is true a stage $\mathfrak{J}^m$ then B is true a stage $\mathfrak{J}^{\theta_\alpha(m)}$.

7. Negation-as-failure and left-termination

In this section we show how negation-as-failure and left-termination can be expressed by inductive definitions. What does this mean? — Consider a predicate R in a logic program P. From the clauses that define R we construct operator forms for three new relation symbols R^s (success), R^f (finite failure) and R^t (left-termination). Even though in most cases the operator forms for R^s are purely existential formulas, the operator forms for R^f and R^t are more complicated. They are universal formulas in general.

It can be shown that $R^\mathrm{t}(a)$ is derivable in the corresponding system BID if, and only if, the goal $R(a)$ is left-terminating in the sense of Apt and Pedreschi [1]. Moreover, the formula $R^\mathrm{s}(a) \wedge R^\mathrm{t}(a)$ is derivable in BID if, and only if, the goal $R(a)$ succeeds and is left-terminating; $R^\mathrm{f}(a) \wedge R^\mathrm{t}(a)$

is derivable in BID if, and only if, the goal $R(a)$ fails finitely and is left-terminating. These results are proved in [23]. What remains to show in this paper is that the system used in [23] can be embedded into BID. In order to do this we need some terminology of logic programming. For a more detailed presentation we refer the reader to [23].

The syntactic objects in logic programming that correspond to formulas are called *goals*. We use different connectives for negation, conjunction and disjunction of goals to point out that they have a different meaning in logic programming. Negation means negation-as-failure; conjunction is a non-commutative operation; disjunction means alternatives in the sense of Prolog. *Goals* (denoted by G, H) are expressions of the following form:

$$\mathit{true} \mid \mathit{fail} \mid s = t \mid R(\mathbf{t}) \mid \mathit{not}\, G \mid G \,\&\, H \mid G \;\mathit{or}\; H \mid \exists x\, G.$$

A goal of the form $G_1 \,\&\, \ldots \,\&\, G_n \,\&\, \mathit{true}$ is called a *query* and we abbreviate it by $[G_1, \ldots, G_n]$. A *program clause* is an expression of the form $R(\mathbf{t}) \leftarrow G$, where G is a goal. $R(\mathbf{t})$ is called the *head* of the clause and G its *body*. For a clause C of the form

$$R(t_1[\mathbf{y}], \ldots, t_n[\mathbf{y}]) \leftarrow G[\mathbf{y}]$$

we define its *definition form* $D_C[\mathbf{x}]$ to be the goal

$$D_C[x_1, \ldots, x_n] :\equiv \exists \mathbf{y}\, (x_1 = t_1[\mathbf{y}] \,\&\, \ldots \,\&\, x_n = t_n[\mathbf{y}] \,\&\, G[\mathbf{y}]),$$

where $\mathbf{y}$ are new variables. A *logic program* P is a finite list of program clauses. For every relation symbol R and logic program P we define a goal $D_R^P[\mathbf{x}]$ called the *definition form* of R with respect to P.

Let $C_1, \ldots, C_m$ be the list of program clauses in P the head of which is of the form $R(\ldots)$. Then we set

$$D_R^P[\mathbf{x}] :\equiv D_{C_1}[\mathbf{x}] \;\mathit{or}\; \ldots \;\mathit{or}\; D_{C_m}[\mathbf{x}].$$

One could as well define a logic program to be a function that assigns to every relation symbol R a goal $D_R^P[\mathbf{x}]$ with distinguished variables $\mathbf{x}$.

Given a logic program P the user can ask queries. In the following we describe a simple query evaluation procedure which directly reflects the stack based memory management of most implementations of logic programming systems.

An *environment* is a finite set of bindings $\{t_1/x_1, \ldots, t_n/x_n\}$ such that the x_i's are pairwise different variables. It is not required that $t_i \not\equiv x_i$. A *frame* consists of a query G and an idempotent environment η. Idempotent means that if $t_i \not\equiv x_i$ then x_i does not occur in $t_1, \ldots, t_n$. Remember that a query is a list of goals. A *frame stack* consists of a (possibly empty) sequence $\langle G_1, \eta_1; \ldots; G_n, \eta_n \rangle$ of frames. The query G_n together with the environment η_n is called the *topmost frame* of the stack. Capital greek letters Φ, Ψ and Θ denote finite, possibly empty, sequences of the form $G_1, \eta_1; \ldots; G_n, \eta_n$. Thus $\langle \Phi; G, \eta \rangle$ denotes a stack with topmost frame G, η. A *state* of a computation is a finite sequence $\langle \Phi_1 \rangle \;\ldots\; \langle \Phi_n \rangle$ of frame stacks. $\langle \Phi_n \rangle$ is called the topmost stack of the state. States are denoted by the capital greek letter Σ. For a query G with free variables $x_1, \ldots, x_n$ let $init(G)$ be the state $\langle G, \{x_1/x_1, \ldots, x_n/x_n\}\rangle$. There are three kinds of final states: $yes(\eta)$, *no* and *error*.

Definition 7.1. The transition rules of the query evaluation procedure are:

1. $\Sigma\ \langle\Phi; true\ \&\ G, \eta\rangle \longrightarrow \Sigma\ \langle\Phi; G, \eta\rangle$
2. $\Sigma\ \langle\Phi; fail\ \&\ G, \eta\rangle \longrightarrow \Sigma\ \langle\Phi\rangle$
3. $\Sigma\ \langle\Phi; s = t\ \&\ G, \eta\rangle \longrightarrow \Sigma\ \langle\Phi; G, \eta\tau\rangle$ [if $\tau = \mathrm{mgu}(s\eta, t\eta)$]
4. $\Sigma\ \langle\Phi; s = t\ \&\ G, \eta\rangle \longrightarrow \Sigma\ \langle\Phi\rangle$ [if $s\eta$ and $t\eta$ are not unifiable]
5. $\Sigma\ \langle\Phi; R(\mathrm{t})\ \&\ G, \eta\rangle \longrightarrow \Sigma\ \langle\Phi; D_R^P[\mathrm{t}]\ \&\ G, \eta\rangle$
6. $\Sigma\ \langle\Phi; (E\ \&\ F)\ \&\ G, \eta\rangle \longrightarrow \Sigma\ \langle\Phi; E\ \&\ (F\ \&\ G), \eta\rangle$
7. $\Sigma\ \langle\Phi; (E\ or\ F)\ \&\ G, \eta\rangle \longrightarrow \Sigma\ \langle\Phi; F\ \&\ G, \eta; E\ \&\ G, \eta\rangle$
8. $\Sigma\ \langle\Phi; (E\ or\ F)\ \&\ G, \eta\rangle \longrightarrow \Sigma\ \langle\Phi; E\ \&\ G, \eta; F\ \&\ G, \eta\rangle$
9. $\Sigma\ \langle\Phi; (\exists x\, F)\ \&\ G, \eta\rangle \longrightarrow \Sigma\ \langle\Phi; F\{y/x\}\ \&\ G, \eta \cup \{y/y\}\rangle$ [where y is new]
10. $\Sigma\ \langle\Phi; (not\ F)\ \&\ G, \eta\rangle \longrightarrow \Sigma\ \langle\Phi; (not\ F)\ \&\ G, \eta\rangle\ \langle[F], \eta\rangle$ [if $F\eta$ is ground]
11. $\Sigma\ \langle\Phi; (not\ F)\ \&\ G, \eta\rangle \longrightarrow error$ [if $F\eta$ is not ground]
12. $\Sigma\ \langle\Phi; (not\ F)\ \&\ G, \eta\rangle\ \langle\Psi; true, \tau\rangle \longrightarrow \Sigma\ \langle\Phi\rangle$
13. $\Sigma\ \langle\Phi; (not\ F)\ \&\ G, \eta\rangle\ \langle\rangle \longrightarrow \Sigma\ \langle\Phi; G, \eta\rangle$
14. $\langle\Phi; true, \eta\rangle \longrightarrow yes(\eta)$
15. $\langle\rangle \longrightarrow no$

Without Rule (8) this definition describes the operational model of pure Prolog with occurs check. We say that

1. a query G *succeeds with answer* σ, if there exists a computation with initial state $init(G)$ and final state $yes(\eta)$ such that σ is the restriction of η to the variables of G;
2. a query G *succeeds with answer including* σ, if there exist substitutions τ and θ such that G succeeds with answer τ and $G\tau\theta \equiv G\sigma$;
3. a query G *fails*, if there exists a computation with initial state $init(G)$ and final state *no*;
4. a query G is *left-terminating*, if all computations with initial state $init(G)$ are finite and do not end in *error*.

Note, that termination means universal termination. For Prolog-like systems this means that one can hit the semicolon key a finite number of times until one finally obtains the message *no more solutions*.

For defining operator forms for success, finite failure and left-termination we need a language with new predicate symbols R^{s}, R^{f} and R^{t}. We also need a special unary predicate *gr* which is used to express that a term is ground. The formulas A, B of the extended language are

$$\top \mid \bot \mid s = t \mid gr(t) \mid R^{\mathrm{s}}(\mathrm{t}) \mid R^{\mathrm{f}}(\mathrm{t}) \mid R^{\mathrm{t}}(\mathrm{t}) \mid$$

$$\neg A \mid A \wedge B \mid A \vee B \mid A \to B \mid \forall x\, A \mid \exists x\, A.$$